Protein Immobilization

Bioprocess Technology

Series Editor

W. Courtney McGregor

Xoma Corporation
Berkeley, California

Volume 1 Membrane Separations in Biotechnology, *edited by W. Courtney McGregor*

Volume 2 Commercial Production of Monoclonal Antibodies: A Guide for Scale-Up, *edited by Sally S. Seaver*

Volume 3 Handbook on Anaerobic Fermentations, *edited by Larry E. Erickson and Daniel Yee-Chak Fung*

Volume 4 Fermentation Process Development of Industrial Organisms, *edited by Justin O. Neway*

Volume 5 Yeast: Biotechnology and Biocatalysis, *edited by Hubert Verachtert and René De Mot*

Volume 6 Sensors in Bioprocess Control, *edited by John V. Twork and Alexander M. Yacynych*

Volume 7 Fundamentals of Protein Biotechnology, *edited by Stanley Stein*

Volume 8 Yeast Strain Selection, *edited by Chandra J. Panchal*

Volume 9 Separation Processes in Biotechnology, *edited by Juan A. Asenjo*

Volume 10 Large-Scale Mammalian Cell Culture Technology, *edited by Anthony S. Lubiniecki*

Volume 11 Extractive Bioconversions, *edited by Bo Mattiasson and Olle Holst*

Volume 12 Purification and Analysis of Recombinant Proteins, *edited by Ramnath Seetharam and Satish K. Sharma*

Volume 13 Drug Biotechnology Regulation: Scientific Basis and Practices, *edited by Yuan-yuan H. Chiu and John L. Gueriguian*

Volume 14 Protein Immobilization: Fundamentals and Applications, *edited by Richard F. Taylor*

Additional Volumes in Preparation

Protein Immobilization

FUNDAMENTALS AND APPLICATIONS

edited by

Richard F. Taylor

Arthur D. Little, Inc.
Cambridge, Massachusetts

MARCEL DEKKER, INC. New York • Basel • Hong Kong

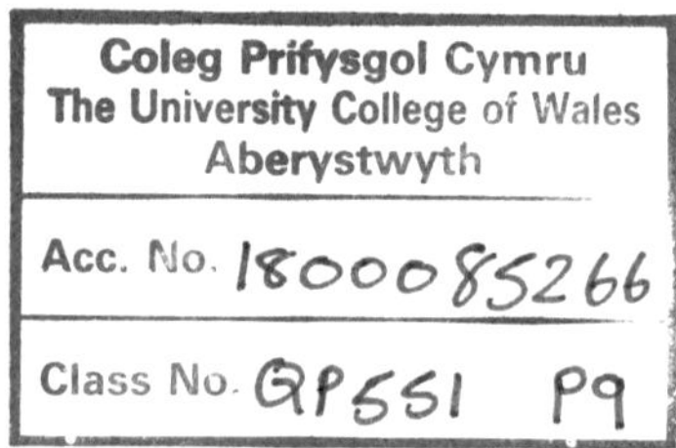

ISBN 0-8247-8271-2

This book is printed on acid-free paper.

Marcel Dekker, Inc.
270 Madison Avenue, New York, New York 10016

Current printing (last digit):
10 9 8 7 6 5 4 3 2 1

PRINTED IN THE UNITED STATES OF AMERICA

Series Introduction

Bioprocess technology encompasses all of the basic and applied sciences as well as the engineering required to fully exploit living systems and bring their products to the marketplace. The technology that develops is eventually expressed in various methodologies and types of equipment and instruments built up along a bioprocess stream. Typically in commercial production, the stream begins at the bioreactor, which can be a classical fermentor, a cell culture perfusion system, or an enzyme bioreactor. Then comes separation of the product from the living systems and/or their components followed by an appropriate number of purification steps. The stream ends with bio-product finishing, formulation, and packaging. A given bioprocess stream may have some tributaries or outlets and may be overlaid with a variety of monitoring devices and control systems. As with any stream, it will both shape and be shaped with time. Documenting the evolutionary shaping of bioprocess technology is the purpose of this series.

Now that several products from recombinant DNA and cell fusion techniques are on the market, the new era of bioprocess technology is well established and validated. Books of this series represent developments in various segments of bioprocessing that have paralleled progress in the life sciences. For obvious proprietary reasons, some developments in industry, although validated, may be published only later, if at all. Therefore, our continuing series will follow the growth of this field as it is available from both academia and industry.

W. Courtney McGregor

Preface

This book presents a practical overview of the technologies used to immobilize proteins, together with examples of specific applications for immobilized proteins.

Rather than being limited to immobilized *enzymes,* this book focuses on immobilized *proteins.* This emphasis is aimed at drawing attention to the increasing importance of immobilized antibodies, structural proteins, and macromolecular complexes (such as cellular receptors) to the development of new immobilized, protein-based products.

Immobilized enzymes (and cells containing desired enzymes), however, continue to command the most attention in the development and commercialization of immobilized proteins. From the first adsorption immobilization of invertase onto charcoal and alumina by Nelson and Griffin in 1916, through the intense research activities on enzyme immobilization in the 1960s, immobilized enzymes emerged in the 1970s as a major, commercially viable technology for food and drug processing. Today, commercial processes routinely utilize immobilized enzymes such as glucose isomerase, glucoamylase, penicillin amidase and acylase, invertase, lactase, fumarase, amino acylase, aspartase, and hydantoinase. In one case—immobilized glucose isomerase—the production of high-fructose syrups from dextrose syrups almost exclusively utilizes the immobilized enzyme.

From the base technology derived from enzyme immobilization, other immobilized proteins are being developed and applied in commerce. Immobilized antibodies form the basis of modern clinical diagnostics and are being applied to new detection and diagnostic devices, such as biosensors, for applications in medicine, food and drug processing, and environmental monitoring. Immobilized peptides and proteins that bind toxic substances are being developed for detoxification applications. New therapeutics are being developed that utilize antibody–toxin conjugates to target and destroy tumor cells. Immobilized cellular receptors, such as receptors isolated from nerve tissue, are being used as the basis for a new generation of diagnostic biosensors. Immobilized antibodies and binding proteins also serve as the basis for biospecific separations, such as affinity chromatography and filtration, which are being applied to the production of new genetically engineered products such as drugs and hormones.

These examples illustrate the increasing uses for immobilized proteins. Key to such use is our better understanding of protein immobilization technology and the application of this technology to useful products and processes. In a world environment where basic research and development is more than ever driven by final, commercializable products, protein immobilization still provides the researcher and inventor with research challenges as well as product development rewards.

This book is aimed at readers who carry out basic research and development in protein immobilization, as well as those who utilize protein immobilization technologies in their processes and products. No text can be inclusive of an area as diverse as protein immobilization. Our aim was to present discussions of both basic and applied aspects of protein immobilization. If, at the very least, the information presented in this text stimulates the reader to further studies in development of protein immobilization methods or in the application of immobilized proteins to new processes or products, it will have achieved its goal.

Richard F. Taylor

Contents

Contributors

Joaquim M. S. Cabral Associate Professor, Department of Chemical Engineering, Instituto Superior Técnico, Universidade Técnica de Lisboa, Lisbon, Portugal

Thomas M. S. Chang Director, Artificial Cells and Organs Research Centre, and Professor, Departments of Physiology and Medicine, McGill University, Montreal, Quebec, Canada

Ichiro Chibata President and Representative Director, Tanabe Seiyaku Co., Ltd., Osaka, Japan

Michael P. Coughlan Professor, Department of Biochemistry, University College, Galway, Ireland

George G. Guilbault Research Professor, Department of Chemistry, University of New Orleans, New Orleans, Louisiana

Subhash B. Karkare Senior Scientist, Cell Culture Process Development, AMGEN, Inc., Thousand Oaks, California

J-Michel Kauffmann Pharmaceutical Institute, Free University of Brussels (U.L.B.), Brussels, Belgium

Rajni Kaul Research Scientist, Department of Biotechnology, Chemical Center, University of Lund, Lund, Sweden

Takuo Kawamoto Instructor, Department of Industrial Chemistry, Faculty of Engineering, Kyoto University, Kyoto, Japan

John F. Kennedy Director, Research Laboratory for the Chemistry of Bioactive Carbohydrates and Proteins, School of Chemistry, University of Birmingham, Birmingham, England

Marek P. J. Kierstan Director of Research, Dalgety plc., Cambridge, England

Bo Mattiasson Department of Biotechnology, Chemical Center, University of Lund, Lund, Sweden

Gaston J. Patriarche Pharmaceutical Institute, Free University of Brussels (U.L.B.), Brussels, Belgium

Tadashi Sato Manager, Biochemistry Department, Research Laboratory of Applied Biochemistry, Tanabe Seiyaku Co., Ltd., Osaka, Japan

Atsuo Tanaka Professor, Department of Industrial Chemistry, Faculty of Engineering, Kyoto University, Kyoto, Japan

Richard F. Taylor Manager, Applied Biotechnology Laboratory, Arthur D. Little, Inc., Cambridge, Massachusetts

Tetsuya Tosa Director, Research Laboratory of Applied Biochemistry, Tanabe Seiyaku Co., Ltd., Osaka, Japan

1

Introduction: The Current Status of Immobilized Protein Technology

Richard F. Taylor

Arthur D. Little, Inc.
Cambridge, Massachusetts

I. INTRODUCTION

The last half of the twentieth century has witnessed a revolution in the practical application of the biological and physical sciences. The merging of disciplines grounded in chemistry and biology with others in physics and electronics has spawned new technologies and products. Of such interdisciplinary technologies, the new biotechnology (as distinct from "old" biotechnology grounded in fermentation processes stretching back to ancient times) stands out as a primary example. To practice the new biotechnology, expertise may be required in disciplines previously thought revolutionary in their own right, such as biochemistry, molecular biology, genetics, biophysics, and microelectronics. The primary difference which has brought biotechnology into our daily lives is that its practice is leading to practical products.

Just as the practical application of physics in the first half of this century resulted in controllable nuclear power and the electronics revolution, biotechnology is being used today to change the fabric and quality of our daily lives. Genetically engineered drugs such as insulin, human growth hormone, α-interferon, and tissue plasminogen activator (TPA) are available and promise to increase the quality of health care. New vaccines such as those for hepatitis B, scours, rabies, and coccidiosis are being produced in recombinant microorganisms or by solid-state peptide synthesis. Trans-

genic plants and animals are being developed to impart herbicide resistance and improve meat production, respectively. Improved methods are being developed for the production of specialty chemicals for the food and processing industries including recombinant enzymes (e.g., α-amylase, rennin, and lipase) and synthetic sweeteners (e.g., aspartame).

The art and science of immoblizing proteins is a biotechnology, and the oldest of the new biotechnologies. The use of immobilized enzymes for production of products such as sugars, amino acids, and drugs predates by nearly two decades biotechnologies such as genetic engineering and monoclonal antibodies. In addition to use in processing, immobilization techniques represent a base technology which is necessary for many biotechnology products. These include immobilized antibodies and enzymes used in diagnostic products; immobilized antibodies and binding proteins used in separation products such as those for bioaffinity chromatography; and immobilized proteins (enzymes, antibodies), protein complexes (receptors), and cells used for biosensors. Immobilization technology, then, is a key technology for the successful application of biotechnology to new products.

II. APPLICATIONS AND MARKETS FOR IMMOBILIZED PROTEINS

A. Applications

The primary application for immobilized proteins is the use of immobilized enzymes as catalysts in industrial processes and products. The high specificity, high rates of reaction, nontoxicity, water solubility, biodegradability, and use under mild conditions of pH, temperature, and pressure are major advantages over inorganic catalysts. For this reason, the use of immobilized enzymes is firmly established as an effective and economically favorable approach for the production of products such as fructose, synthetic penicillins, and amino acids. Specific examples of these processes are presented throughout this text.

While many texts have been written on immobilized enzymes, the focus of this text includes not only immobilized enzymes but other proteins which function through means other than catalytic activity, e.g., antibodies, receptors, and binding proteins such as proteins A and G, lectins, and avidin. These noncatalytic proteins are the focus of the most rapidly growing area of immobilized protein research, development and application.

As shown in Table 1, modern biotechnology needs have stimulated new applications of immobilized proteins. Of these, the areas being most affected by improvements in immobilized protein technology are bioseparations, diagnostics, bioprocessing, and new disease therapies.

Table 1 Immobilized Protein Application Examples

Application	Immobilized Protein(s)	Support	Ref.
Research Applications			
Artificial photosystem	*Desulfovibrio vulgaris* hydrogenase	Nylon gel	18
Photoreaction systems	Spinach photosystem II submembrane fraction	Cross-linked albumin membrane	19
Drug metabolism studies	Various metabolism enzymes	Agarose, sepharose, acrylamide, etc.	20
Sialic acid assay	*N*-Acetylneuraminic acid aldolase and lactic dehydrogenase	Gelatin membrane	21
Diagnostic Applications			
Analytical bioassays	Bacterial luciferase	Nylon, glass and agarose beads, etc.	22
DNA analysis	Restriction endonucleases	Nylon, cellulose, gelatin, and agarose	23
Hydrolysis of nerve gases	Acetylcholinesterase	Agarose beads	24
Detection of cholinergics	Acetylcholine receptor	Polymeric membrane	25
Medical Applications			
Fibrinolytic polymers and vascular prosthesis devices	Urokinase, trypsin, and streptokinase	Dacron-reinforced collagenous tubes, polyvinylidene fluoride films, nylon, etc.	26–28
Removal of drugs, toxins, antibodies, etc., from blood	Appropriate antibodies and binding proteins	Agarose beads, etc.	29
Intracorporeal treatment of gout	Uricase	Polyethylene glycol	29
Extracorporeal removal of asparagine in cancer patients	Aspariginase	Methacrylate plates	30
Inhibition of platelet adhesion to medical prosthesis devices	Albumin–heparin complex	Polyvinyl chloride and other polymers	31
Extracorporeal heparin removal	Heparinase	Agarose beads	32
Plasmapheresis immunotherapy	Antibodies to IgE	Porous cellulose, agarose, and glass beads	33

Table 1 (Continued)

Application	Immobilized Protein(s)	Support	Ref.
Industrial Applications			
Sterilization of dairy products	Catalase	Collagen spiral membrane reactor	34
Tallow hydrolysis	Lipase	Microporous acrylic membrane reactor	35
Aglycone production from glycosides	Naringinase	Controlled pore glass	36
Nucleotide production	T4 Polynucleotide kinase	Sepharose beads	37
Oxygen extraction from seawater	Hemoglobin	Polyurethane	38
Purification of immunoglobulins	Proteins A and G	Vinyl-cellulose spiral membrane cartridge	39
Sugar production from hemicelluloses	Cellulases	Controlled-pore glass, alumina, and titania	40
Juice clarification	Endo-polygalacturonase	Trimethylchitosan and others	41
Milk lactose hydrolysis	β-Galactosidase	Polyvinyl chloride-silica spiral flow reactor	42

Immobilized antibodies and binding proteins (such as concanavalin A and protein A) are widely used for separation and purification of materials by affinity chromatography and filtration (1,2). Improvements in the stability of support matrix materials and coupling methods have led to more rapid affinity chromatography methods, i.e., high performance affinity chromatography (3–5) and new membranes for fast flow affinity filtration (6). The use of immobilized protein affinity supports in two-phase aqueous systems is also being developed for application to process scale separations (7). While this method has, to date, primarily utilized affinity dyes bound to polyethylene glycol for purification of enzymes (such as dehydrogenases and kinases)(8,9) from salt or dextran second phases, other workers are developing two-phase systems for immunopartitioning using immobilized antibodies to purify enzymes, cells, and hormones (10,11). Examples of support materials used for affinity separations and their application to enzyme immobilization will be discussed in Chapter 4 of this book.

Diagnostics rely heavily on immobilized proteins. Most nonisotopic immunoassays utilize either immobilized antibody or enzymes or both (12–14). The proteins may be used in a variety of immobilization modes in such assays including adsorption onto solid supports, entrapment into membranes and films, and covalently bonded to carriers such as polymeric beads. Immobilized enzymes, antibodies, and receptors also form the basis for biosensors: emerging detection devices applicable not only to clinical diagnostics but also to chemical, food, and drug process control; environmental monitoring; and robotics. Chapters 7 and 8 of this text discuss biosensor technology and applications.

New membrane-based bioreactors promise to increase the number of bioprocess products in the next 5 years. By utilizing new polymeric membranes as the base for protein and cell immobilization, systems are being developed which take advantage of not only increased protein stability on the membranes, but also the built-in separation and concentration capabilities of the membranes (15). These membrane reactors, together with new bioreactor technologies for microencapsulation of cells and enzymes and hollow fiber reactors for cell cultivation, are already impacting production of specialty chemicals and pharmaceuticals.

Immobilized proteins are also being developed to improve the quality of health care. A number of these applications are cited in Table 1. The successful development of extracorporeal immobilized protein devices for removal of toxic substances from blood and of intracorporeal immobilized protein therapeutics will significantly impact the treatment of a wide range of diseases, from AIDS to gout, within the next 5 to 8 years.

Future uses of immobilized proteins may also include application as protection products against environmental hazards. For example, the specificity of binding proteins and the catalytic activity of enzymes can be exploited to remove toxic substances from individuals, machines, and the environment. Figure 1 illustrates the principle of an immoobilized protein dressing which was developed in our laboratories for the detoxification of organophosphorus and carbamate nerve toxins (16,17). Co-immobilization of a binding enzyme (AChE, acetylcholinesterase) and an organophosphorous agent hydrolyzing enzyme (from either squid hepatopancreas or rat liver) resulted in a bifunctional dressing in which binding to the AChE provides initial detoxification followed by longer-term (and higher capacity) hydrolysis action on the toxic agent. This approach of using binding and hydrolysis proteins (or their active-site peptides) for removal and destruction of toxic agents has potential for widespread application in defense, agriculture, and industry.

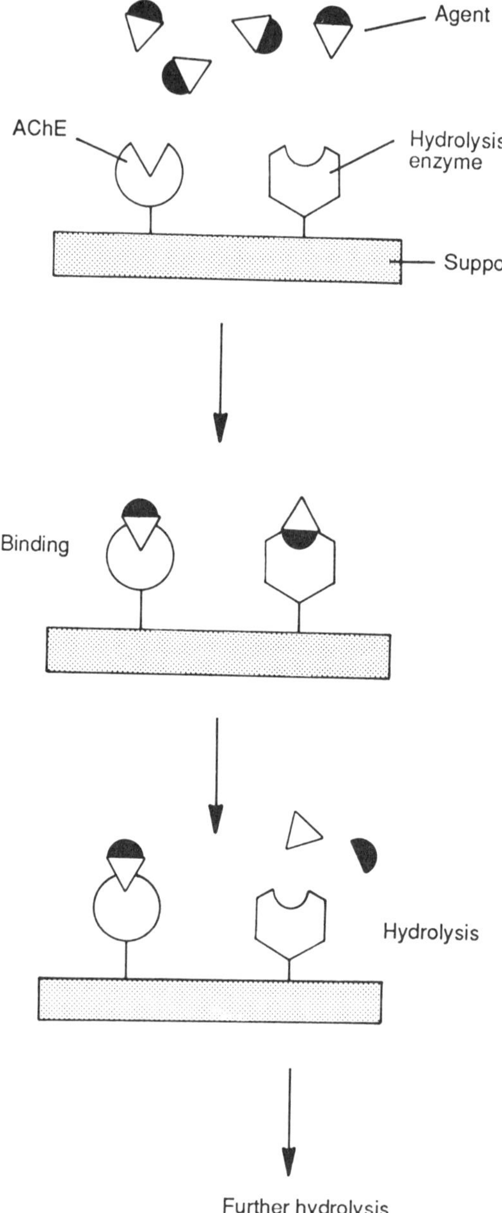

Figure 1 Detoxification of toxic organophosphorus agents by an immobilized protein wound dressing.

B. Markets

The market for immobilized proteins is difficult to estimate due to the diversity of their applications. However, since immobilized enzymes currently represent the bulk of the immobilized protein market, estimates on the total world enzyme market can be used to establish markets and market growth rates for immobilized proteins.

Worldwide sales of enzymes in 1987 totaled approximately $445 million and are expected to reach of $800 million by 1997, thus realizing a 5–7% annual average growth rate (43). The food industry accounted for over half the world market for enzymes in 1987, with detergent applications the second largest market (Fig. 2). Specialty enzyme products currently repre-

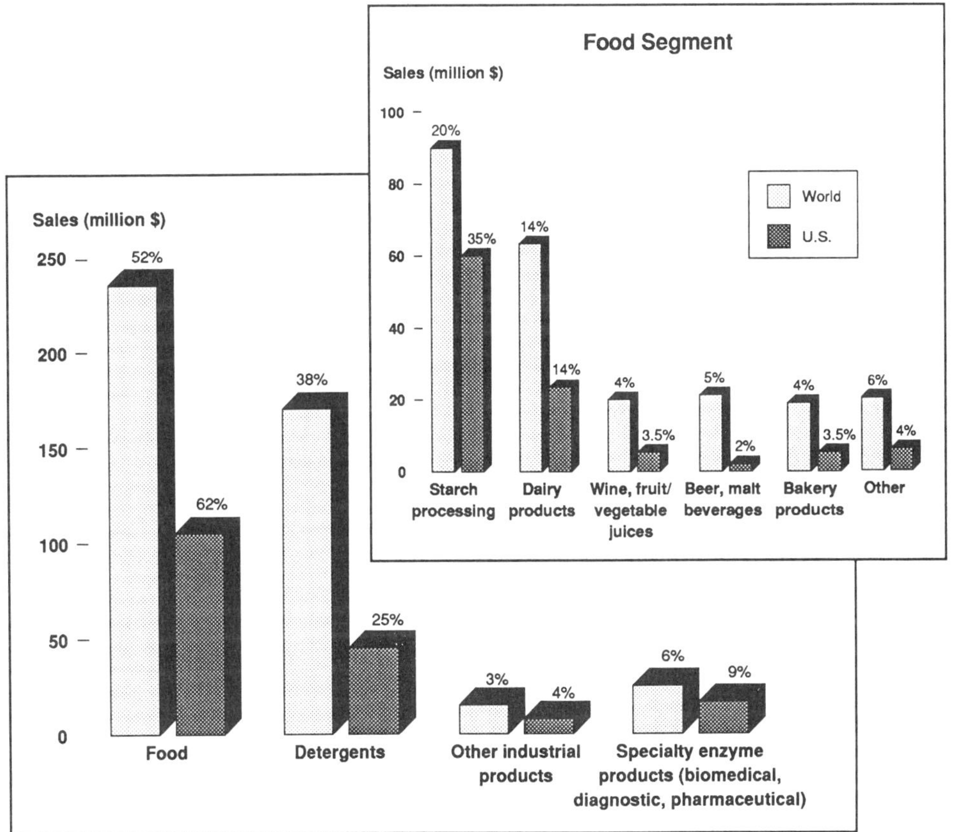

Figure 2 The 1987 world market for enzymes. (From Ref. 43.)

sent only a small percentage of total enzyme sales but are projected to grow significantly (to perhaps 15% of the total world market) by 1995.

Immobilized proteins are estimated to have commanded approximately 35% of the total enzyme market or $165 million in sales in 1987 and are expected to achieve sales of $325 million in 1997 (4–9% annual average growth rate). A primary factor in this growth will be increased application of immobilized enzymes and other immobilized proteins to new diagnostics, biomedical, pharmaceutical, and environmental products resulting from applications such as those described in Table 1.

III. PROTEIN IMMOBILIZATION: BASIC AND APPLIED ASPECTS

The growing applications and markets for immobilized proteins are providing a positive environment for development of new and improved immobilization technologies. The chapters in this book address a wide spectrum of activities in protein immobilization in order to provide a text containing not only basic immobilization principles and methods, but practical information and applications as well.

The first section of the text addresses basic principles and methods for protein immobilization. Protein immobilization is accomplished by two basic methods, either noncovalent or covalent attachment to an appropriate carrier. Chapter 2 addresses noncovalent protein immobilization methods including adsorption, entrapment, microencapsulation, and aggregation. Chapter 3 addresses covalent and coordination protein immobilization. Chapter 4 addresses commercially available supports for protein immobilization which may be utilized quickly by workers with minimal preparation. Chapter 5 completes the section on basic immobilization principles by addressing practical aspects of handling proteins during immobilization and means to determine immobilization coupling yields.

The second section of the text is directed at practical applications of immobilized proteins. Chapter 6 addresses the use of immobilized enzymes in organic solvents, a method which promises to expand the commercial application base for immobilized enzymes. Such organic solvent–based systems may be used in either mono- or biphasic systems to allow reactions not possible with traditional immobilized enzyme systems. Chapters 7 and 8 address a newly emerged application for immobilized proteins: as the basis for biosensors. Biosensors are electronic detection devices which utilize an immobilized biomolecule (enzyme, antibody, or receptor) or cell as the basic detection component. Chapter 7 addresses the oldest biosensor technology, that based on enzyme electrodes, while Chapter 8 presents a discussion of a new type of blosensor, bioaffinity sensors. Chapter 9 presents a discussion on advances in the use of immobilized proteins for medical applications. The use of immobilized proteins in medicine appears poised for rapid growth in the 1990s as practical devices are developed for removal of toxicants from blood and urine, for correcting congenital and disease-caused enzyme defects, and for detection and diagnosis.

Chapters 10 and 11 discuss industrial applications of immobilized enzymes. Chapter 10 addresses immobilized cell reactors and recent advances which have significantly increased interest and development of this method for the production of new products. Chapter 11 presents an overview of the more classical use of immobilized enzymes in food and pharmaceutical processing.

The purpose of this book is to provide the reader with a basic understanding of the principles and methodologies of immobilized proteins, and to illustrate the wide diversity of applications for these materials. While no text can present all the information available in such a diverse area, it is the hope of the chapter authors and the editor that the information provided can stimulate further productive development and use of immobilized proteins.

REFERENCES

1. T. C. J. Gribnau, J. Visser, and R. J. F. Nivard (Eds.), *Affinity Chromatography and Related Techniques,* Elsevier Scientific Publishing, Amsterdam, 1982.
2. I. M. Chaiken, M. Wilchek, and I. Parikh, *Affinity Chromatography and Biological Recognition,* Academic Press, New York, 1983.
3. P. O. Larsson, M. Glad, L. Hansson, M. O. Mansson, S. Ohlson, and K. Mosbach, *Adv. Chromatogr. 21:*41 (1983).
4. R. F. Taylor, in *Proceedings of Biotech USA '85,* Online International, New york, 1985, p. 381.
5. T. M. Phillips, *LC 3:*962 (1985).
6. S. Brandt, R. A. Goffe, S. B. Kessler, J. L. O'Connor, and S. E. Zale, *Bio/technology 6:*779 (1988).
7. J. C. Janson, *Trends Biotechnol. 2:*35 (1984).
8. G. Johansson and M. Joelsson, *Biotechnol. Bioeng. 27:*621 (1985).
9. H. Hustedt, K. H. Kroner, U. Menge, and M. R. Kula, *Trends Biotechnol. 3:*54 (1985).
10. H. Walter and G. Johannson, *Analyt. Biochem. 155:*215 (1986).
11. L. J. Karr, S. G. Shafer, J. M. Harris, J. N. Van Alstine, and R. S. Snyder, *J. Chromatogr. 354:*269 (1986).
12. A. Voller, D. E. Bidwell, and A. Bartlett, *Enzyme Linked Immunosorbent Assay (ELISA),* Dynatech Laboratories, Alexandria, VA, 1979.
13. Maggio, E. T. (Ed.), *Enzyme-Immunoassay,* CRC Press, Boca Raton, FL, 1980.
14. D. Moore, *Analyt. Chem 56:*920A (1984).
15. G. Belfort, *Biotechnol. Bioeng. 33:*1047 (1989).

16. R. F. Taylor, *Immobilized Protein Dressings for Wound Decontamination from Nerve Agents,* Final Report USAFSAM-TR-88-38, USAF School of Aerospace Medicine, San Antonio, TX, 1989.
17. R. F. Taylor, *Article for Inactivation of Toxic Materials and Novel Peptides for Use Therein,* U.S. Patent Appl. 07/105,312, 1987.
18. Y. Nosaka, A. Kuwabara, T. Kobayashi, and H. Miyama, *Biotechnol. Bioeng. 28:*456 (1986).
19. R. Carpentier and S. Lemieux, *Appl. Biochem. Biotechnol. 15:*107 (1987).
20. D. M. Dukik and C. Fenselau, *FASEB J. 2:*2235 (1988).
21. F. N. Kolisis, *Biotechnol. Appl. Biochem. 8:*148 (1986).
22. N. N. Ugarova and O. V. Lebedeva, *Appl. Biochem. Biotechnol. 15:*35 (1987).
23. M. Nasri and D. Thomas, *Appl. Biochem. Biotechnol. 15:*119 (1987).
24. F. C. G. Hoskin and A. H. Roush, *Science 215:*1255 (1982).
25. R. F. Taylor, I. G. Marenchic, and E. J. Cook, *Analyt. Chim. Acta 213:*131 (1988).
26. L. C. Mercer, K. E. Everse, A. W. Holmes, and J. Everse, *Thromb. Res. 13:*931 (1978).
27. F. F. Senatore and F. R. Bernath, *Res. Comm. Chem. Pathol. Pharmac. 49:*295 (1985).
28. J. E. Wilson, *Polym. Plast. Technol. Eng. 25:*233 (1986).
29. S. Margel and L. Marcus, *Appl. Biochem. Biotechnol. 12:*37 (1986).
30. M. D. Klein and R. Langer, *Trends Biotechnol 4:*179 (1986).
31. S. W. Kim, *Artificial Org. 11:*228 (1987).
32. H. Bernstein, V. C. Yang, and R. Langer, *Appl. Biochem. Biotechnol. 16:*129 (1987).
33. H. Sato, T. Kidaka, and M. Hori, *Appl. Biochem. Biotechnol. 15:*145 (1987).
34. H. D. Chu, J. G. Leeder, and S. G. Gilbert, *J. Food Sci. 40:*641 (1975).
35. F. Taylor, C. C. Panzer, J. C. Craig, Jr., and D. J. O'Brien, *Biotechnol. Bioeng. 28:*1318 (1986).
36. P. Turecek and F. Pittner, *Appl. Biochem. Biotechnol. 13:*1 (1986).
37. D. W. Hutchinson and R. Collier, *Biotechnol. Bioeng. 29:*793 (1987)
38. T. H. Maugh, *Science 223:*474 (1984).
39. R. M. Mandaro, S. Roy, and K. C. Hou, *Bio/technology 5:*928 (1987).
40. K. Shimizu and M. Ishihara, *Biotechnol. Bioeng. 29:*236 (1987).
41. P. G. Pifferi, M. Tramontini, and A. Malacarne, *Biotechnol. Bioeng. 33:*1258 (1989).
42. A. P. Bakken, C. G. Hill Jr., and C. H. Amundson, *Biotechnol. Bioeng. 33:*1249 (1989).
43. C. G. Greenwald and J. M. Nystrom, *Outlook for Food Enzymes,* Arthur D. Little Decision Resources, Cambridge, MA, 1988.

I

BASIC PRINCIPLES AND METHODS FOR PROTEIN IMMOBILIZATION

2

Immobilization of Proteins by Noncovalent Procedures: Principles and Applications

Marek P. J. Kierstan

Dalgety plc.
Cambridge, England

Michael P. Coughlan

University College
Galway, Ireland

I. INTRODUCTION

Seventy-one years ago Nelson and Griffin (1) found, quite fortuitously, that invertase retained its catalytic ability when adsorbed to activated charcoal. This, the first demonstration of the general phenomenon of noncovalent attachment of proteins to insoluble supports, may be considered to be the cornerstone of what has since become an immense area of study and application. While an established use for immobilized enzymes did not then exist, the observations of Nelson and his colleagues (1,2) stimulated others to investigate a wide range of materials as possible immobilizing supports. Not surprisingly, many materials specifically designed to adsorb proteins, albeit for chromatographic purposes, were found to be suitable. Synthetic ion exchange resins, in particular, such as Dowex 50 (3), DEAE- and CM-celluloses (4–6), have been used extensively. Indeed, some of the early large-scale processes for "high fructose" corn syrup production were possible because of the availability, in quantity, of such resins and the ease of enzyme immobilization thereon. For example, the procedure used by the Clinton Corn Processing Company involved the use of glucose isomerase immobilized on DEAE-cellulose (7).

Other noncovalent immobilization techniques that have been extensively reported, especially in recent years, include entrapment and encapsulation. Here again one finds an example of the transfer of large-scale technology from one industry to another. Thus, high fructose corn syrups have been produced using glucose isomerase entrapped in cellulose acetate fibers (8).

A review of the pertinent literature shows that quite a wide range of techniques for noncovalent immobilization of proteins is available (see Table 1). Most of the developments in this area took place between 1965 and 1975. Regrettably, few of the proposed or hoped-for applications of such preparations have been realized. Those that have have been restricted to industrial-scale single enzyme processes. Moreover, apart from microencapsulated preparations, there has been relatively little interest in using noncovalent techniques for immobilization of proteins other than enzymes.

That the use of noncovalent techniques has been limited to large-scale applications is due mainly to the comparative simplicity and cheapness of the procedures involved and to the fact that the supports (e.g., ion-exchange resins) used retain good flow characteristics under the high compression forces generated in large-scale processes. In small-scale systems, by contrast, the disadvantages of noncovalent immobilization (see below) have dictated that the more expensive and involved covalent techniques be used. This would be true, for example, in the immobilization of high value proteins such as antibodies. Despite its shortcomings, noncovalent immobilization is the preferred technique in some applications. Thus, entrapment or encapsulation is the norm for immobilization of living and dead cells while physical adsorption to appropriate supports is attracting considerable attention as a means of immobilizing enzymes for use in nonaqueous systems (see Ref. 9).

As stated above, examples of important industrial, commercial, and medical applications of noncovalently immobilized proteins are few. Nevertheless, a wide range of techniques have been developed, and the systems involved have been thoroughly investigated. Unfortunately, man's ingenuity in this context has scant regard for the work of the cataloguer. Thus, while the bulk of this review will, in keeping with the title, be concerned solely with strictly noncovalent techniques, other processes must necessarily be included. Although purists might disagree, we are of the opinion that these follow naturally from the core material and that their omission would lead to an unbalanced view of the true range of techniques available for immobilization of proteins generally and enzymes specifically. Accordingly, the text will cover the following: (1) adsorption, including physical adsorption and ionic binding; (2) entrapment and related techniques; (3) microencapsulation, including liposome and hollow fiber systems; (4) inorganic bridge formation; and (5) aggregation. Included in the text are systems involving purified or partially purified enzymes or proteins. Passing mention will be made of

Table 1 Noncovalently Immobilized Enzymes, the Supports Used, and the Mode of Immobilization

Enzyme	Support	Interaction	Ref.
Acetylcholinesterase	Silastic resin	ENT	115,312
Acid phosphatase	Carbon (cephalin-coated)	PA	308
	Polyacrylamide-Separan AP273	ENT	205
	Silica	PA	308
	Silica (cephalin-coated)	PA	308
Acid protease	Konjak powder	ENT	374
AMP deaminase	DEAE-cellulose	I	366,367
	Silica gel	PA	324
ATPase	Millipore filter	PA	310
	Silastic resin	ENT	106
ATP deaminase	Carbon	PA	366
	DEAE-cellulose	I	366
	TEAE-cellulose	I	366
Albumin	Cation-exchangers CM-2p,SG-1m	I	279
Alcohol dehydrogenase	Aluminas (various)	PA	214
	Collodion	ENC	126
	Collagen membrane	ENT	178
	Hexadecyl silica	PA,H	267
	Polyacrylamide	ENT	402
	Polyaminomethyl styrene	PA	299
	Silica or protein-modified silica	PA	197
Alcohol oxidase	Microporous membrane	ENT	217
Aldehyde dehydrogenase	*n*-Octyl-Sepharose via hydrophobic membrane proteins	PA,H	270
Aldehyde oxidase	DEAE-Sepharose 6B	PA,H	254
	n-Hexyl-Sepharose 4B	PA,H	254
	n-Octyl-Sepharose 4B	PA,H	254
Aldolase	Polyacrylamide	ENT	117,311,430,436
	Poly(vinyl alcohol)	ENT	188
Alkaline phosphatase	Antigen-labelled liposomes	ENC	195
	n-Octylamino-Sephadex	I	20
	Polyacrylamide gel	ENT	283,312
	T_4-labeled liposomes	ENC	220
Alkaline protease	Polyacrylamide	ENT	423
α-Amino-ϵ-caprolactam hydrolase	DEAE-Sephadex	I	369

Table 1 (Continued)

Enzyme	Support	Interaction	Ref.
α-Amino-ε-caprolactam racemase	DEAE-Sephadex	I	369
D-Amino acid oxidase	Polyacrylamide	ENT	174,175,185,423
L-Amino acid oxidase	Polyacrylamide	ENT	173,185,428
Aminoacylase	Aluminium oxide (acid)	PA	172
	Aluminium oxide (neutral)	PA	172
	κ-Carrageenan	ENT	389
	DEAE-cellulose	I	21,23,172,373, 375,376
	DEAE-Sephadex	I	23,172,313,314, 375,377,378,380
	Diaion SA-11A and SA-21A (Cl^-)	I	23
	ECTEOLA-cellulose	I	23,172
	Ethylcellulose	ENC	172
	HPMCP-DEAE	ENT	172
	Nitrocellulose	ENT	127
	Nylon	ENC	172
	Polyacrylamide	ENT	113,423
	Polyurea	ENC	172
	Tannin-aminohexyl cellulose	PA,H	370,371
	TEAE-cellulose	I	23,172
α-Amylase	Activated carbon	PA	10
	Bentonite	PA	315
	Biocarb	I	242
	Clay (bleaching)	PA	10
	Collagen	PA	316
	Diaflo PM 10 membrane/UFC	ENT	128,129
	Diaflo PM 30 membrane/UFC	ENT	130
	Kaolinite	PA	327
	Polyacrylamide	ENT	423,426,429
	Starch	PA	382
α-Amylase + phosphorylase	UFC	ENT	232
β-Amylase	Activated carbon	PA	383
	Bentonite	PA	327
	Collagen	PA	316
	Diaflo PM 30 membrane	ENT	130

Table 1 (Continued)

Enzyme	Support	Interaction	Ref.
	Gluten	PA	384
	Kaolinite	PA	327
	Polyacrylamide	ENT	423
	Polyiminoethylene-polystyrene	ENC	285
β-Amylase + phosphorylase	UFC	ENT	232
Amyloglucosidase	Activated carbon	PA	317,383,393–401
	Alumina	PA	317,403
	Amberlite IR-45	I	4,404
	Amberlite CG-50(CG-4B)	I	317
	Amicon HP-10	PA	188
	Calcium alginate	ENT	241
	Calcium alginate-polyethyleneimine	ENT	239
	Carbon	PA	24,400
	Cation-exchange resins (various)	I	198,317
	Clay (acid)	PA	24,317,398,400,405
	CM-cellulose	I	317
	Con A in calcium alginate	PA/ENT	262
	DEAE-cellulose	I	4,223,317,406,407
	Diaflo PM 10 membrane/UFC	ENT	128,155
	Diaflo PM 30 membrane/UFC	ENT	130
	Diaflo UM 2 membrane/UFC	ENT	318
	Diatomaceous earth	PA	24
	Dowex-1-X10 (Cl^-)	I	4
	Gelatin	ENT	246
	Hydrophobic cotton	PA	226
	Liposome	ENC	156
	Palmityl-Sepharose	PA,H	209
	Polyacrylamide	ENT	273,423,439–442
	Polyacrylic acid	ENT	122
	Polystyrene	PA	238
	Polyvinyl alcohol	ENT	441
	Polyvinylpyrrolidone	ENT	124
	Silanized alumina	PA	233
	Tannin-coated Ti-activated alkylamine-CPG	PA	221

Table 1 (Continued)

Enzyme	Support	Interaction	Ref.
Apyrase	Polyacrylamide	ENT	106
	Silica resin	ENT	106
Arginase	Epoxy	ENC	184
	Nylon	ENC	184
	Poly(phthaloyl piperazine)	ENC	184
Arylsulfatase	Concanavalin A	PA	385
Asparaginase	Amicon HP-10	PA	188
	CM-cellulose	I	5,6
	Cellulose nitrate	ENC	184
	Collodion	ENC	138,319,320
	DEAE-cellulose	I	387
	Fibrin	ENT	131
	Liposomes	PA,H	307
	Nitrocellulose	ENC	132
	Nylon	ENC	133–136,138, 319,320
	Polyacrylamide	ENT	111,137,175,185
	Polyurea	ENC	136
Aspartase	Calcium phosphate gel	PA	119
	κ-Carrageenan	ENT	389
	Collodion	ENC	138
	DEAE-cellulose	I	119
	DEAE-Sephadex	I	119
	ECTEOLA-cellulose	I	119
	β-Naphthyl-cotton	PA,H	278
	Polyacrylamide	ENT	119
	Silica gel	PA	119
	TEAE-cellulose	I	119
Carbonic anhydrase	Collodion	ENC	166,319–321
	Nylon	ENC	139,140,142,166, 319,321
Carboxypeptidase	Amicon HP-10	PA	188
Carboxypeptidase A	Monoclonal antibody bound covalently to Eupergit	BS	302
	Monoclonal antibody bound noncovalently to Sepharose-Protein A	BS	302
Catalase	Amberlite XE-97	I	391
	Bentonite	PA	327
	Butyl rubber	ENC	184
	Calcium carbonate	PA	327

Table 1 (Continued)

Enzyme	Support	Interaction	Ref.
	Carbon (lauric acid- or cephalin-coated)	PA	328
	CM-cellulose	I	13
	Cellulose nitrate	ENC	184
	Cellulose triacetate	ENT	47
	Collagen	ENT	141,201
	Collodion	ENC	64,138,142,143, 145,319–321
	DEAE-cellulose	I	13,388
	Glass (barium stearate-, thorium nitrate-, or sodium deoxycholate-treated)	PA,H	329,330
	Metal (barium stearate-, thorium nitrate-, or sodium deoxycholate-treated)	PA,H	329,330
	Nickel-silica-alumina	PA	390
	Phenylsiloxane	ENC	146
	Polyaminopolystyrene	I	391
	Polyacrylamide	ENT	147,177,185,206
	Polyion complex membrane	ENT	293
	Polystyrene	ENC	146
	Silica gel	ENT	324
	Silica (lauric acid-, cephalin-, or tridodecylamine-treated)	PA,H	328
Cellulase system	BM100 membrane/UFC	ENT	266
	Calcium phosphate gel	PA	192
	Cellulose	BS	335
	Collagen	PA	316
	HFA 300 membrane/UFC	ENT	148
	PM 30 membrane/UFC	ENT	148
Cephalosporin acetyl-transferase	Bentonite	PA	392
Cholesterol oxidase	Bicarb-D	I	212
	Collagen membrane	ENT	179
Cholinesterase	Polyacrylamide	ENT	309,437,438
	Silastic resin	ENT	309
	Starch	ENT	309,331,379, 381,386

Table 1 (Continued)

Enzyme	Support	Interaction	Ref.
Chymotrypsin	CM-cellulose	I	393
	Cellulose-citrate	I	393
	Diaflo XM 100 membrane/UFC	ENT	88
	Glass (treated with metal then coated with barium stearate and treated with uranyl acetate)	PA,H	333,334
	Hexadecyl silica	PA,H	267
	Kaolinite	PA	16–18,32,426
	Palmityl-Sepharose	PA,H	209
	Phosphocellulose	I	393
	Polyacrylamide	ENT	106,206,436
	Polyethyleneimine-nylon	ENC	184
	Silastic resin	ENT	106
Chymotrypsinogen-A	Carboxyl cation-exchangers CM-2p and SG-lm	I	279
Citrate synthase	Polyacrylamide	ENT	116,406
Creatinine deiminase + glutamate dehydrogenase	Poly(vinyl chloride)	PA	298
Cyclomaltodextrin glucotransferase	DIAION HP-20	I	231
dCMP aminohydrolase	Polymeric membrane	ENT	304
Cytochrome D	Polyionic microcapsules	ENC	213
Cytosine deaminase	Cellulose tubes	ENC	249
Deoxyribonuclease I	Cellulose	PA	315,343
Dextranase	Bentonite	PA	230
Dextransucrase	DEAE-Sephadex	I	394
	Modified silochrome	PA	306
Endo-β-1,4-glucanase	Alumina	PA	251
	Concanavalin A-Sepharose	PA	15,225
	Polyurethane	ENT	222
	Silica	PA	251,274
	Titanium	PA	251
Enniatin synthase	Propyl agarose	PA	236
Enolase	Polyacrylamide	ENT	118
Exo-β-1,4-cellobiohydrolase	Alumina	PA	251
	Concanavalin A-Sepharose	PA	15,225

Table 1 (Continued)

Enzyme	Support	Interaction	Ref.
	Silica	PA	251
	Titanium	PA	251
β-D-Fructofuranosidase (invertase)	Activated carbon	PA	1,2,102,338
	Alumina	PA	417
	Aluminum hydroxide gel	PA	1,2,338
	Amberlite (various types)	I	418
	Amicon HP-10	PA	188
	Amicon XM-50	PA	188
	Bentonite	PA	263,417
	Cellulose triacetate	ENT	150,159
	Con A in Ca alginate	PA/ENT	262
	Concanavalin A-Sepharose	PA	215
	Collagen	PA	316,337
	Collagen membrane	ENT	180
	DEAA-cellulose	I	397,415
	DEAE-cellulose	I	102,419,420
	DEAE-Sephadex-A-50	I	186
	Polyacrylamide	ENT	102–104,170,423, 441,442
	Polyacrylamide-Separan AP273	ENT	205
	Polyacrylic acid	ENT	122
	Polyethyleneglycol dimethacrylate	ENT	123
	Poly(ethylene-vinyl alcohol) modified	I	277
	Polyion complex membrane	ENT	294
	Polystyrene	ENC	102
	Polysulphone membrane/ UFC	ENT	189
	Polyvinyl alcohol	ENT	127,275,441
	Polyvinyl pyrrolidone	ENT	124,125
Fumarase	κ-Carrageenan	ENT	389
Galactose oxidase	Polyacrylamide	ENT	280
α-Galactosidase	Amicon HP-10	PA	188
	Romicon P10	PA	188
β-Galactosidase	Amicon HP-10	PA	188
	Cellulose nitrate	ENC	66
	Cellulose triacetate	ENT	47
	Collagen	ENT	151

Table 1 (Continued)

Enzyme	Support	Interaction	Ref.
	Collodion	ENC	152
	DEAE-cellulose	I	339
	Duolite S-761 ion-exchange resin	I	193,261
	Hydrophobic cotton	PA,H	226
	Nylon	ENC	154
	Phenyl-Sepharose CL 4B	PA	272
	Polyacrylamide	ENT	372,442,449
	Polyacrylamide-Separan AP273	ENT	205
	Polyacrylic acid	ENT	122
	Polygalacturonate	I	291
	Poly (vinyl alcohol)	ENT	188
	Polyvinyl pyrrolidone	ENT	124
	Romicon XM-50	PA	188
	Silyl undecanoic acid-modified inorganic carrier	PA,H	265
	Tannin-aminohexyl cellulose	PA,H	23
	Titanium oxide-steel	PA	395
Glucose dehydrogenase	Carbon electrode	PA	284
Glucose isomerase	Alumina (large pore)	PA	204
	Amicon P-10	PA	188
	Amicon XM-50	PA	188
	Cellulose triacetate	ENT	157
	Collagen	PA	316
	Polyacrylamide	ENT	120,121
	Romicon PM-10	PA	188
	Tannin-aminohexyl cellulose	PA,H	23
	Tannin-silochrome	PA	257
Glucose oxidase	Activated carbon	PA	317
	Alumina	PA	317
	Amberlite CG-50	I	317,411
	Carbon electrode	PA	296
	Cellulose acetate	PA	188
	Cellulose triacetate	ENT	47
	Collagen	ENT	180
	ConA in calcium alginate	PA/ENT	262
	Concanavalin A-Sepharose	PA	215

Table 1 (Continued)

Enzyme	Support	Interaction	Ref.
	Clay (acid)	PA	317
	Clay (montmorillonite)	PA,I	228
	Cuprophane	ENT	187
	Glass	PA	412
	Glassy carbon	PA	240
	Graphite	PA	216,237,240
	Nitrocellulose	ENT	127,297
	Polyacrylamide	ENT	147,185,340,402,409, 410,416,421,423
	Poly (vinyl alcohol)	PA	258
	Starch	ENT	22
	Titania	PA	413
Glucose oxidase + catalase	Alumina	PA	317
	Amberlite CG-50 Type II	PA	317
	Calcium alginate	ENT	235
	Carbon	PA	317
	Cellulose membrane	ENT	199
	Clay	PA	317
	ConA-cellulose	PA	200
Glucose oxidase + invertase	Polymethacrylate	ENC	211
Glucose oxidase + peroxidase	Silastic resin	ENT	312
Glucose-6-phosphate dehydrogenase	Collodion	PA	341
	Polyacrylamide	ENT	116
	Silica gel	PA	342
	Silica gel (lecithin-coated)	PA/H	342
Glucose phosphate isomerase	DEAE-Sephadex	I	414
	Polyacrylamide	ENT	430
β-Glucosidase	Calcium alginate	ENC	194
	Ceramic support	PA	292
	ConA-Sepharose	PA	15,186
	ConA-Sepharose in Ca alginate	PA/ENC	38
	Hydrophobic cotton	PA,H	226
	Hydroxyethyl cellulose-Tylose T4000	ENT	205
	Polyacrylamide	ENT	185,443
	Poly (vinyl alcohol)	ENT	190
	Radiation-induced polymer	ENT	295
	Visking tubing	ENT	259

Table 1 (Continued)

Enzyme	Support	Interaction	Ref.
Glutamate dehydrogenase	Hexadecyl Fractosil	PA,H	300
	Hexadecyl silica	PA,H	267
	Palmityl-Sepharose	PA,H	209
	Polyacrylamide	ENT	409
	Ultrafiltration membrane	ENT	182,183
Glutamate-pyruvate aminotransferase	Nitrocellulose	ENT	127
Glutaminase	Polyacrylamide	ENT	174
Hemoglobin	Polyionic microcapsules	ENC	213
Hexokinase	Collodion	ENC	158
	Polyacrylamide	ENT	421,430
	Silica gel	PA	344,345
	Silica gel (lauric acid- or cephalin-coated)	PA,H	345
Hexokinase + glucose-6-P dehydrogenase	Polyacrylamide	ENT	255
Hydrogenase	Polyacrylamide	PA,ENT	234
Laccase	Ion-exchange polyacrylonitrile fibre	I	281
Lactate dehydrogenase	Albumen-coated carbon fibre electrode	PA	288
	Collagen membrane	ENT	178
	Nitrocellulose filter	PA	346
	n-Octylamino-Sephadex	I	368
	Polyacrylamide	ENT	402,406,409
	Ultrafiltration membrane	ENT	183
Leucine aminopeptidase	Calcium phosphate gel	PA	347,422
	Hydroxylapatite	PA	422
Lipase	Amicon	PA	188
	Amberlite XE-97	I	391
	Bis(2-ethylhexyl) sodium sulfosuccinate reverse micelles	ENC	303
	Collagen	ENT	160
	Polystyrene/silicone/ethylcellulose	ENC	146
	Hydrophobic membrane	PA,H	282
	Oil-impregnated polypropylene	PA,H	271
	Organic and inorganic supports (various)	PA	196

Table 1 (Continued)

Enzyme	Support	Interaction	Ref.
	Phenylsiloxane	ENC	146
	Polyaminostyrene	PA	391
	Polystyrene	ENC	146
	Polyurethane/photo-cross-linkable copolymer	ENT	196
	Porous glass	PA	424
Lipoamide dehydrogenase	Butyl-Sepharose	PA,H	425
	Hexyl-Sepharose	PA,H	425
Lipoprotein lipase	Heparin-Sepharose	PA	207
Lipoxygenase	Glucose-calcium alginate	ENC	248
Lysine decarboxylase	Alumina	PA	348
Lysine oxidase	Gelatin-polypropylene on oxygen electrode	PA	208
Lysozyme	Collagen	PA	32,316,337
	Kaolinite	PA	426
Malate dehydrogenase	Collodion	ENC	126
	Polyacrylamide	ENT	406
	Silica gel	PA	349
	Silica gel (lecithin-, cephalin- or cholesterol-coated)	PA,H	349
Mutarotase	Collagen membrane	ENT	180
NAD pyrophosphorylase	Hydroxylapatite	PA	427
Naringinase	Tannin-aminohexyl cellulose	PA,H	370
Natural protease	Polyacrylamide	ENT	423
Orsellinate decarboxylase	Polyacrylamide	ENT	107
Oxalate decarboxylase	Polyacrylamide	ENT	244
D-Oxynitrilase	ECTEOLA-cellulose	PA	353
Pancreatin	Pectin-based carrier	I	268
Papain	Calcium-pectin gel	ENT	219
	Collagen	ENT	161
	Polyacrylamide	ENT	436
	Porous glass	PA	412,429
	Tannin-aminohexyl cellulose	PA,H	370
Pectinesterase	Polyethylene terephthalate	PA	269
Penicillin acylase	Poly (vinyl alcohol)	ENT	188
	Ultrafiltration membrane	ENT	210,252

Table 1 (Continued)

Enzyme	Support	Interaction	Ref.
Penicillinase	Polyacrylamide	ENT	114,185
Pepsin	DEAE-cellulose	I	393
	Metal or glass plates (treated with barium stearate, thorium nitrate or sodium deoxycholate)	PA,H	330,336
	Sand	PA	256
Peptidase	Alumina	PA	276
Peroxidase	Nitrocellulose	ENT	127
	Polyacrylamide	ENT	173,350
	Polyacrylonitrile ion-exchange fiber	I	281
Phenylalanine ammonia lyase	Amicon HP-10	PA	188
	Collodion cells	ENC	243
	Microcapsules	ENC	253
Phosphodiesterase	Concanavalin A-Sepharose	PA	14
Phosphofructokinase	Polyacrylamide	ENT	430
Phosphoglucomutase	Carbon (cephalin-coated)	PA,H	308
	Silica	PA	308
	Silica (cephalin-coated)	PA,H	308
Phosphoglycerate mutase	Polyacrylamide	ENT	351
Phosphomonoesterase	CM-cellulose	I	352
Phospholipase	Polyamine-modified phosphatidylethanolamine	BS	260
Phosphorylase	Silica (and glass)	PA	431
Polygalacturonase	Alumina	PA	251
	Silica	PA	251
	Titanium	PA	251
Polynucleotide phosphorylase	Nitrocellulose fiber	PA	246
Polyphenol oxidase	Ion-exchange polyacrylonitrile fiber	I	281
Protease	DEAE-cellulose	I	393
Protosubtilin	Pectin-based carrier	I	268
Pullulanase	Diaflo PM 30 membrane/UFC	ENT	130
Pyrophosphatase	DEAE-cellulose	I	427
Pyruvate kinase	Collodion	ENC	133
Ribonuclease	Cationic resin SBS 4 (H)	I	354

Table 1 (Continued)

Enzyme	Support	Interaction	Ref.
	Dowex-2 anion-exchanger	I	3
	Dowex 50 cation-exchanger	I	3
	n-Octylamino-Sephadex	I	368
	Polyacrylamide	ENT	436
	Porous glass	PA	3,433,434
Rulactine	Liposomes	ENC	286
Salicylate hydroxylase	Dialysis membrane	ENT	244
Steroid Δ^1 dehydrogenase	Polyacrylamide	ENT	355
Succinate dehydrogenase	Carbon coated with a monolayer of cephalin or heparin	PA,H	3
	Polyacrylamide	ENT	185
	Silica gel coated with a monolayer of cephalin or heparin	PA,H	3
Thermolysin	Radiation-induced polymers	ENT	218
Transglucosidase	Clay (acid)	PA	435
Transglutaminase	Monoclonal antibody-coated agarose	BS	247
Triacylglcerol acylhydrolase	Celite 560	PA	244
	Filtercel	PA	244
	Hyflosupercel	PA	244
	Silica	PA	244
Trypsin	CM-cellulose	I	393
	Cellulose-citrate	I	393
	Collodion	ENC	142,321
	Gelatin-modified phenolic resin	PA	250
	Glycidyl methacrylate-ethylene dimethacrylate copolymer with trypsin inhibitor	BS	264
	Hexadecyl silica	PA,H	267
	Kaolinite	PA	426
	Methyl methacrylate-styrene sulfonate copolymer	PA	229
	Nylon	ENC	142,321
	Palmityl-Sepharose	PA,H	209

Table 1 (Continued)

Enzyme	Support	Interaction	Ref.
	P-cellulose	I	391
	Polyacrylamide	ENT	105–110,147, 421,436
	Polymethyl methacrylate	PA	203
	Polystyrene	PA	203
	Silastic resin	ENT	106
	Silica gel	ENT	356
	Silica with immob. trypsin inhibitor	BS	264
	Styrene-glycidyl methacrylate copolymer	PA	290
	Styrene-2-hydroxyethyl copolymer	PA	203
Tryptophan synthase	Cellulose triacetate	ENT	47
β-Tyrosinase	Cellulose triacetate	ENT	47
L-Tyrosine decarboxylase	Polyacrylamide	ENT	185
	Ultrafiltration membrane	ENT	181
Urate oxidase	Cellulose acetate	PA	188
	Collodion	ENC	321
	Nylon	ENC	321
	Poly(vinyl alcohol) HFM	PA	188
	Protamine-aminohexyl CPG500	PA	287
Urease	Amicon PM-30 membrane/UFC	ENT	191
	Benzalkonium-heparin-collodion	ENT	332
	Butyl acetate cellulose	ENC	162,169
	Cellulose acetate HFM	PA	188
	Cellulose nitrate	ENC	184,305
	Collagen	ENT	150,163–165
		PA	316,361
	Collodion	ENC	64,138,142,143, 158,166,319–321, 323,364,365
	Gelatin	ENT	167
	Kaolinite	PA	362,363
	Lipid-polyamide	ENC	184
	Metal or glass plates (treated with barium stearate, thorium nitrate or sodium deoxycholate)	PA,H	330,336

Table 1 (Continued)

Enzyme	Support	Interaction	Ref.
	Nylon	ENC	138,139,142,154, 166,319–323,325, 326,364,365
	n-Octylamino-Sephadex	I	368
	Organosmectite	PA,H	289
	Phenylsiloxane	ENC	146
	Polyacrylamide	ENT	112,147,176,185, 309,357–360,421
	Polyacrylonitrile	PA	301
	Polyionic microcapsules	ENC	213
	Polystyrene	PA	301
	Polystyrene/silicone/ ethylcellulose	ENC	146
	Polyvinyl alcohol	ENT/PA	168,188
	Silastic resin	ENT	309
	Silica gel (hydrogel)	ENT	324
	Starch	ENT	309
	Ultrafiltration membrane	ENT	181
	Urea liquid membrane	ENT	171
Urease + hemoglobin	Cellulose nitrate	ENC	227
Xanthine oxidase	*n*-Octylamino-Sephadex	I	368

Abbreviations used: PA = physical adsorption; I = ionic binding; BS = biospecific sorption; ENT = entrapment; ENC = encapsulation; H = hydrophobic; HFM = hollow fiber membrane; UFC = ultrafiltration cell.
Acetylcholinesterase and cholinesterase may be same enzyme.

immobilized cell systems in which the activity of but a single enzyme is exploited. (See Chapters 10 and 11.)

II. ADSORPTION

Adsorption, the longest established technique for the noncovalent immobilization of enzymes, is based on the physical adsorption or ionic binding, or both, of the enzyme to the surface of the support. An early example of immobilization by physical adsorption is that provided by Stone in 1955 (10). He found that α-amylase bound to activated carbon when the two were mixed in solution and stirred at 10°C for 1 h. Moreover, the immobilized enzyme when packed into a column effected the conversion of starch to glucose. In 1956 Mitz (11) reported that some of the catalase passed through a column of DEAE-cellulose bound to the support. The resultant ionically bound preparation efficiently catalyzed the conversion of hydrogen peroxide in solutions subsequently fed through the column.

As one might expect, the nature of the enzyme and of the carrier influence both the amount of enzyme bound and the activity of the final preparation (12). The greater the surface area of the carrier and, in general, the higher its hydrophilic/hydrophobic group ratio, the greater is the amount of enzyme that may be bound. Immobilization by physical adsorption or by ionic binding, both low-cost procedures, are simple and effective and, unlike covalent procedures, usually bring about little change in the overall conformation of the protein or of the active site (12). The activity of enzyme preparations immobilized by ionic binding may be quite high, and such preparations have been used over the years in certain large-scale applications. Nevertheless, the general applicability of physically adsorbed enzymes is low. During operational use noncovalently immobilized enzymes, whether physically adsorbed or ionically bound, may leak from the carrier because the binding force(s) between the protein and the carrier is weak. This has frequently necessitated the inclusion of a chemical cross-linking step to allow such preparations to be used effectively in industrial systems (13). It should be mentioned, however, that the amounts of protein that may be immobilized by physical adsorption or by ionic binding may be increased by recycling the protein through the reactor containing such preparations. This fact also underscores one advantage of such preparations over encapsulated proteins, i.e., reactors containing adsorbed enzymes can be regenerated. A selection of the organic and inorganic supports that have been used, the proteins that have been noncovalently immobilized thereon, and the nature of the immobilization procedure are listed in Table 1.

In the case of physical adsorption the forces responsible for immobilization include hydrogen bonding, Van der Waals forces, and, importantly, hydrophobic interactions. Immobilization via ionic binding, as the term implies, involves salt-linkage formation between charged groups on the protein and opposite charges on the carrier. In general, interactions between the support and the enzyme involves only the protein moiety of the latter. However, as a number of enzymes are glycoproteins, adsorption via interactions involving the carbohydrate portion may also be possible. Immobilization of phosphodiesterase (14) and all three components, namely, endoglucanase, exocellobiohydrolase, and β-glucosidase, of the cellulase system of *Trichoderma reesei* (15) on concanavalin A-Sepharose are cases in point.

Because of the nature of the forces involved in the noncovalent immobilization of proteins, it is relatively simple to reverse the immobilization process and so to effect desorption of the protein from the support. Since both adsorption and desorption would be affected by conditions that influence the strength of the interaction between support and protein, parameters such as pH, ionic strength, and temperature must be controlled. Manipulation of these parameters allows of a simple and cost-effective method for immobilization. However, operational systems have to be engineered with care so as to minimize desorption losses and prevent inadvertant total loss of enzyme. This is an especially significant engineering problem in large-scale operations. Even if total desorption is avoided, the dynamic equilibrium between adsorbed and desorbed species invariably leads to a low but continuous loss of enzyme from the system. This has two consequences. First, the inherent activity of the immobilized preparation is gradually diminished. Second, the product stream contains active enzyme that may continue to operate on the original or other substrates. This can be an especially serious problem in large-scale food processes.

In order to ensure mild conditions for the noncovalent immobilization of protein, the supports most often used are polyfunctional. As a result, immobilization can lead to dramatic changes in the microenvironment at the support surface, the environment experienced by the protein. These changes can lead to apparent alterations in stability and pH/activity profiles. Moreover, when the substrates and/or products are themselves charged, their interaction with the support or the microenvironment may lead to diffusional or kinetic alterations. The nature of the interaction with the support can also be critical to the demonstrated characteristics of the immobilized enzyme if metal ion cofactors are involved. Chibata (12) and others (16–18) have provided a thorough discussion with many examples of the effects of charged supports on the pH optima or pH/activity profiles of ionically bound enzymes. In general, the apparent pH for optimal activity shifts towards the acid side when an enzyme is bound to a polycationic support and to the alkaline side if bound to a polyanionic support. Such shifts in the apparent optimum pH may be as much as 2 units. One may explain these observations as follows. When an enzyme binds to a polycationic support the positive charge on the enzyme increases and the microenvironment of the enzyme/support becomes proton deficient, i.e., more alkaline than the bulk external solution. One might say that the external solution compensates by becoming more acid. Thus, the apparent optimum pH for the reaction will be on the acid side of that catalyzed by the free enzyme. Similar reasoning may be used to explain the shift in apparent optimum pH to the alkaline side when an enzyme is bound to a polyanionic support. Indeed, the fact that immobilization on polyanionic and polycationic supports has opposite effects on pH optima has been demonstrated experimentally (19). Shifts in pH optima may be minimized by carrying out the catalytic reactions at higher ionic strengths. In such cases the charges on the carrier are neutralized by Na^+ and Cl^- rather than by H^+ and OH^- with the result that the difference in pH between the internal and external environments is lessened (12).

By using covalent techniques proteins may be immobilized at nonessential residues. By contrast, immobilization by adsorption is less specific and a real or apparent decrease in activity will ensue if adsorption involves groups at the active site. Covalent techniques also allow of immobilization to supports via spacer arms thereby obviating steric hindrance by the support. This is not possible in immobilization by adsorption and support overloading can be a problem. This results from protein–protein interaction and leads to apparent loss in catalytic activity (20). Although immobilization by adsorption is mild compared with the chemical interactions required for covalent binding, the final environment of covalently immobilized preparations can be engineered to be essentially neutral by the use of minimally charged base media. By contrast, the microenvironment of an adsorbed protein will almost certainly be different than the macroenvironment. This, in association with the gross interactions that can take place between a charged support and the charged protein may lead to alterations in quaternary structure and a destabilizing microenvironment. Chibata (12) has given comprehensive coverage to comparison of the effects of different methods of immobilization on such parameters as yield of immobilized enzyme, optimum pH, optimum temperature, activation energy, K_m and V_{max} values, as well as thermal, storage, and operational stability.

The following discussion and examples address specific methods for preparing adsorbed proteins.

A. DEAE-Sephadex

As mentioned earlier this technique involves electrostatic bonding of the protein to the polymer. The example described is that for the immobilization of amyloglucosidase (21).

Materials: Amyloglucosidase solution (5 g/100 ml of 20 mM acetate buffer, pH 4.2); DEAE-Sephadex Type A25 (10 g); 20 mM acetate buffer, pH 4.2; 0.5 M HCl; 0.5 M NaOH.

Method: Suspend 10 g DEAE-Sephadex in 50 ml of 0.5 M HCl and stir for 20 min. Filter, resuspend in 50 ml of 0.5 M NaOH and stir for 20 min. Filter and wash with distilled water until the washings are neutral. Resuspend in 100 ml of 20 mM buffer, pH 4.2, and stir for 10 min. Filter, resuspend in 100 ml of the same buffer and leave to equilibrate overnight. Filter to dryness. Add the prepared DEAE-Sephadex to 100 ml of the amyloglucosidase solution, stir at room temperature for 2 h and then filter. Reslurry the enzyme-Sephadex complex in 200 ml of acetate buffer, pH

4.2, stir for 15 min, and filter. Repeat the last step until, as judged by absorbance at 280 nm, the filtrate is free of protein. Resuspend the immobilized enzyme preparation in 50 ml of 20 mM acetate buffer, pH 4.2, and store at 4°C until required for use.

The level of observed enzyme activity of such preparations is dependent on the binding capacity of the immobilizing media and on the diffusional and steric restrictions on the immobilized enzyme. Yields as low as 1% of soluble activity have been reported (22). However, the DEAE-Sephadex immobilized aminoacylase preparation above exhibited 86% of the original activity (23). Immobilization on DEAE-Sephadex would, as discussed earlier, change the microenvironment of the enzyme with respect to pH with a consequence shift in the apparent pH activity curve towards acid pH values. This has been demonstrated experimentally for amyloglucosidase adsorbed on DEAE-Sephadex (24).

B. Alumina

Techniques for immobilization of protein on alumina can be used for immobilization onto other inorganic carriers. Messing (25) recommends that glass and silica be used as supports for systems operating below pH 7.0. Titania is suitable for immobilized preparations working in the pH range 3–9 while alumina should be chosen for alkaline operating conditions. However, alumina is effective as a carrier in the pH range 5–11. It has been noted that alumina carriers adsorb considerable quantities of cobalt and magnesium. Since removal of these metals from cobalt- or magnesium-activated enzyme systems by the support would reduce activity, the recommended procedure outlined below, which has been designed for improved maintenance of enzyme activity, includes preconditioning with cobalt and magnesium.

Materials: Controlled-pore alumina (Corning Glass Works product) 2 × 60 mesh, average pore diameter 140 Å, maximum pore diameter 230 Å, surface area 100 m^2/g and porosity 63.5%. Glucose isomerase (590 IU/ml in 60% saturated ammonium sulfate). Buffer, pH 7.5, consisting of 50 mM magnesium acetate and 100 mM cobalt acetate. Sodium bicarbonate solution (0.5 M). Sodium chloride solution (0.5 M).

Method: Equilibrate an 11-g batch of the carrier by repeated incubation with 100 ml of 50 mM magnesium acetate/100 mM cobalt acetate buffer, pH 7.5. The first incubation is carried out in a stoppered cylinder at 60°C (in a water bath) for 10 min. Decant the supernatant and resuspend the alumina in the same volume of preconditioning buffer. Mix and allow to stand at room temperature for 2 h. The solution is then decanted and the carrier is ready for use as an immobilization support.

To 40 ml of the glucose isomerase solution add 1.4 g of ammonium sulfate and stir the slurry for 20 min at room temperature. The sample is then centrifuged and the supernatant discarded. Redissolve the precipitate in 12 ml of 50 mM magnesium acetate/100 mM cobalt acetate, pH 7.5. Add an additional 3 ml of 0.5 M sodium bicarbonate to the solution and mix. Heat the mixture for 15 min in a water bath set at 60°C. After centrifugation the precipitate is discarded and the supernatant (vol 28 ml, pH 7.5) containing the equilibrated enzyme is ready for immobilization.

Add the 28 ml of enzyme solution to the 11 g of preconditioned alumina carrier in a stoppered 10-ml cylinder. Allow the enzyme to react with the carrier for 2.5 h at 60°C in a water bath. During this period the cylinder is inverted every 15 min to mix the alumina with the enzyme solution. The mixture is then allowed to react for a further 2 h at room temperature with inversion at 30-min intervals and finally to continue to completion by overnight incubation at room temperature without inversion. The liquid is then decanted from the carrier and tested for residual enzyme activity. Wash the immobilized enzyme system with 60 ml of distilled water, then with 40 ml of 0.5 M sodium chloride, and finally with 40 ml of 0.05 M magnesium acetate/0.01M cobalt acetate buffer, pH 7.5. The system is then ready for use.

In the original report (25) about 18% of the original activity was exhibited by the immobilized system. As 19% of the original enzyme was recovered unbound, one may calculate that the immobilized enzyme activity was only 22% of that of an equivalent amount of free enzyme. Moreover, binding to the carrier altered the apparent pH activity profile. The enzyme in solution has a pH optimum of 6.8 while the optimum for the immobilized enzyme was about 7.5. The immobilized enzyme system had a half life of 42 days. However, plugging of the pores with consequent depression of observed activity was noted. This could be alleviated by using the washing procedure outlined in the immobilization protocol.

In general the advantage of using alumina or ceramics for enzyme immobilization is that they are robust materials that can be used repeatedly for efficient immobilization. Regeneration of the carrier can be effected by using pyrolysis. Temperatures above 400°C in the presence of air or oxygen are recommended.

C. Collagen

Vieth and Venkatsubramanian (26) have reviewed the techniques that have been developed for the use of collagen as an immobilizing suport (see below). It has been established that the bonds between protein and support for several collagen immobilization procedures are noncovalent (27). However, in the case of one variant technique, namely that of complexation, the authors (27) claim that immobilization does not take place via simple adsorption. Unlike adsorption, which involves just surface interactions, complexation involves bulk-phase interactions. The basis of this claim is as follows. In adsorbed systems total desorption can be achieved. By contrast, in the complexation system, although some bound enzyme is desorbed by washing with solutions of high electrolyte concentration, a limit is reached beyond which no further desorption occurs. This was interpreted to mean that a stable complexed network, stabilized by quasi-reversible covalent bonds between the collagen and the enzyme, had been formed (28). The use of collagen allows for preparation of a range of reactor types, the most commonly cited system being a spirally wound collagen–enzyme membrane system (29). Collagen–protein complexes can be prepared by a number of methods classified as (1) membrane impregnation, (2) macromolecular complexation, and (3) electrodeposition (26). These are represented diagrammatically below.

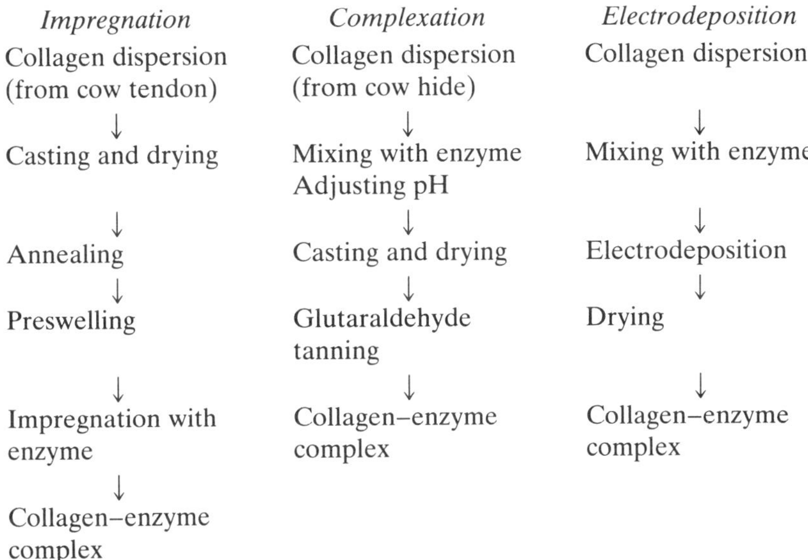

1. Membrane Impregnation Method

In this method a solid collagen membrane is impregnated with enzyme. Efficient penetration of the enzyme into the membrane is achieved as a consequence of the concentration and pH gradients set up between the enzyme bath and the membrane during impregnation. The gradients facilitate molecular diffusion processes. The subsequent drying process removes water clusters around charged groups on both the collagen and enzyme, thereby reducing the intermolecular distances between the participating species and promoting the completion of complex formation. The membrane so formed can be used directly or following treatment with a range of tanning agents of which glutaraldehyde is the most commonly used. Glutaraldehyde covalently cross-links protein molecules. By acting on both the immobilizing matrix and the adsorbed proteins, it provides a membrane of improved mechanical strength.

Materials: Collagen (30% w/v solids). Enzyme (10–20 mg/ml in the example cited). Buffer solution, the composition being dependent on the enzyme to be immobilized. Lactic acid (95% w/v). Glutaraldehyde solution (0.1–5% w/v). Sodium bicarbonate.

Method: Disperse 60 g of hide collagen (30% solids) in 4 liters of water. Adjust the pH to 3.0 by dropwise addition of lactic acid. Homogenize in a Waring blender using 15 × 1-min treatments. During this operation the temperature should be maintained below 25°C by cooling of the blender and homogenate in ice between treatments. The dispersion is then degassed and a membrane is cast either by using specially designed equipment or more simply by pouring the dispersion onto glass plates and spreading uniformly with a glass rod. After the casting operation, the dispersion is left to dry at room temperature. The membrane is allowed to swell for 2 h in buffer, the composition and pH of which is optimal for immobilization of the enzyme in question. The cast membrane is then immersed fully in the solution of enzyme (10–20 mg/ml). The impregnation step is carried out at room temperature for 24–36 h. Lower temperatures can be used for thermosensitive enzymes. After impregnation the membranes are allowed to dry at room temperature prior to tanning. Alternatively, the membrane-bound enzyme preparation can be used directly.

Tanning is carried out by adjusting the glutaraldehyde solution to the pH value at which the immobilized enzyme is most stable. This is achieved by addition of solid sodium bicarbonate. The enzyme membrane is immersed in the tanning solution at room temperature for 0.5–2 min followed immediately by washing in running cold tap water for 1 h. If required, membrane preparation can be carried out under alkaline conditions. This is achieved by using 1 N sodium hydroxide so as to effect collagen dispersion at pH 10.5–11.

2. Macromolecular Complexation

This procedure, a simplification of the impregnation technique, involves inclusion of the enzyme to be immobilized at the dispersion step. This is dependent on the enzyme having significant stability at the acid (2.5–4.5) or alkaline (10.0–11.5) pH values required for dispersion. Glutaraldehyde may also be added at the time of dispersion so as to achieve simultaneous tanning.

Materials: As for impregnation above.

Method: A collagen dispersion is prepared as previously described. To this a quantity of enzyme is added and the mixture is homogenized. The ratio of enzyme dry weight to collagen is typically between 0.1–1 and 0.3–1. The enzyme is dissolved in distilled water prior to addition. To this dispersion a quantity of glutaraldehyde (without pH adjustment) can also be added to achieve simultaneous tanning. Alternatively, the dispersion is first cast to form membranes, which can then be tanned if required.

3. Electrodeposition

This technique involves the formation via dispersion of a collagen–enzyme macromolecular complex that under the influence of an external electrical field gradient is caused to migrate and be deposited on a solid electrode surface. By judicious adjustment of experimental parameters membranes of different thicknesses can be produced (30).

Materials: As for impregnation above.

Methods: An acid or alkaline dispersion of collagen is prepared as previously described. The dispersion is then placed into an electrodeposition vessel containing a stainless steel plate of suitable dimensions as the anode onto which the complex is deposited. Following deposition the film can be peeled off and tanned or deposited and tanned directly onto structures to be used in a final reactor. A number of different types of apparatus for electrodeposition have been devised (31,32). In these techniques recovered enzyme activities of 20–30% have been recorded with a further decrease of 30% as a result of tanning (20,33,34).

Studies on the microenvironment of enzymes immobilized in collagen indicate that the bulk and microenvironmental pH are approximately equivalent (20). However, it is possible to adjust the microenvironment by the introduction of charged groups into the system or by chemical blocking of amino or carboxyl groups. The use of collagen allows of the production of a variety of structures suitable for incorporation into bioreactors. The simple tanning operation, by introducing covalent bonds, strengthens the membrane structure. Indeed, the simplicity of the tanning process and the advantages it imparts are such that it has become standard practice in immobilization using collagen.

D. Charcoal

The properties of charcoal with respect to adsorption of protein have been exploited in industrial processes. In the example cited below, solvent systems have been utilized in order to increase the yield of adsorbed enzyme. In order to achieve stability in aqueous operating media glutaraldehyde treatment has also been incorporated (13,35). The immobilized preparations are obtained by coating the support material with an aqueous solution of the free enzyme. A hydrophobic organic solvent is then used to facilitate adsorption of the protein or enzyme onto the support with a final cross-linking to gel the coating.

Material: Acid-washed bone char (mesh size 500–1000 μm). Acetone. Glutaraldehyde (80% w/v). Amyloglucoside solution (40%, w/v).

Method: Mix 100 ml of acid-washed bone char with 60 ml of a 40% solution of amyloglucosidase. To this mixture add, dropwise, 150 ml of solution consisting of 124 ml of acetone and 26 ml of 80% (w/v) aqueous glutaraldehyde. The formed gel is then broken up by passage through a 2000-μm sieve. Following washing, the immobilized enzyme is then ready for use.

E. Concanavalin A

Concanavalin A, a lectin, is a protein that binds reversibly with mannose, glucose, and sterically related sugars or sugar residues. Thus, the use of concanavalin A allows of gentle and specific immobilization of a range of glycoenzymes and glycoproteins. As examples we may note that invertase and cellobiase have been immobilized by adsorption to concanavalin A-Sepharose (36,37). Binding of glycoproteins to concanavalin A is dependent on Mn^{2+} and Ca^{2+}. Once formed, the complex is stable at neutral pH in the absence of these cations. However, at pH values below 5 they must be present so as to preserve the binding ability of the concanavalin A. For these reasons concanavalin A-Sepharose is normally sold as a slurry containing Mn^{2+}, Mg^{2+}, and Ca^{2+} cations.

Materials: Concanavalin A-Sepharose. Buffer, pH 7.4, consisting of 10 mM sodium phosphate, 0.5 M NaCl, and 1 mM $CaCl_2$. Buffer, pH 4.8, consisting of 10 mM sodium acetate, 1 mM $CaCl_2$, and 1 mM $MnCl_2$. Industrial grade liquid cellobiase (β-glucosidase).

Method: Prepare about 10 ml of ConA-Sepharose by washing thoroughly with the pH 7.4 buffer. Dilute the commercial cellobiase 100-fold with the same buffer. Stir 5 ml of diluted enzyme with 5 ml of ConA-Sepharose slurry for 30 min at room temperature. Filter the ConA-Sepharose–enzyme complex through a medium-sized porous glass filter and then wash with buffer, pH 4.8, until enzyme activity in the washings is no longer detectable. Resuspend the complex in 10 ml of buffer pH 4.8 for assay. It is claimed that no enzyme activity is lost by immobilization using this technique. This is attributed to the fact that the adsorption mechanism, unlike adsorption onto ion-exchange resins, does not involve the protein moiety of the enzyme. No desorption of enzyme from the complex occurred between pH 3.5 and 7.4 (38,39).

F. Protein Adsorption for use in Nonaqueous Systems

Once a protein is adsorbed to a support, from an aqueous system, substantial binding forces are not required to prevent migration from the immobilized state if the bulk medium to be used is a solvent (e.g., various nonpolar organic solvents) for which the protein has little affinity. One specific interest in these systems has been in the use of lipases. The commonest immobilization methods used for lipases have been the coating of the surfaces of macroporous inorganic particles such as flux calcined diatomaceous earths with a layer of enzyme. Solvent precipitation has been used to achieve coating (40), as has controlled drying of an excess of support particles with concentrated enzyme solution (41). Alternatively, lipases can be adsorbed onto weak anion-exchange resins (42). A procedure for the preparation of lipase for use in nonaqueous systems (42b) is outlined below.

Materials: A solution of lipase containing approximately 700 IU/ml. Celite. Acetone.

Methods: To 40 ml of enzyme solution at 0°C add 10 g of Celite. Mix and add, over a period of 5 min, 60 ml of ice-cold acetone. Stir for 30 min, during which time the temperature is maintained at 0°C. The solid product that forms, i.e., the immobilized lipase, is removed by filtration and dried at 20°C under reduced pressure. The preparation is then ready for use.

III. ENTRAPMENT

While entrapment of enzymes and proteins in various polymers has received some attention, the technique has limited application for immobilizing macromolecules. The matrices commonly used have lattice dimensions that allow of molecular migration relatively easily. Thus, to maximize enzyme entrapment and minimize leakage, a high degree of polymer cross-linking is required. This extensive cross-linking limits the diffusion of substrates and products. For example, enzymes so entrapped cannot interact with subtrates such as cellulose or DNA. Nevertheless, while diffusional problems are very evident in gel entrapped systems, the use of fiber-spinning technology has enabled the use of cellulose triacetate fiber as an entrapping medium. Indeed, glucose isomerase so immobilized has had industrial application (8).

A. Entrapment in Polyacrylamide Gel

Polyacrylamide is the most commonly used gel matrix for entrapment of enzymes. Since it is uncharged, the pH profile characteristics of the immobilized enzyme are much the same as those of the free enzyme. Charged substrates and products do not accumulate, nor are they depleted in the matrix. On the other hand, the failure of the matrix to interact with the entrapped protein does little to prevent leakage. Consequently, generation of a high degree of cross-linking is necessary to obviate this problem. Thus, polymerization must be efficient (an oxygen-free medium is the most important factor in this context). Unfortunately, the use of highly cross-linked matrices may lead to diffusional problems, especially when higher molecular weight substrates are concerned. Two alternate strategies can be used to minimize this problem. If a single enzyme activity is involved it may be possible to immobilize nonviable cells exhibiting the required activity (43). This would allow the use of less highly cross-linked matrices. Alternatively, the size of the immobilized protein can be enlarged by attachment to a support such as Sepharose. This technique has been used for co-immobilization of enzymes and cells in macroporous alginate systems (44,45). A number of procedures for the entrapment of enzymes in polyacrylamide gels have been published. The process described below was designed for immobilization of urease (46). To immobilize other enzymes and proteins it may be necessary to modify conditions to meet specific requirements.

Materials: Enzyme powder or concentrated solution. Buffer, pH 7.0, consisting of 0.1 mM EDTA and 0.1 M Tris-HCl. Buffer, pH 7.0, consisting of 0.1 mM EDTA, 0.5 M NaCl, and 0.1 M Tris-HCl. N,N^1-methylene-bis-acrylamide. Acrylamide. Dimethylaminopropionitrile. Freshly prepared potassium persulfate (10 mg/ml).

Method: Dissolve 110 mg of methylene-bis-acrylamide and 10 mg of acrylamide in 10 ml of 0.1 mM EDTA, 0.1 M Tris-HCl buffer, pH 7.0, and keep in a stoppered bottle. Dissolve 15 mg of enzyme in this solution. Purge solution with nitrogen for at least 20 min. To the oxygen-free solution add 0.2 ml of dimethylaminopropionitrile and mix gently. Then add 0.5 ml of potassium persulfate (10 mg/ml) and mix again. Leave in a stoppered flask at room temperature for 20 min. An opaque gel will form. Disrupt the gel by shaking the flask and then by passing the contents through a fine-bore syringe needle. Wash the gel with 50 ml of 0.1 mM EDTA, 0.5 M NaCl, 0.1 M Tris-HCl buffer, pH 7.0. The washing procedure is repeated at least three times or until significant amounts of enzyme activity are no longer detectable in the wash solution. The enzyme-containing gel is resuspended in 15 ml of 0.1 mM EDTA, 0.1 M Tris-HCl buffer, pH 7.0, and is ready for use.

B. Fiber Entrapment

Various techniques have been developed for the physical entrapment of enzymes within the microcavities of porous fibers (47–49). Enzyme entrapment is achieved by using a water-immiscible organic solvent to dissolve the fiber-forming polymer. These solutions are emulsified with an aqueous solution containing the enzyme, and the final emulsion is extruded through fine holes into a liquid coagulant that precipitates the polymer into a filamentous form. These processes are similar to those used in the manufacture of manmade fibers. The fibers which can be of any length may be collected into bundles. This gives considerable flexibility to the design of reactors that can be employed. The individual filaments consist of a macroscopically homogeneous dispersion of small droplets of aqueous solution in a porous polymer gel. Pore size can be varied so as to allow of the free diffusion of low molecular weight compounds, while preventing the escape of high molecular weight enzyme molecules. A range of polymers, including cellulose, ethyl cellulose, nitrocellulose, and poly-γ-methyl-L-glutamate, have been used for these purposes. However, greatest interest is currently being shown in cellulose triacetate because of its low cost, good biological resistance, and the fact that the fibers are unaffected by dilute solutions of weak acids and are resistant to many common solvents. The process for production of cellulose triacetate fiber entrapped enzyme is described. Other fiber-forming materials can be used in a similar manner.

Materials: Methylene chloride. Cellulose triacetate. Concentrated protein solution. Toluene.

Method: Mix 2.1 ml of methylene chloride and 200 mg of cellulose triacetate in a round-bottom flask. Stir at room temperature until complete solution is obtained. Cool to 0°C and add, dropwise, 0.4 g of concentrated enzyme (e.g., invertase) solution. Increase the rate of stirring to approx. 900 rpm and continue stirring for about 5 min so as to form an emulsion of an aqueous enzyme solution dispersed in the polymer phase. The emulsion should consist of a uniform distribution of aqueous droplets not greater than 2 μm in diameter. This can be checked by microscopic examination. Allow the emulsion to stand for 20 min at 0°C to permit air bubbles trapped during emulsification to escape. Then, using nitrogen pressure (0.7 atms), the emulsion is pumped to a spinneret (internal diameter 80 μm) immersed in a coagulating bath containing toluene. As soon as the emulsion flows from the spinneret, the coagulating polymer is wound onto a driving roll rotating at 60 rpm. The filament is wound continuously until all of the emulsion is extruded. The fibers are collected from the roll and allowed to stand in air to remove organic solvents. After spinning, the spinneret and other parts of the apparatus are washed in methylene chloride. A number of reactors for use of fiber-entrapped enzymes have been devised (50). For industrial application slurry reactors, in which the fibers are cut into small pieces 1–3 cm in length and maintained in the reaction mixture by mechanical stirring, are considered most convenient. Hydrophilic and hydrophobic fibers can be prepared. For example, the ability of cellulosic fibers to absorb water is gradually reduced as the degree of acetylation of the hydroxyl groups on each anhydroglucose unit is increased.

The catalytic properties of fiber-entrapped enzymes generally differ from those of the corresponding free enzyme. These differences are primarily ascribed to the restricted diffusion of substrates and products inside the porous enzyme support (50). Immobilized enzyme activities of 30% of those of the free enzyme have been observed. However, the density of enzyme loading affects observed efficiency. A decrease in efficiency on increased loading results from limitation of substrate and product diffusion. This effect is highlighted at the pH value at which activity is maximal. pH profiles are not affected unless the reaction products include H^+ or OH^- whose diffusion from the microenvironment of the enzyme is limited.

C. Entrapment in Alginate

The formation of calcium alginate gels allows of immobilization of biological materials under mild conditions. This technology has been used extensively for the immobilization of living cells (51), of nonviable cells in which the activity of a single enzyme is desired (52), and of enzymes (53). Leakage of enzyme from gels can be minimized by increasing alginate and calcium concentrations (53). The effects of pore size on the properties of calcium alginate beads and on the diffusion characteristics of substrates in such beads have been investigated (54,55). While calcium is the cation most commonly used with alginates, other di- and trivalent cations may be used and, it has been claimed, gels with altered properties produced (56,57). Entrapment within alginate gels may also be used for the co-immobilization of living cells and enzymes (38,44,45). In such cases it is essential that the enzyme be "enlarged" by covalent linkage to a support such as Sepharose so as to prevent its loss through pores that readily prevent leakage of cells. The technique described below represents a standard procedure for the entrapment of enzyme within calcium alginate. Before use, however, the beads should be washed repeatedly so as to remove enzyme bound loosely to the surface of the beads (the washings should be assayed for the relevant activity). Furthermore, while the conditions described efficiently entrapped an enzyme with a molecular weight in excess of 100,000, higher or lower concentrations of both calcium and alginate should be tried so as to optimize the system for different enzymes.

It should be noted that conditions that facilitate the effectiveness of enzyme immobilization may also decrease the diffusion of substrates and products. It should also be noted that enzymes that are inactivated or inhibited by calcium cannot be immobilized by this technique and that the presence of phosphates would disrupt calcium alginate gels.

Materials: Sodium alginate solution (4%, w/v). Calcium chloride (0.5 M). Enzyme solution (20 mg/ml).

Method: Mix equal volumes of the sodium alginate and enzyme solutions. Extrude the mixture, dropwise, into an excess of 0.5 M calcium chloride. Leave the beads of calcium alginate-entrapped enzyme to harden in the calcium chloride solution for 20 min. Remove the beads and wash twice with 0.5 M calcium chloride. Assay the washings and the beads for the enzyme activity in question. Various labor-saving procedures for the preparation of beads of calcium alginate-entrapped biological materials in large quantities and of uniform defined sizes have been published recently (58–61). Modification of beads, e.g., by the inclusion of magnetite (62), are options that may be worth trying.

IV. MICROENCAPSULATION AND ALLIED TECHNIQUES

Microencapsulation, entrapment in liposomes, and in hollow fibers are considered together since proteins immobilized by these techniques are confined in an environment that is often comparable in volume to the bulk phase being studied.

A. Microencapsulation within Semi-Permeable Membranes

Microencapsulated proteins may be thought of as being enveloped within a semi-permeable membrane from which they cannot leak. By contrast, small molecules may readily cross into or out of the microscapsule and interact with the encapsulated protein. A range of membrane materials including nylon, epoxy resins, butyl rubber, and cellulose nitrate have been used (63–65) (see Table 1). Moreover, the technique has the potential for immobilization of mixtures of active substances in an environment in which they can act in a predetermined sequence. Materials that have been microencapsulated range from hemoglobin to the entire cell contents of erythrocytes. However, practical applications of these systems have not been utilized. Microencapsulation is nevertheless used for the slow release of oils, perfumes, dyes, and hormones.

By comparison with the conditions required for encapsulation of nonbiological materials, those for microencapsulation of enzymes etc. must be carefully controlled (12). There are in fact three basic procedures for the encapsulation of enzymes. These are coacervation (or phase separation), interfacial polymerization, and liquid drying. In the coacervation or phase separation process, an aqueous solution of enzyme is dispersed in a water-immiscible organic solvent containing the polymer in question. To this mixture a second water-immiscible organic solvent is added with stirring. Phase separation of colloidal particles of the polymer which then takes place causes them to associate around the small aqueous droplets. A continuous membrane is formed by coalescence of the droplets. This occurs when cellulose nitrate, cellulose acetate, or butyl rubber is used as the polymer. Chibata (12) points out that it is important that the polymer should separate as a phase of concentrated solution but should not precipitate. If precipitation occurs, the enzyme will not be encapsulated.

A common example of interfacial polymerization is the formation of nylon. In the context of enzyme immobilization by this technique, an aqueous solution of enzyme is emulsified with a hydrophilic monomer in a water-immiscible organic solvent. A hydrophobic monomer solution dissolved in the same water-immiscible organic solvent is then added and the mixture is stirred. The polymerization reaction that occurs at the interface between the aqueous and organic phases results in enzyme being encapsulated by a membrane of polymer. In this procedure the size of the capsules can be varied by adjustment of stirring rates. However, while the time required for encapsulation is short, the combination of monomers used should be those that effect least inactivation of the enzyme and that are most suited to the intended application of the final preparation.

In the liquid drying procedure for encapsulation, a suitable polymer (e.g., polystyrene) is dissolved in a water-immiscible solvent (e.g., benzene, chloroform) with a boiling point lower than water. An aqueous solution of the protein to be encapsulated is then dispersed in the organic phase to form an emulsion. This emulsion is then dispersed in an aqueous phase containing protective colloidal substances, such as gelatin, to form a second emulsion. While the emulsion is being stirred, the organic solvent is removed by warming in vacuo. During this process the enzyme is encapsulated by a membrane of the polymer. The size of the capsules can be varied by altering the concentration of polymer or the type of protective colloid used or by adjusting stirrer speeds. A procedure for the encapsulation of catalase within butyl rubber by the liquid drying method is described below.

Materials: Mixture A, 20 g Span 85 (ICI) in 25 ml cyclohexane. Mixture B, 1 ml (650 IU) of catalase in 50 mM phosphate buffer, pH 7.0. Mixture C, 0.1 g of butyl rubber in 50 ml of cyclohexane—stir for 1–2 h for complete dissolution. Mixture D, 15 ml cyclohexane plus 10 ml of vegetable oil. Tween 20 and buffer (in this case 50 mM phosphate, pH 7) appropriate to the protein to be immobilized are also required.

Method: Add Mixture A to Mixture B with continuous stirring. Then, while continuing with stirring, add Mixture C. Finally add Mixture D. While continuing to stir the mixture the cyclohexane is slowly evaporated in vacuo (26 mmHg). Collect the microcapsules by centrifuging the oil phase. Resuspend in 15 ml of 50 mM phosphate buffer, pH 7. Centrifuge again and discard the supernatant. Repeat the washing procedure several times. Finally, resuspend the microcapsules in the buffer solution containing Tween 20 so as to keep the particles separate.

The techniques involved in microencapsulation are relatively mild. However, low recoveries of activity have been recorded. The major factor affecting the observed activity is that of mass transfer resistance around the membrane. This includes limitation of diffusion rates across the membrane and through stagnant boundary layers, inside of and external to the microcapsule. In addition, mixing within the microcapsule can also be important (66–70). All of the mass transfer resistances have been combined into a unified model for microcapsules and applied to microencapsulated urease (71).

B. Encapsulation within Liposomes

Liposomes are assemblages of phospholipids and other lipids sustaining a biomolecular configuration and not requiring mechanical support for their stability (72). They are formed when a water-insoluble polar lipid, such as phosphatidylcholine, is confronted with water and undergoes a sequence of conglomerations. In fact, the highly ordered structures that emerge and persist in the presence of excess of water lead to further arrangement of the lipid molecules. The final structure is a system of concentric closed membranes, each of which is an unbroken sheet of molecules. During this period of arrangement of the lipid molecules, protein or other materials in the aqueous solution can be entrapped within the liposomes. However, it is also possible that hydrophobic regions of the protein finally reside in a lipid phase of the liposome or are adsorbed on the exterior of the structure. Multilamellar, monolamellar, or macro-vesicular liposomes can be formed with central aqueous cores, the dimensions of which are variable. Aqueous cores of multilamellar liposomes have a diameter of about 0.15 μm. The volume of the entrapped water is governed by the equation, $V_{H_2O} = k$, where V_{H_2O} is the volume of the aqueous phase within the liposomes and k is the ratio of the charge of the inner or outer surfaces of the lipid bilayer to the ionic strength of the medium used in the formation of the liposomes (73). Thus, in order to increase the volume of entrapped water, media of low ionic strength and/or incorporation of charge into the lipid structure is necessary. Liposome-entrapped proteins have enjoyed considerable success with respect to therapeutic use. In some cases the protein in these preparations exerts its effect in the circulatory system. However, suitably designed liposome preparations have also been used for delivery of proteins to specific tissues (see Refs. 74–78). Experimental conditions need to be controlled so as to ensure the protein is suitably entrapped, that it is able to act as required, and that immunological reactivity is minimized (74–76).

The general process for preparation of protein-containing liposomes consists of dissolving lipid in organic solvent with subsequent elimination of solvent under reduced pressure. This leaves a thin lipid layer on the walls of the container. This is dispersed with the protein solution. Liposomes containing the protein form spontaneously. Prior to their separation from nonentrapped protein, liposomes can be reduced in size to monolamellar structures by sonication. In order to avoid oxidation of unsaturated lipids oxygen-free argon is used for flushing and purging the operating system, and the lipids are stored in organic solvents in sealed ampules at −30°C under argon. Glassware should be cleaned with detergents and thoroughly washed with tap and distilled water before use.

Materials: Chloroform. Phospholipid (e.g., egg lecithin). Sterol (e.g., cholesterol). Charged amphiphile (e.g., diacetyl phosphate). Protein solution (e.g., neuraminidase in 3 mM sodium phosphate buffer, pH 7.2). 3 mM sodium phosphate buffer, pH 7.2.

Method: Add to an argon-purged flask 40 μmol of phospholipid, 11.4 μmol of sterol, and 5.7 μmol of charged amphiphile in 2.5 ml of chloroform. Remove the solvent in a rotary evaporator under reduced pressure at about 37°C until a thin film is visible on the walls of the flask. As it is essential to remove all traces of solvent, continue evaporation for a further 10 min. Remove the flask and flush with argon. Add 1.5 ml of protein solution. Stopper the flask and shake gently until the lipid film no longer adheres to the walls of the flask (if necessary add clean glass beads to facilitate this process). Keep the milky suspension at room temperature (or at 4°C if required for protein stability) for 1 h for completion of liposomal formation. The size of the liposomes so formed may, if necessary, be reduced by sonication. Separate the liposomes from the nonentrapped enzyme by centrifugation and repeated washing in buffer. A wash with 1% (w/v) saline should be interspersed between buffer washes.

Lecithin is the most common lipid component used, but synthetic lecithins, because of their higher purity, have been used for work with human patients (77). Charged amphiphiles are included to improve the water content of the liposomes (78). Liposome systems have been recently been used for the delayed release of enzymes in cheese-ripening processes (79).

C. Encapsulation within Hollow Fiber Membranes

Hollow fiber semi-permeable membranes provide a ready means for the physical immobilization of protein. The technique is simple and chemical reactions are not required. The membranes confine macromolecules but allow small molecules to pass through. The major problem of such systems is membrane fouling. Laboratory scale hollow fiber devices are available. They consist of bundles of hollow fibers with their open ends fitted into plugs and their surfaces immersed in a reaction solution. While a range of membrane materials have been used, the most typical include polysulfones, cellulose acetate, and acrylic copolymers (67). These membranes can be prepared with a range of molecular weight cut-off values from as low as 200 to as high as 100,000 daltons. The systems are used by the introduction of the enzyme or protein solution into the lumen of the fibers while the reactants are introduced into the reaction container. Various enzymes have been immobilized in such systems and alternative procedures include circulation of the enzyme through a closed loop which includes the hollow fibers (81–85) (see Table 1).

Two problems are commonly associated with hollow fiber techniques. In the first place the fibers may break during handling. A single fiber, if broken in a recycling system, can result in substantial enzyme loss. Second, concentration polarization occurs at the membrane surface. This results in decreased apparent activity (86,87). Operations have been varied to include immobilization of enzymes onto polymeric support for circulation through a fiber reactor. This allows the use of membranes of greater pore size and may confer greater stability on the immobilized preparation (80,88).

V. TRANSITION METAL CHELATION

Methods of protein immobilization in which transition metals are used have been developed (89). A wide range of such metals may be used, but titanium and zirconium are favored because their oxides are not toxic. The process, using these transition metal salts, is based on the formation of a link between a protein and a carrier such as cellulose or glass. Linkage involves initial chelation of vicinal diol groups in support media by transition metal ions, the chelate being formed by replacement of two ligands of the transition metal ion by hydroxyl groups of the supports. For steric reasons it is impossible that all the water or chloride ligands on the transision metal ions be replaced by hydroxyl groups of the support media. Therefore, chelation of any molecule containing groups appropriate for replacement of the remaining ligands of the support-linked transition metal bound can be achieved. The incoming molecule should be in aqueous solution. At neutral pH coupling will occur. The groups on proteins that can act as ligands are the free side chain groups, carboxyl groups of acidic amino acid residues, the phenolic hydroxyl groups of tyrosyl residues, the alcoholic hydroxy groups of seryl and threonyl residues, free sulfhydryl groups of cysteine residues, and the ε-amino groups of lysyl residues. The formation of the immobilized complexes depends on the availability of the required groups at sterically favorable positions. With respect to enzyme activity, it is obviously essential that such sterically available groups not be at the active site. The bonds formed in these complexes are not ionic as successful immobilization can be achieved in ionic media. It should also be mentioned that while this technique is relatively simple, results obtained are not always reproducible (90,91). Immobilization via metal link chelation is achieved in two stages: (1) activation of the support and (2) coupling of the enzyme.

Materials: Microcrystalline cellulose. Titanium (IV) chloride. HCl. Enzyme solution. Buffer appropriate to the protein to be immobilized and the same buffer solution containing 1 M NaCl.

Method: To 10 g of microcrystalline cellulose add 50 ml of a 15% (w/v) titanium (IV) chloride solution in 15% (w/v) HCl and mix. Place the mixture in a desiccator containing sodium hydroxide and evacuate. Place the evacuated mixture in an air oven at 45°C until completely dry (24 h). Wash the solid three times with suitable buffer (the pH of which is usually optimum for enzyme activity) until the solid is off-white in color. To the washed activated support add a solution of the protein to be immobilized (1 g of protein per 10 g of cellulose) and stir at 4°C for 18 h. Centrifuge the suspension and wash the immobilized protein with two cycles of buffer and buffer/NaCl solution. Finally wash twice with buffer. Resuspend the immobilized protein in buffer and store at 4°C.

The retention of specific activity of transition metal-chelated enzymes is reported to range from 4% to 80%, although typically about 50% is obtained. The pH optimum is usually largely unaltered, though shifts of 0.5–1.0 pH units in both acid and alkaline directions have been noted (92–94). Alternative supports have been used and to enhance stability glutaraldehyde cross-linking has been applied (89).

VI. AGGREGATION AND OTHER TECHNIQUES

Proteins may also be immobilized by aggregation. This is usually achieved by chemical means using glutaraldehyde or other cross-linking reagents (95). Noncovalently aggregated systems are less well known and usually involve nonviable cells containing the desired enzyme activity. In these cases the active species is assumed to be physically entrapped within or adsorbed onto the cell surface. An example in point is the aggregation using polyelectrolyte flocculating agents of *Arthrobacter* sp. containing glucose isomerase (96,97). Spray drying of cells has also been used to immobilize glucose isomerase activity for industrial purposes (7,98,99). While large-scale applications of glucose isomerase have attracted considerable attention, it is of interest to note that a major part of the current market for such preparations is supplied by a pelletizing operation using a range of general binding agents (100a,b, 101).

VII. APPLICATIONS OF NONCOVALENTLY IMMOBILIZED PROTEINS

Immobilized enzyme technology may be described as technological answers in search of commercial problems. The volume of relevant published material more than adequately supports the view that any enzyme may be immobilized by any one of a range of methods and on any one of a variety of supports. Initially the use of immobilized enzymes was expected to provide financial savings because of their potential for repeated utilization. This reasoning does not necessarily hold true any longer. What changes have taken place? First, enzyme production procedures have become more efficient so that the unit cost of enzymes has fallen considerably. Second, immobilization procedures have not dramatically improved the stability of enzymes generally. There is little virtue in immobilizing labile enzymes if the immobilized preparation is also labile in operation. This is not to say that immobilized enzyme technology has been without success (see below). Among the advantages of such preparations over soluble enzymes is that their action, whether in industrial processes or clinical analyses, is more readily controlled, cleaner products are obtained and smaller volume reactors are needed. A specially commissioned working party within the European Federation of Biotechnology has provided guidelines for the characterization of immobilized biological preparations (444). In contemplating any new process, one may have to choose between traditional chemical or modern enzyme-catalyzed technologies, and whether, if the latter is chosen, the use of immobilized preparations can be justified. Not easy decisions, perhaps, but ones that might be made more easily if the strategies recommended by Boyce (445) are considered.

Applications of immobilized biological preparations generally are dealt with in detail elsewhere in this volume. In this chapter we are restricted to immobilization by noncovalent means. It must be borne in mind, as stated in the introduction, that important applications of such preparations are few, having being superceded by covalently immobilized proteins on the one hand and by entrapped or encapsulated whole cells on the other. However, in some areas the use of adsorbed, entrapped, and encapsulated proteins has made its mark.

Clinical analysis is one such area that immediately comes to mind. The most simple analytical devices utilizing noncovalently immobilized enzymes are the dipsticks used by physicians throughout the world for determination of analyte concentrations in serum or urine samples. These are essentially strips of filter paper containing adsorbed enzymes and other reagents pertinent to the analyte in question. They are dipped in the serum or urine sample and the color that develops is compared with those on the standard chart provided. One such dipstick contains adsorbed glucose oxidase, peroxidase, and a suitable chromogen. Glucose present in the sample is determined by the amount of color produced as a result of oxidation of the chromogen as indicated in Eq. 1 below.

$$\text{Glucose} + O_2 \xrightarrow{\text{glucose oxidase}} \text{gluconic acid} + H_2O_2$$

$$H_2O_2 + \text{reduced chromogen} \xrightarrow{\text{peroxidase}} H_2O + \text{oxidized chromogen} \quad (1)$$

Enzymes have also been used in the form of immobilized enzyme reactors for automated analysis. The ready acceptance of such devices stemmed from the fact that they could readily be incorporated into existing instrumentation. While it is true that most such preparations are comprised of covalently linked enzymes, reactors consisting of polyacrylamide-entrapped glucose oxidase and lactate dehydrogenase have been used for the automated determination of the concentrations of glucose and lactate, respectively, in biological fluids (409). Similarly, acetylcholinesterase (also known as cholinesterase) entrapped in polyacrylamide (309,437,438), in silastic resin (115,309,312), or in starch (309,331,379,381,386) has been used for determination of acetylcholine; urate oxidase adsorbed to protamine-aminohexyl controlled-pore glass (287) has been used for measurement of uric acid concentration.

In recent years, biosensors have become fashionable. These devices have been defined as "analytical tools comprised of an immobilised biological material in intimate contact with a suitable transducer that converts the biochemical signal into a quantifiable electrical signal" (446). The biological component of these devices include enzymes, proteins such as receptors and antibodies, whole cells, and tissue slices. The combination of a biocomponent that reacts specifically with the analyte in question and any one of several possible transducers allows the detection and measurement of a wide variety of analytes (447). Hence such systems are finding, or are expected to find, considerable use in clincial analysis and in analyses carried out in the food, drink, dairy, and environmental sectors. The best known biosensors are the so-called "enzyme-electrodes" (see Ref. 185). Examples of enzyme electrodes, in which the enzyme or combination of enzymes (see Table 1) is noncovalently immobilized adjacent to or, more efficiently, on the transducer include those for the measurement of amino acids (175), asparagine (175), cholesterol (179), creatinine (298), ethanol (217), galactose (280), glucose (284,296,297), glutamic acid (182), glutamine (174), lysine (208), peroxide (178), penicillin (114), oxalate (244), phenylalanine (173), pyruvate (288), salicylate (224), sucrose (180), tyrosine (181), urea (176,181,358,360), and uric acid (287).

Immunosensors, in which the biocomponent is an antibody or antigen, may also be used for analytical purposes. Indeed, the principle of antibody/antigen reactions has for some time been used in straightforward ELISA techniques. One example of the latter is that for the measurement of progesterone (448). In this case immunoglobulin G was noncovalently adsorbed to the walls of wells in polystyrene microtiter plates. nonadsorbed protein was then removed by washing and horseradish peroxidase-labeled progesterone and samples of saliva or plasma were then added, mixed, and incubated. After incubation the wells were emptied and washed three times. Substrate and chromogen, appropriate to the enzyme label (in this case peroxidase), were then added and the plates were again incubated. The reaction was stopped by the addition of 4 N sulfuric acid and absorbance at 490 nm determined. Comparison with appropriate standards gives the progesterone concentration in the samples used. Since the assay is based on competition between peroxidase-labeled progesterone and unlabeled sample progesterone, the greater the concentration of the latter, the less of the former will be bound.

Immobilized enzymes have considerable potential for therapeutic use (see Ref. 52). In theory, such preparations, when implanted in the recipient, could replace a genetically defective or missing enzyme and by so doing convert the substrate in question to the required product. Unfortunately, it is not always possible to get the enzyme to the site at which its activity is required. Gregoriadis and colleagues (74–78, 156), by judicious choice of lipid composition, have had some success in preparing site-specific or "homing" liposome-entrapped preparations. Among the other problems facing the therapeutic use of immobilized enzyme preparations are the fact that activity may be lost as a result of proteolysis or antigen/antibody reaction. The first problem can be minimized by using microencapsulated preparations in which the pore size allows substrates and products to diffuse freely but prevents leakage of the encapsulated enzyme or access by proteases. Stimulation of antibody production or clot formation, a serious problem (to the patient and the preparation), may be minimized by using nonimmunogenic and nonthrombogenic materials (heparin-coated membranes) (332) to form microcapsules or by using extracorporeal shunt systems (319–323) instead of implants. Microencapsulated enzyme preparations, whether for implantation or extracorporeal shunt use, have been used with some success, at least in the short term. Asparaginase, encapsulated in nylon, has been used for the treatment of asparagine-requiring lymphosarcoma (134), while encapsulated phenylalanine ammonia lyase has been used to reduce circulating phenylalanine concentrations in cases of phenylketonuria (243,253). The use of immobilized arginase in the in vivo treatment of murine leukemia has been evaluated (54). Implantable cytosine deaminase encapsulated in cellulose tubes may be useful for deamination of fluorocytosine (249), and catalase encapsulated in collodion was of benefit to acatalasemic mice (143).

Cellulose, the major constituent of biomass, is available in vast quantities in primary and waste sources, and about 10% of the total is replenished each year as a result of photosynthesis (432). Biomass (hemicellulose and lignin are the other major constituents) is therefore a huge, renewable potential source of food, fuel, and chemical feedstocks—and one that is in receipt of increasing attention as fossil fuels are being depleted and politicization of oil supplies drives prices to unrealistic levels. The problem with respect fo realization of this potential is that cellulose is crystalline and is coated with lignin. Both of these factors restrict acid or enzymic hydrolysis of the substrate, and until they are solved no real industry based on biomass utilization will be realized. Nevertheless, many investigators continue to examine ways and means of improving the yield of cellulases by suitable organisms and to increase the efficiency of the cellulolytic process. In this context, one notes that cellulases immobilized on various soluble supports (225), entrapped in ultrafiltration reactors (148), and/or biospecifically adsorbed to the substrate itself (148,335) have been used for the continuous conversion of cellulose to glucose. A feature of these reactors is that they allow of the continuous removal of glucose and so obviate product inhibition. The effectiveness of noncovalently immobilized β-glucosidase in converting cellobiose, a product of cellulose hydrolysis, to glucose has been demonstrated (194,292). Moreover, by coentrapment of *Zymomonas mobilis* or *Saccharomyces cerevisiae* with β-glucosidase in calcium alginate gels (38,44,45), the cellobiose may be converted to ethanol as it is formed during cellulolysis.

For those in the dairy industry disposal of whey can be a problem. Dumping it in rivers, unforunately an all-too-common practice, is not designed to win environmentalist-of-the-year awards. In any event, it wastes a valuable byproduct. it would make much more sense economically to hydrolyze the lactose in the whey to the sweeter and more valuable mixture of glucose and galactose. While β-galactosidase adsorbed to ion-exchange resins (193,261) or to polygalacturonate (291) or entrapped in polyacrylamide (149) or cellulose triacetate (153) have been found to be effective in this regard, any of the immobilized β-galactosidase preparations listed in Table 1 could theoretically be used for this process. Indeed, a completely automated pilot plant utilizing β-galactosidase adsorbed to ion-exchange resin in fixed bed reactors capable of processing 1000 liters of raw cheese whey per day is in operation (193). The Tate and Lyle process currently in operation consists of lactase adsorbed to charcoal and cross-linked with glutaraldehyde (202). The same preparations could also be used for converting milk lactose and so rendering milk or milk products tolerable to Asiatic or other lactase-deficient adults. It may be noted that hydrolyzed wheys are used as food syrups, brewing substrates, fermentation substrates, silage additives, and animal feeds. In these cases increasing sweetness is not the only reason for carrying out hydrolysis. Due to the relatively poor solubility of lactose, it is not possible to concentrate whey or whey permeates above 30% w/w. At this concentration lactose crystallizes and there are substantial storage problems in terms of microbial contamination. Hydrolysis allows concentration to >65% w/w. This greatly improves bacteriostatic properties and reduces the volume of material to be transported. Hydrolysis also minimizes the lactose intolerance evident in pigs fed "high-whey" diets. Substantial quantities of whey permeate are currently being hydrolyzed using soluble enzyme in one industrial operation in Ireland. The use of soluble enzyme is possible because of the efficiency of lactase production by continuous fermentation of permeate at its natural concentration. The operation of the plant is substantially more economical than if immobilized preparations were used and highlights the recent advances in enzyme production technologies that have eroded the cost advantages commonly associated with immobilized systems.

Cheese ripening, i.e., development of the appropriate flavor and texture, typically requires 4–15 months depending on the type of cheese and the market for which it is intended. The ripening process involves the selective but restricted hydrolysis of constituent protein. A recent report (79) valued British cheddar cheese production at £300 million and pointed out that storage (while ripening) costs amount to a significant £25 million. However, the AFRC Institute of Food research has shown that the addition of liposome-encapsulated endo- and exopeptidases to cheesemilk prior to curd formation can reduce markedly the time required for ripening. The liposomal membrane protects the casein from premature proteolysis and minimizes losses of enzyme to the whey fraction on curd separation. It appears that while the encapsulated enzymes are released in the curd matrix during the early days of storage, the inherent instability of the peptidases used prevented overripening. The bottom line of this endeavor was that the flavor and texture of 6- to 12-month-old cheese could be reproduced in 2–5 months by enzyme treatment and the product met with consumer approval. Similar work on cheese ripening using liposome-entrapped rulactine, a metalloproteinase from *Micrococcus caseolyticus,* is being carried out by INRA in France (286). Immobilized proteolytic enzymes have also found potential use in other areas of the food and drink industries. For example, trypsin has been used to prevent development of oxidized flavor in milk and papain has been used in chill-proofing of beers. In a different context, one notes that immobilized catalase may be useful for peroxide-sterilization of dairy products (141).

Investigations of the use of noncovalently immobilized lipases for the hydrolysis of olive oil triglycerides are of sufficient promise (196,271, 282,303) to suggest that such preparations may soon be used on a large scale. The use of enzyme technology has long been accepted by those in the food industry engaged in the production of fructose-rich syrups. Indeed, one might say that a whole new industry, involving many companies in many countries, devoted to this end has developed. The increasing cost of sugar, whether from sugar cane or sugar beet, and the abundance (in some countries) of D-glucose obtained by hydrolysis of corn starch provided the impetus for these endeavours. Glucose isomerase catalyzes the conversion of D-glucose to an equimolar mixture of D-glucose and D-fructose (see Eq. 2) with a concomitant increase in sweetness. In theory, noncovalently immobilized glucose isomerase preparations (see Ref. 82 and Table 1) could be used for the conversion process. In fact, glucose isomerase adsorbed to ion-exchange resins has been used in industrial processes in the past (7,245). In current practice, covalently bound preparations or entrapped whole cells exhibiting the required activity have proven to be more effective. One of the major reasons for the use of immobilized whole cells is that glucose isomerase is a heat-stable enzyme. Consequently, whole cell preparations exhibiting only the required activity are readily prepared by heat treatment.

$$\text{D-Glucose} \xrightarrow{\text{glucose isomerase}} \text{D-fructose} \tag{2}$$

$$\text{Sucrose} + H_2O \xrightarrow{\text{invertase}} \text{glucose} + \text{fructose} \tag{3}$$

It would seem that any of the noncovalently immobilized amyloglucosidase preparations listed in Table 1 could be used for starch hydrolysis and so provide the substrate for the glucose isomerase process. However, we are not aware of the use of such preparations on a large scale. Another route for the production of fructose-rich syrups is via hydrolysis of sucrose as outlined in Eq. 3. Indeed, as mentioned in the introduction, the first demonstration of the phenomenon of retention of enzyme activity on immobilization involved invertase and many laboratory-scale preparations have been investigated in the interim (see Table 1). However, the only large-scale operation of which we have knowledge is the 44.5 m^3 packed-bed reactor containing invertase (β-fructofuranosidase) adsorbed to granular bone charcoal (quoted in Ref. 144).

We end this chapter on a high note for noncovalently immobilized enzyme preparations with two examples of the large-scale and very successful large-scale applications. The first of these is the process for the production of L-amino acids developed by Chibata and colleagues (12) and operated by Tanabe Seiyaku Co. in Japan. One is reminded that chemically synthesized amino acids are optically inactive racemic mixtures of D- and L-isomers and that the L-isomer is the physiologically active form required for human and animal consumption and for medical use. To obtain the L-form, optical resolution of the mixture is necessary. This can be accomplished, as outlined in Eq. 4, by using aminoacylase. Various types of preparations of immobilized aminoacylase, covalently and noncovalently bound, were tested for the efficiency in this regard. Aminoacylase adsorbed to DEAE-Sephadex was found to be the most suitable because of ease of preparation. The residence time in the reactor was only 20–30 min, and conversion was close to 100% and the typical operational half-life was about 65 days. Moreover, regeneration was possible. L-lysine can be produced in a similar fashion by using caprolactam hydrolase and racemase adsorbed to DEAE-Sephadex (369).

$$\text{Acyl-DL-amino acid} + H_2O \xrightarrow{\text{aminoacylase}} \text{L-amino acid} + \text{D-amino acid} \quad (4)$$

(D-amino acid → racemization → Acyl-DL-amino acid)

The second example is that of penicillin acylase, which catalyzes the reaction shown in Eq. 5. This is of particular significance to the antibiotic producers, since one of the products, 6-amino penicillanic acid, is the substrate for the chemical synthesis of a host of modified penicillins.

$$\text{Benzyl penicillin} + H_2O \xrightarrow{\text{penicillin acylase}} \text{6-amino penicillanic acid} + \text{phenyl acetate} \quad (5)$$

While many companies are understandably loathe to divulge their secrets, it would seem that some, if not many, use penicillin acylase adsorbed to bentonite for this process. Indeed, Vandamme (396) states that, whereas the productivity of the soluble enzyme is 0.5–1 kilo of 6-APA/kilo of enzyme, that of the immobilized preparation is 100–200 kilos/kilo of enzyme. In 1980 the production of 6-APA amounted to 3000 tons. From this figure he calculated that approximately 15 to 30 tons of penicillin acylase were used. No doubt this pleases the enzyme producers.

ACKNOWLEDGMENT

We acknowledge with thanks the assistance of R. J. Fensom in the preparation of this manuscript.

REFERENCES

1. J. M. Nelson and E. G. Griffin, *J. Amer. Chem. Soc. 38:*1109 (1916).
2. J. M. Nelson and D. I. Hitchcock, *J. Amer. Chem. Soc. 43:*1956 (1921).
3. L. B. Barnett and H. B. Bull, *Biochim. Biophys. Acta 36:*244 (1959).
4. M. J. Bachler, G. W. Strandberg, and K. L. Smiley, *Biotechnol. Bioeng. 12:*85 (1970).
5. A. Y. Nikolaev and S. R. Mardashev, *Biokhimiya* (English trans.) *26:*565 (1961).
6. A. Y. Nikolaev and S. R. Mardashev, *Biokhimiya* (English trans.) *26:*641 (1961).
7. B. J. Schnyder, *Stärke 26:*409 (1974).
8. D. Dinelli, F. Morisi, S. Giovenco, and P. Pansolli, UK Patent No. 1,361,963 (1974).
9. A. R. Macrae, in *International Conference on Enzyme Engineering,* IBC Technical Services Ltd., Byfleet, U.K., 1986. ISBN 0-907822-96-7. See also *Biochem. Soc. Trans. 17:*1146 (1989).
10. I. Stone, U.S. Patent 2,717,852 (1955).
11. M. A. Mitz, *Science 123:*1076 (1956).
12. I. Chibata, (Ed.), *Immobilised Enzymes,* Wiley, New York, 1978.
13. J. Daniels and D. M. Farmer, UK Patent No. 2,070,022A (1981).
14. E. Sulkowski and M. Laskowski, Sr., *Biochem. Biophys. Res. Comm. 57:*463 (1974).
15. J. Woodward and G. S. Zachry, *Enzyme Microbiol. Technol. 4:*245 (1982).
16. A. D. McLaren, *Science 125:*697 (1957).
17. A. D. McLaren and E. F. Esterman, *Arch. Biochem. Biophys. 68:*157 (1957).
18. A. D. McLaren, *Enzymologia 21:*356 (1960).
19. L. Goldstein, *Biochemistry 11:*4072 (1972).
20. F. R. Bernath and W. R. Vieth, in *Immobilised Enzymes in Food and Micro-*

bial Processes, A. C. Olson and C. L. Cooney (Eds.), Plenum Press, New York, 1974, pp. 157–185.
21. T. Tosa, T. Mori, N. Fuse, and I. Chibata, *Enzymologia 32:*153 (1967).
22. O. R. Zaborsky, *Immobilized Enzymes,* CRC Press, Boca Raton, FL, 1973.
23. T. Tosa, T. Mori, N. Fuse, and I. Chibata, *Enzymologia 31:*214 (1966).
24. S. Usami and H. Shirasaki, *J. Ferment. Technol. 48:*506 (1970).
25. R. A. Messing, *Meth. Enzymol. 44:*148 (1976).
26. W. R. Vieth and K. Venkatsubramanian, *Meth. Enzymol. 44:*243 (1976).
27. K. Venkatsubramanian, R. Saini and W. R. Vieth, *J. Ferment. Technol. 52:*268 (1974).
28. M. L. Tanzer, *Science 180:*561 (1973).
29. W. R. Vieth and K. Venkatsubramanian, *Chem. Technol. 4(7):*434 (1974).
30. W. R. Vieth, S. S. Wang, S. G. Gilbert, and R. Saini, U.S. Patent No. 3,758,396 (1973).
31. A. Constantinides, W. R. Vieth, and P. M. Fernandes, *Mol. Cell. Biochem. 1:*127 (1973).
32. K. Venkatsubramanian, W. R. Vieth, and S. S. Wang, *J. Ferment. Technol. 50:*600 (1972).
33. S. S. Wang and W. R. Vieth, *Biotechnol. Bioeng. 15:*93 (1973).
34. K. Venkatsubramanian, W. R. Vieth, and F. R. Bernath, in *Enzyme Engineeering* Vol. 2, (E. K. Pye and L. B. Wingard, Jr., Eds.), Plenum Press, New York, 1974, p. 439.
35. Tate & Lyle Ltd., European Patent Application, EP 34,933, 2 Sept., 1981.
36. J. Woodward and A. Wiseman, in *Developments in Food Carbohydrate* Vol. 3, (C. K. Lee and M. G. Lindley, Eds.), Applied Science Publishers, Barking, U.K., 1982, p. 1.
37. J. Woodward and A. Wiseman, *Enzyme Microb. Technol. 4:*73 (1982).
38. J. M. Lee and J. Woodward, *Biotechnol. Bioeng. 25:*2441 (1983).
39. J. Woodward and D. L. Wohlpart, *J. Chem. Technol. Biotechnol. 32:*547 (1982).
40. R. A. Wisdom, P. Dunnill, M. D. Lilly and A. R. Macrae, *Enzyme Microb. Technol.* 6:443 (1984).
41. T. Matsuo, N. Sawamura, Y. Hashimoto, and W. Hashida, European Application No. EP 0,035,883 (1981).
42. T. T. Hansen and P. Eigtved in *Proc. World Conference on Emerging Technologies in the Fats and Oils Industry* (A. R. Baldwin, Ed.), American Oil Chemists, Champaign, IL, 1986, p. 365.
42b. M. H. Coleman and A. R. Macrae, UK Patent No. 1,577,933 (1966).
43. I. Chibata, T. Tosa, and T. Sato, Ger. Offen. 2,414,128 (1974).
44. M. P. J. Kierstan, A. McHale, and M. P. Coughlan, *Biotechnol. Bioeng. 24:*1461 (1982).
45. B. Hahn-Hägerdal, *Biotechnol. Bioeng. 26:*771 (1984).
46. M. D. Trevan and S. Grover, *Biochem. Soc. Trans. 7:*28 (1979).
47. D. Dinelli, *Process Biochem. 7:*9 (1973)

48. W. Marconi, S. Gulinelli, and F. Morisi, in *Insolubilised Enzymes* (M. Salmona, C. Saroni, and S. Garattini, Eds.), Raven Press, New York, 1974, p. 51.
49. D. Dinelli and F. Morisi, in *Enzyme Engineering,* Vol. 2, (E. K. Pye and L. B. Wingard, Jr., Eds.), Plenum Press, New York, 1974, p. 293.
50. D. Dinelli, W. Marconi, and F. Morisi, *Meth. Enzymol. 44:*227 (1976).
51. B. Mattiasson, in *Immobilized Cells and Organelles,* Vol. 1, (B. Mattiasson, Ed.), CRC Press Inc., Boca Raton, FL, 1983, p. 3.
52. A. Johansen and J. M. Flink, *Enzyme Microb. Technol. 8:*485 (1986).
53. M. P. J. Kierstan and C. Bucke, *Biotechnol. Bioeng. 19:*387 (1977).
54. J. Klein, J. Stock, and K.-D. Vorlop, *Eur. J. Appl. Microbiol. Biotechnol. 18:*86 (1983).
55. H. Tanaka, M. Matsumura, and I. A. Veliky, *Biotechnol. Bioeng. 26:*53 (1984).
56. F. Paul and P. M. Vignais, *Enzyme Microb. Technol. 2:*281 (1980).
57. W. E. Rochefort, T. Rehg, and P. C. Chau, *Biotechnol. Lett. 8:*115 (1986).
58. A. C. Hulst, J. Tramper, K. Van't Riet, and J. M. M. Westerbeek, *Biotechnol. Bioeng. 27:*870 (1985).
59. T. Rehg, C. Dorger, and P. C. Chau, *Biotechnol. Lett. 8:*111 (1986).
60. B. Larsen, PCT Patent Appl., WO 8603781.
61. U. Matulovic, D. Rasch, and F. Wagner, *Biotechnol. Lett. 8:*485 (1986).
62. G. I. Kvesitadze, D. J. Graves, and M. A. Burns, *Biotechnol. Bioeng. 27:*137 (1985).
63. T. M. S. Chang, in *Biomedical Applications of Immobilized Enzymes and Proteins* Vol. I, (T. M. S. Chang, Ed.), Plenum Press, New York, 1977, p. 69.
64. A. O. Mogensen and W. R. Vieth, *Biotechnol. Bioeng. 15:*467 (1973).
65. P. O'Grady and P. Joyce, *Enzyme Microb. Technol. 3:*149 (1981).
66. D. T. Wadiak and R. G. Carbonnell, *Biotechnol. Bioeng. 17:*1157 (1975).
67. A. Artmains, J. Neufeld, and T. M. S. Chang, *Enzyme Microb. Technol. 6:*135 (1984).
68. L. S. Nang and P. F. Carlier, in *Microencapsulation* (J. R. Nixon, Ed.), Marcel Dekker, New York, 1976, p. 185.
69. M. Takenaka, N. Kawashima, M. Saitoh, and R. Ishibashi in *Microencapsulation* (K. Tamutsu, Ed.), Techno Inc., Tokyo, 1979, p. 35.
70. D. N. Dean, M. J. Fuchs, J. M. Schaffer, and R. G. Carbonnell, *Ind. Eng. Chem. Fund. 16:*452 (1977).
71. W. R. Vieth, A. K. Mendiratta, A. O. Mogensen, R. Saini, and K. Venkatasubramanian, *Chem. Eng. Sci. 28:*1013 (1973).
72. A. D. Bangham, M. W. Hill, and N. G. A. Miller, in *Methods in Membrane Biology,* Vol. 1 (E. D. Korn, Ed.), Plenum Press, New York, 1974, p. 1.
73. D. Papahadjopoulos and N. Miller, *Biochim. Biophys. Acta 135:*624 (1967).
74. G. Gregoriadis, *FEBS Lett. 36:*292 (1973).
75. G. Gregoriadis and A. C. Allinson, *FEBS Lett. 45:*71 (1974).
76. G. Gregoriadis, D. Putman, L. Louis, and G. D. Neerunjun, *Biochem. J. 140:*323 (1974).

77. G. Gregoriadis, P. C. Swain, E. J. Wills, and A. S. Tavill, *Lancet 1*:1313 (1974).
78. G. Gregoriadis and E. D. Neerunjun, *Eur. J. Biochem. 47*:179 (1974).
79. Anon., *AFRC News,* Jan./Feb, 4 (1987).
80. R. P. Chambers, W. Cohen, and W. H. Baricos, *Meth. Enzymol. 44*:291 (1976).
81. J. R. Ford, Enzyme engineering. I. Characterisation of immobilized enzymes. II. Multienzyme reaction system, Ph.D. dissertation, Tulane University, New Orleans, LA *Diss. Abstr. Int. B. 33*:3624 (1973).
82. R. P. Chambers, J. R. Ford, J. H. Allender, W. H. Baricos, and W. Cohen, in *Enzyme Engineering,* Vol. 2, (E. K. Pye and L. B. Wingard, Eds.), Plenum Press, New York, 1974, p. 195.
83. P. R. Rony, *J. Amer. Chem. Soc. 94*:8247 (1972).
84. J. R. Ford, W. Cohen, and R. P. Chambers, *Meet. Am. Inst. Chem. Eng. Mich.* Paper 24a (1973).
85. J. C. Davis, *Biotechnol. Bioeng. 16*:1113 (1974).
86. M. C. Porter, in *Enzyme Engineering,* Vol. 1 (L. B. Wingard, Jr., Ed.), Wiley, New York, 1972, p. 115.
87. M. C. Porter and L. Nilsen, in *Recent Developments in Separation Science,* Vol. II (N. N. Li, Ed.), CRC Press, Cleveland, OH, 1972, p. 227.
88. S. P. O'Neill, J. R. Wykes, P. Dunnill, and M. D. Lilly, *Biotechnol. Bioeng. 13*:319 (1971).
89. J. F. Kennedy and J. M. S. Cabral, in *Immobilized Cells and Enzymes* (J. Woodward, Ed.), IRL Press, Oxford, 1985, p. 19.
90. J. P. Cardoso, M. F. Chaplin, A. N. Emery, J. F. Kennedy, and L. P. Revel-Chion, *J. Appl. Chem. Biotechnol. 28*:775 (1978).
91. J. F. Kennedy and P. M. Watts, *Carbohydr. Res. 32*:155 (1974).
92. S. A. Barker, A. N. Emery, and J. M. Novais, *Process Biochem. 6*(10):11 (1971).
93. A. N. Emery, J. S. Hough, J. M. Novais, and T. P. Lyons, *Chem. Eng. 258*:71 (1972).
94. C. A. Kent and A. N. Emery, *J. Appl. Chem. Biotechnol. 24*:663 (1974).
95. G. B. Broun, *Meth. Enzymol. 44*:263 (1976).
96. R. J. Reynolds Tobacco Co., UK Patent No., 1,368,650 (1974)
97. N. E. Lloyd, L. T. Lewis, R. M. Logan, and D. N. Patel, U.S. Patent No., 3,817,832 (1974).
98. Y. Takasaki, Y. Kasugi, and A. Kanbayashi, in *Fermentation Advances* (D. Perlman, Ed.), Academic Press, New York, 1969, p. 561.
99. B. L. Scallet, K. Shieh, I. Ehrenthal, and L. Slapshak, *Stärke 26*:405 (1974).
100a. T. K. Neilsen and E. K. Markussen, British Patent No. 1,361,387 (1974).
100b. T. K. Neilsen and E. K. Markussen, British Patent No. 1,362,365 (1974).
101. L. Zittan, P. B. Poulsen, and S. H. Hemmingsen, *Stärke 27*:236 (1975).
102. S. Usami, E. Hasegawa, and M. Karasawa, *Hakko Kyokaishi 33*:44 (1975).
103. H. Nakagawa, T. Arao, T., Matsuzawa, S. Ito, N. Ogura, and H. Takehana, *Agr. Biol. Chem. 39*:1 (1975).

104. S. Usami and Y. Kuratsu, *J. Ferm. Tech. 51:*789 (1973).
105. H. Ozawa, *J. Biochem. 62:*531 (1967).
106. H. D. Brown, A. B. Patel, and S. K. Chattopadhyay, *J. Biomed. Mater. Res. 2:*231 (1968).
107. K. Mosbach and R. Mosbach, *Acta Chem. Scand. 20:*2807 (1966).
108. J. Dobo, *Izotoptechnika* (Hung.) *16,* 215 (1973).
109. Å. Johansson, J. Lundberg, B. Mattiasson, and K. Mosbach, *Biochim. Biophys. Acta 304:*217 (1973).
110. G. J. Bartling, H. D. Brown, and S. K. Chattopadhyay, *Nature 243:*342 (1973).
111. T. Mori, T. Tosa, and I. Chibata, *Cancer Res. 34:*3066 (1974).
112. W. F. Shipe, G. Senyk, and H. H. Weetall, *J. Dairy Sci. 55:*647 (1972).
113. T. Mori, T. Sato, T. Tosa, and I. Chibata, *Enzymologia 43:*213 (1972).
114. G. J. Papariello, A. K. Mukherji, and C. M. Shearer, *Anal. Chem. 45:*790 (1973).
115. S. N. Pennington, H. D. Brown, A. B. Patel, and C. O. Knowles, *Biochim. Biophys. Acta 167:*479 (1968).
116. K. Mosbach and B. Mattiasson, *Acta Chem. Scand. 24:*2084 (1970).
117. P. Bernfeld, R. E. Bieber, and P. C. Macdonnell, *Arch. Biochem. Biophys. 127:*779 (1968).
118. P. Bernfeld and R. E. Bieber, *Arch. Biochem. Biophys. 131:*587 (1969).
119. T. Tosa, T. Sato, T. Mori, Y. Matsuo, and I. Chibata, *Biotechnol. Bioeng. 15:*69 (1973).
120. G. W. Strandberg and K. L. Smiley, *Appl. Microbiol. 21:*588 (1971).
121. T. Kasumi, K. Kawashima, and N. Tsumura, *J. Ferment. Technol. 52:*321 (1974).
122. H. Maeda, H. Suzuki, A. Yamauchi, and A. Sakimae, *Biotechnol. Bioeng. 17:*119 (1975).
123. S. Fukui, A. Tanaka, T. Iida, and E. Hasegawa, *FEBS Lett. 66:*179 (1976).
124. H. Maeda, H. Suzuki, A. Yamauchi, and A. Sakimae, *Biotechnol. Bioeng. 16:*1517 (1975).
125. H. Maeda, *Biotechnol. Bioeng. 17:*1571 (1975).
126. J. Campbell and T. M. S. Chang, *Biochem. Biophys. Res. Comm. 69:*562 (1976).
127. F. Leuschner, German Patent 1,227,855 (1966).
128. T. A. Butterworth, D. I. C. Wang, and A. J. Sinskey, *Biotechnol. Bioeng. 12:*615 (1970).
129. J. R. Wykes, P. Dunnill, and M. D. Lilly, *Biochim. Biophys. Acta 250:*522 (1971).
130. J. J. Marshall and W. J. Whelan, *Chem. Ind. 25:*701 (1971).
131. Y. Inada, S. Hirose, M. Okada, and H. Mihama, *Enzyme 20:*188 (1975).
132. T. M. S. Chang, *Enzyme 14:*95 (1973).
133. T. M. S. Chang, *Proc. Can. Fed. Biol. Soc. 12:*62 (1969).
134. T. M. S. Chang, *Nature 229:*117 (1971).
135. T. Mori, T. Sato, T. Matuo, T. Tosa, and I. Chibata, *Biotechnol. Bioeng. 14:*663 (1972).

136. T. Mori, T. Tosa, and I. Chibata, *Biochim. Biophys. Acta 321:*653 (1973).
137. T. Mori, R. Sano, Y. Iwasawa, T. Tosa, and I. Chibata, *J. Solid-Phase Biochem. 1:*15 (1976).
138. T. M. S. Chang, *Biochem. Biophys. Res. Comm. 44:*1531 (1971); correction *ibid. 45:*1367 (1971).
139. T. M. S. Chang and F. C. MacIntosh, *Pharmacologist 6* (Abstr. No. 190):198 (1964).
140. R. C. Bogulaski and A. M. Janik, *Biochim. Biophys. Acta 250:*266 (1971).
141. H. D. Chu, J. G. Leeder, and S. G. Gilbert, *J. Food Sci. 40:*641 (1975).
142. T. M. S. Chang, *Science 146:*524 (1964).
143. T. M. S. Chang and M. J. Poznansky, *Nature 218:*243 (1968).
144. P. Monson and D. Coombes, *Biotechnol. Bioeng. 26:*347 (1984).
145. M. J. Poznansky and T. M. S. Chang, *Proc. Can. Fed. Biol. Soc. 12:*54 (1969).
146. M. Kitajima, S. Miyano, and A. Kondo, *Kogyo Kagaku Zasshi 72:*493 (1969).
147. G. Broun, D. Thomas, G. Gellf, D. Domurado, A. M. Berjonneau, and C. Guillon, *Biotechnol. Bioeng. 15:*359 (1973).
148. T. K. Ghose and J. A. Kostick, *Biotechnol. Bioeng. 12:*921 (1970).
149. A. Dahlqvist, B. Mattiasson, and K. Mosbach, *Biotechnol. Bioeng. 15:*395 (1973).
150. W. R. Vieth, S. G. Gilbert, and S. S. Wang, *Biotechnol. Bioeng. Symp. 3:*285 (1972).
151. J. Jakubowski, J. R. Giacin, D. H. Kleyn, S. G. Gilbert, and J. G. Leeder, *J. Food Sci. 40:*467 (1975).
152. M. A. Paine and R. G. Carbonell, *Biotechnol. Bioeng. 17:*617 (1975).
153. L. E. Wierzbicki, V. H. Edwards, and F. V. Kosikowsky, *J. Food Sci. 38:*1070 (1973).
154. J. C. W. Østergaard and S. C. Martiny, *Biotechnol. Bioeng. 15:*561 (1973).
155. D. I. C. Wang and A. E. Humphrey, *Chem. Eng. 76:*108 (1969).
156. G. Gregoriadis, P. D. Leathwood, and B. E. Ryman, *FEBS Lett. 14:*95 (1971).
157. S. Giovenco, F. Morisi, and P. Pansolli, *FEBS Lett. 36:*57 (1973).
158. D. F. Tallon, J. P. Gosling, P. M. Buckley, M. M. Dooley, W. F. Cleere, E. M. O'Dwyer, and P. F. Fottrell *Clin. Chem. 30:*1507 (1984).
159. W. Marconi, S. Gulinelli, and F. Morisi, *Biotechnol. Bioeng. 16:*501 (1974).
160. M. Kubo, I. Karube, and S. Suzuki, *Biochem. Biophys. Res. Comm. 69:*731 (1976).
161. K. Venkatsubramanian, R. Saini, and W. R. Vieth, *J. Food. Sci. 40:*109 (1975).
162. Anon., *Chem. Eng. News 49*(39):31 (1971); see also D. L. Gardner and D. Emmerling, *Abstr. 162nd Meet. Am. Chem. Soc. Div. Organic Coatings Plastics Chem. 31:*366 (1971).
163. Y. Nakamoto, I. Karube, and S. Suzuki, *J. Ferment. Technol. 53:*595 (1975).
164. K. Venkatasubramanian and W. R. Vieth, *Biotechnol. Bioeng. 15:*583 (1973).

165. I. Karube, Y. Nakamoto, K. Namba, and S. Suzuki, *Biochim. Biophys. Acta 429:*975 (1976).
166. T. M. S. Chang, F. C. MacIntosh, and S. G. Mason, *Can. J. Physiol. Pharmacol. 44:*115 (1966)
167. J. P. Bollmeier and S. Middleman, *Biotechnol. Bioeng. 16:*859 (1974).
168. W. J. Blaedel, T. R. Kissel, and R. C. Boguslaski, *Anal. Chem. 44:*2030 (1972).
169. D. L. Gardner, R. D. Falb, B. C. Kim, and D. C. Emmerling, *Trans. Amer. Soc. Artif. Int. Organs 17:*239 (1971).
170. K. Kawashima and K. Umeda, *Agr. Biol. Chem. 40:*1151 (1976).
171. S. W. May and N. N. Li, *Biochem. Biophys. Res. Comm. 47:*1179 (1972).
172. I. Chibata and T. Tosa, in *Applied Biochemistry and Bioengineering,* Vol. 1, Academic Press, New York, 1976, p. 329.
173. G. G. Guilbault and G. Nagy, *Anal. Lett. 6:*301 (1973).
174. G. G. Guilbault and F. R. Shu, *Anal. Chim. Acta 56:*333 (1971).
175. G. G. Guilbault and E. Hrabankova, *Anal. Chim. Acta 56:*285 (1971).
176. G. G. Guilbault and J. G. Montalvo, *J. Am. Chem. Soc. 92:*2533 (1970).
177. D. R. Senn, P. W. Carr, and L. N. Klatt, *Anal. Chem. 48:*954 (1976).
178. M. Aizawa, I. Karube, and S. Suzuki, *Anal. Chim. Acta 69:*431 (1974).
179. I. Satoh, I. Karube, and S. Suzuki, *Biotechnol. Bioeng. 19:*1095 (1977).
180. I. Satoh, I. Karube, and S. Suzuki, *Biotechnol. Bioeng. 18:*269 (1976).
181. G. G. Guilbault and F. R. Shu, *Anal. Chem. 44:*2161 (1972).
182. B. K. Ahn, S. K. Wolfson, Jr., and S. J. Yao, *Bioelectrochem. Bioenerget. 2:*142 (1975).
183. P. Davies and K. Mosbach, *Biochim. Biophys. Acta 370:*329 (1974).
184. H. E. Klei, D. W. Sundstrom, and D. Shim, in *Immobilised Cells and Enzymes. A Practical Approach* (J. Woodward, Ed.), IRL Press, Oxford, 1985, p. 49.
185. G. G. Guilbault and G. de Olivera Neto, in *Immobilised Cells and Enzymes. A Practical Approach* (J. Woodward, Ed.), IRL Press, Oxford, 1985, p. 55.
186. J. Woodward, in *Immobilised Cells and Enzymes. A Practical Approach* (J. Woodward, Ed.), IRL Press, Oxford, 1985, p. 3.
187. L. C. Clark, Jr. and C. Lyons, *Ann. N.Y. Acad. Sci. 102:*29 (1962).
188. V. C. Gekas, *Enzyme Microb. Technol. 8:*450 (1986).
189. K. Ima, T. Shiomi, K. Sato, and A. Fujiyama, *Biotechnol. Bioeng. 25:*613 (1983).
190. S. Miyairi, *Biochim. Biophys. Acta 571:*374 (1979).
191. P. Gacesa, R. Eisenthal, and R. England, *Enzyme Microb. Technol. 5:*191 (1983).
192. J. M. Sarkar and R. G. Burns, in *Anaerobic Digestion and Carbohydrate Hydrolysis of Wastes* (G. L. Ferrero, M. P. Ferranti, and H. Naveau, Eds.), Elsevier Applied Science Publishers, London, 1984 p. 168.
193. J. E. Prenosil and H. Pedersen, *Ann. N.Y. Acad. Sci. 469:*289 (1986).
194. J. Woodward, S. R. Krasniak, R. D. Smith, F. Spielberg, and G. S. Zachry, *Biotechnol. Bioeng. Symp. 12:*485 (1982).

195. G. F. Vovis, S. J. Law, D. Bowden, P. MacDonnell, J. Braman, A. Myles, R. Broeze, F. X. Cole, and K. Clough, Biotech 83. Online Publications, Northwood, UK, 1983, p. 397.
196. Y. Kimura, A. Tanaka, K. Sonomoto, T. Nihira, and S. Fukui, *Eur. J. Appl. Microbiol. Biotechnol. 17:*107 (1983).
197. Z. Mikelsone, A. N. Mitrofanova, and O. M. Poltorak, *Zh. Fiz. Khim. 57:*351 (1983).
198. T. A. Kovaleva, T. Klochkora, V. A. Uglyanskaya, and G. A. Chikin, *Zh. Prikl. Spektrosk. 38:*247 (1983).
199. M. Hashimoto, T. Tsukamoto, S. Morita, and J. Okada, *Chem. Pharm. Bull. 31:*1 (1983).
200. J. Iqbal and M. Saleemuddin, *Indian J. Biochem. Biophys. 20:*33 (1983).
201. H. D. Schell, G. Angel, and V. Govoreanu, *Stud. Cercet. Biochim. 26:*68 (1983).
202. M. J. Daniels, *Food Technol. 39:*68 (1985).
203. M. Okubo, S. Kamei, K. Mori, and T. Matsumoto, *Kobunshi Ronbunshu 40:*449 (1983).
204. M. Ying, X. Ma, N. Ye, and X. Li, *Huaxe Tongbao 5(7):*15 (1983).
205. L. Gianfreda, M. Modafferi, and G. Greco, *Enzyme Microb. Technol. 7:*78 (1985).
206. I. I. Nikol'skaya and N. F. Kazanskaya, *Appl. Biochem. Microbiol 19:*162 (1983).
207. I. Posner, C. S. Wang, and W. J. McConathy, *Arch. Biochem. Biophys. 226:*306 (1983).
208. J. L. Romette, J. S. Yang, H. Kusakabe, and D. Thomas, *Biotechnol. Bioeng. 25:*2557 (1983).
209. M. Nemat-Gorgani and K. Karimian, *Biotechnol. Bioeng. 25:*2617 (1983).
210. G. Greco, Jr., F. Veronese, R. Largajolli and L. Gianfreda, *Eur. J. Appl. Microbiol. Biotechnol. 18:*333 (1983).
211. H. Obata, J. Tanishita, and T. Tokuyama, *Technol. Rep. Kansai Univ. 24:*193 (1983).
212. F. A. Eyterkate, *Appl. Environ. Microbiol. 47:*177 (1984).
213. K. Pommering, D. Ristau, H. Rein, H. Dautzenberg, and F. Loth, *Biomed. Biochim. Acta 42:*813 (1983).
214. G. A. Kovalenko and V. D. Sokolovskii, *Biotechnol. Bioeng. 25:*3177 (1983).
215. J. Iqbal and M. Saleemuddin, *Biotechnol. Bioeng. 25:*3191 (1983).
216. P. G. Pifferi, C. Rovito, and E. Petrocchi, *Rev. Ferment. Ind. Aliment 38:*84 (1983).
217. M. Mason, *Am. J. Enol. Vitic. 34:*173 (1983).
218. M. Kumakura, I. Kaetsu, and T. Kobayashi, *Enzyme Microb. Technol. 6:*23 (1984).
219. S. Kawabe and S. Usami, *Nippon Shokuhin Kogyo Gakkaishi 30:*501 (1983).
220. J. C. Braman, R. J. Broeze, D. W. Bowden, A. Myles, T. R. Fulton, M. Rising, J. Thurston, F. X. Cole, and G. F. Vovis *Bio/Technology 2:*349 (1984).

221. J. M. S. Cabral, J. M. Novais, and J. P. Cardoso, *Biotechnol. Bioeng. 26:*386 (1984).
222. M. Kumakura and I. Kaetsu, *J. Dispersion Sci. Technol. 4:*147 (1983).
223. M. Jach and H. Sugier, *Staerke 35:*427 (1983).
224. T. Fonong and G. A. Rechnitz, *Anal. Chim. Acta 158:*357 (1984).
225. M. B. Fadda, M. R. Dessi, R. Maurici, A. Rinaldi, and G. Satta, *Appl. Microbiol. Biotechnol. 19:*306 (1984).
226. S. Sharma and H. Yamazaki, *Biotechnol. Lett. 6:*301 (1984).
227. R. J. Neufeld and T. M. S. Chang, *Enzyme Microb. Technol. 6:*135 (1984).
228. G. A. Garwood, M. M. Mortland, and T. J. Pinnavaia, *J. Mol. Catal. 22:*153 (1983).
229. M. Okuko, K. Mori, S. Kamei, and T. Matsumoto, *Kobunshi Ronbunshu 40:*809 (1983).
230. H. Ohta, H. Fujiwara, and G.-L. Tsuchihashi, *Agric. Biol. Chem. 48:*317 (1984).
231. T. Kato and K. Horikoshi, *Biotechnol. Bioeng. 26:*595 (1984).
232. H. Nakatani, A. Tanaka, and K. Hiromi, *J. Appl. Biochem. 5:*371 (1983).
233. W. Krakowiak and H. Sugier, *Staerke 36:*60 (1984).
234. E. Ziomek, W. G. Martin, and R. E. Williams, *Appl. Biochem. Biotechnol. 9:*57 (1984).
235. H. N. Chang, I. S. Joo, and Y. S. Chim, *Biotechnol. Lett. 6:*487 (1984).
236. N. Madry, R. Zocher, K. Grodzki, and H. Kleinkauf, *Appl. Microbiol. Biotechnol. 20:*83 (1984).
237. T. Ikeda, I. Katasho, and M. Kamei, *Agric. Biol. Chem. 48:*1969 (1984).
238. K. Makino, H. Tanigami, and T. Takeuchi, *Chem. Lett. 8:*1323 (1984).
239. H. Takaka, H. Kurosawa, E. Kukufuta, and I. A. Veliky, *Biotechnol. Bioeng. 26:*1393 (1984).
240. O. Miyawaki and L. B. Wingard, Jr., *Biotechnol. Bioeng. 26:*1364 (1984).
241. J. E. McGhee, M. E. Carr, and G. St. Julian, *Cereal Chem. 61:*446 (1984).
242. V. V. Belakhov and N. N. Momot, *Khim.-Farm. Zh. 18:*980 (1984).
243. L. Bourget and T. M. S. Chang, *Appl. Biochem. Biotechnol. 10:*57 (1984).
244. P. Vadgama, W. Sheldon, J. M. Guy, and A. K. Covington, *Clin. Chim. Acta 142:*193 (1984).
245. The Baxter Labs. Inc., British Patent No. 1,274,158 (1972).
246. J. F. Kennedy, J. M. S. Cabral, and B. Kalogerakis, *Enzyme Microb. Technol. 7:*22 (1985).
247. K. Ikura, K. Okumura, M. Yoshikawa, R. Sasaki, and H. Chiba, *J. Appl. Biochem. 6:*222 (1984).
248. J. K. Liou, K. C. A. Luyben, and S. Bruin, *Biotechnol. Bioeng. 27:*109 (1985).
249. T. Sakai, T. Katsuragi, K. Tonomura, T. Nishiyama, and Y. Kawamura, *Biotechnol. 2:*13 (1985).
250. Y. Nakamoto and S. Ishida, *Kobunshi Ronbunshu 41:*559 (1984).
251. V. D. Semichaevskii, *Ukr. Bot. Zh. 41:*60 (1984).
252. F. M. Veronese, E. Boccu, O. Schiavon, and G. Greco, *Ann. N.Y. Acad. Sci. 434:*127 (1984).

253. L. Bourget and T. M. S. Chang, *FEBS Lett. 180:*5 (1985).
254. S. A. G. F. Angelino, F. Mueller, and H. C. Van der Plas, *Biotechnol. Bioeng. 27:*447 (1985).
255. K. Mosbach and B. Mattiasson, *Acta Chem. Scand. 24:*2093 (1970).
256. R. Puvanakrishnan and S. M. Bose, *Indian J. Biochem. Biophys. 21:*323 (1984).
257. A. V. Ananichev and I. V. Ulezlo, *Appl. Biochem. Microbiol. 20:*373 (1985).
258. O. A. Belozerova, Y. D. Zytner, L. S. Tikhonova, A. N. Galushkin, and K. A. Makorov, *Appl. Biochem. Microbiol. 20:*379 (1985).
259. T. Tanaka, *Agric. Biol. Chem. 49:*1267 (1985).
260. I. T. Yakubov, R. Akhmedzhanov, and M. U. Tuichibaev, *Uzb. Biol. Zh. 1:*7 (1985).
261. H. Pedersen, L. Furler, K. Venkatasubramanian, J. Prenosil, and E. Stuker, *Biotechnol. Bioeng. 27:*961 (1985).
262. Q. Husain, J. Iqbal, and M. Saleemuddin, *Biotechnol. Bioeng. 27:*1120 (1985).
263. M. Tomar and K. A. Prabhu, *Enzyme Microb. Technol. 7:*454 (1985).
264. N. I. Larionova, N. F. Kazanskaya, and G. V. Mityushina, *Appl. Biochem. Microbiol. 20:*667 (1985).
265. N. M. Samoshina, E. Yu. Lotmentseva, V. N. Borisova, and L. A. Nakhapetyan, *Appl. Biochem. Microbiol. 20:*673 (1985).
266. I. Frennesson, G. Tragardh, and B. Hahn-Hägerdal, *Biotechnol. Bioeng. 27:*1328 (1985).
267. M. Nemat-Gorgani, K. Karimian, and F. Mohanazadeh, *J. Am. Chem. Soc. 107:*4756 (1985).
268. B. M. Manaev, I. A. Yamskov, Z. D. Ashubaeva, and V. A. Davankov, *Appl. Biochem. Microbiol. 21:*47 (1985).
269. O. Markovic and E. Machova, *Collect. Czech. Chem. Commun. 50:*2021 (1985).
270. P. Klossek, M. Kirchner, and J. Kurth, *J. Microbiol. 25:*429 (1985).
271. M. M. Hoq, M. Koike, T. Yamane, and S. Shimizu, *Agric. Biol. Chem. 49:*3171 (1985).
272. A.-K. Frej, J.-G. Gustafsson, and P. Hedman, *Biotechnol. Bioeng. 28:*133 (1986).
273. A. Kumar and H. S. Mainawatee, *Haryana Agric. Univ. J. Res. 15:*26 (1985).
274. V. D. Semichaevskii, *Quel. Ukr. Bot. Zh. 42:*48 (1985).
275. H. Ichijo, H. Uedaira, T. Suehiro, J. Nagasawa, and A. Yamauchi, *Agric. Biol. Chem. 49:*3591 (1985).
276. E. K. Denyakina, A. D. Neklyudov, T. A. Loginova, I. A. Krest'yanova, and Y. E. Bartoshevich, *Appl. Biochem. Microbiol. 21:*140 (1985).
277. K. Imai, T. Shiomi, K. Uchida, and M. Miya, *Biotechnol. Bioeng. 28:*198 (1986).
278. H. Yamazaki, S. Suniti, and S. Joshi, *Biotechnol. Lett. 8:*107 (1986).
279. V. S. Vol'skii, L. K. Shataeva, and G. V. Samsonov, *Zh. Prikl. Khim.* (Leningrad) 58:2074 (1985).
280. A. I. Yaropolov, O. V. Skorobogat'ko, T. T. Buglova, and I. Y. Zakharova, *Appl. Biochem. Microbiol. 21:*336 (1985).

281. N. A. Goncharova, V. P. Gavrilova, I. I. Shamolina, A. B. Lobova, and L. A. Vol'f, *Appl. Biochem. Microbiol. 21:*262 (1985).
282. M. M. Hoq, T. Yamane, and S. Shimizu, *Enzyme Microb. Technol. 8:*236 (1986).
283. S. Kunugi, H. Kodama, H. Yamada, and Y. Nakamura, *Sen'i Gakkaishi 41:*T355 (1985).
284. G. Marko-Varga, R. Appelqvist, and L. Gorton, *Anal. Chim. Acta 179:*371 (1986).
285. E. Kokufuta, T. Sodeyama, and T. Katano, *J. Chem. Soc., Chem. Commun. 9:*641 (1986).
286. J.-C. Piard, M. E. Soda, W. Alkhalaf, M. Rousseau, M. Desmazeaud, et al., *Biotchnol. Lett. 8:*241 (1986).
287. N. Nakamura, K. Murayama, and T. Kinoshita, *Anal. Biochem. 152:*386 (1986).
288. M. F. Suaud-Chagny and F. G. Gonon, *Anal. Chem. 58:*412 (1986).
289. S. A. Boyd and M. M. Mortland, *Experientia 41:*1564 (1985).
290. M. Okubo, Y. Aoki, K. Mori, and S. Kamai, *Kobunshi Ronbunshu 42:*829 (1985).
291. F. Ortega and M. Ortega, *An. R. Acad. Farm. 51:*757 (1985).
292. R. P. Rohrbach, D. S. Scherl, and R. W. Detroy, *Biotechnol. Bioeng. Symp. 15:*363 (1985).
293. S. Kohjiya, K. Maeda, Y. Ikushima, and Y. Ishihara, *Nippon Kagaku Kaishi 12:*2302 (1985).
294. Tj. Uragami, T. Aketa, S. Gobodani, and M. Sugihara, *Polym. Bull. (Berlin) 15:*101 (1986).
295. O. Z. Higa, N. L. Del Mastro, and A. C. Castagnet, *Radiat. Phys. Chem. 27:*311 (1986).
296. L. O. Gorton, F. Scheller, and G. Johansson, *Stud. Biophys. 109:*199 (1986).
297. T. Ikeda, H. Hamada, and M. Senda, *Agric. Biol. Chem. 50:*883 (1986).
298. K. Kihara and E. Yasukawa, *Anal. Chim. Acta 183:*75 (1986).
299. W. Schoepp and M. Grunow, *Appl. Microbiol. Biotechnol. 24:*271 (1986).
300. M. Nemat-Gorgani and K. Karimian, *Biotechnol. Bioeng. 28:*1037 (1986).
301. H. Xu, Y. X. Liu, B. Chen, H. Yang, and G. Yao, *Polym. Prepr. Am. Chem. Soc., Div. Polym. Chem. 27:*466 (1986).
302. B. Solomon, R. Koppel, G. Pines, and E. Katchalski-Katzir, *Biotechnol. Bioeng. 28:*1213 (1986).
303. D. Hans and J. S. Rhee, *Biotechnol. Bioeng. 28:*1250 (1986).
304. G. Iorio, E. Drioli, G. Catapano, R. Molinari, and M. Rossi, *Chem. Eng. Commun. 44:*227 (1986).
305. C. L. Dueck, R. J. Neufeld, and T. M. S. Chang, *Can. J. Chem. Eng. 64:*540 (1986).
306. N. N. Nemtsova, M. N. Birtseva, T. S. Bezyazychnaya, G. S. Parr, O. V. Kondratieva, and B. V. Moskvichev, *Prikl. Biokhim. Mikrobiol. 22:*466 (1986).
307. V. S. Mikhailin, A. A. Kondrashin, and T. T. Berezov, *Vopr. Med. Khim. 32:*68 (1986).

308. E. S. Vorobeva and O. M. Poltorak, *Vestn. Mosk. Univ., Ser. II, 21:*17 (1966).
309. G. G. Guilbault and J. Das, *Anal. Biochem. 33:*341 (1970).
310. K. D. Wheeler, B. A. Edwards, and R. Whittam, *Biochim. Biophys. Acta 191:*187 (1969).
311. P. Bernfeld, R. E. Bieber, and P. C. MacDonnell, *Fedn. Proc. 27:*(Abstr. no. 3185) 782 (1968).
312. S. N. Pennington, H. D. Brown, A. B. Patel, and C. O. Knowles, *J. Biomed. Mater. Res. 2:*443 (1968).
313. T. Tosa, T. Mori, and I. Chibata, *Enzymologia 40:*49 (1971).
314. T. Tosa, T. Mori, and I. Chibata, *J. Ferment. Technol. 49:*522 (1971).
315. G. L. Fletcher and S. Okada, *Nature 176:*882 (1955).
316. Anon., *Chem. Eng. News 49*(26):23 (1971).
317. K. Miyamoto, T. Fujii, and T. Miura, *J. Ferment. Technol. 49:*565 (1971).
318. P. L. Stavenger, *Chem. Eng. Progr. 67*(3):30 (1971).
319. T. M. S. Chang, A. Pont, L. J. Johnson, and N. Malave, *Trans. Am. Soc. Artif. Intern. Organs 14:*163 (1968).
320. T. M. S. Chang, *New Sci 42:*18 (1969).
321. T. M. S. Chang, *Sci. J. 3*(7):62 (1967).
322. T. M. S. Chang, *Trans. Am. Soc. Artif. Intern. Organs 12:*13 (1966).
323. T. M. S. Chang, A. Gonda, J. H. Dirks, and N. Malave, *Trans. Am. Soc. Artif. Intern. Organs 17:*246 (1971).
324. F. H. Dickey, *J. Phys. Chem. 59:*695 (1955).
325. S. N. Levine and W. C. LaCourse, *J. Biomed. Mater. Res. 1:*275 (1967).
326. R. E. Sparks, R. M. Salemme, P. M. Meier, M. H. Litt, and O. Lindan, *Trans. Am. Soc. Artif. Intern. Organs 15:*353 (1969).
327. L. L. Velikanov, N. L. Velikanov, and D. G. Zvyaginstev, *Pochvovedenie 3:*62 (1971).
328. M. G. Goldfeld, E. S. Vorobeva, and O. M. Poltorak, *Zh. Fiz. Khim. 40:*2594 (1966).
329. W. D. Harkins, L. Fourt, and P. C. Fourt, *J. Biol. Chem. 132:*111 (1940).
330. I. Langmuir and V. J. Schaefer, *Chem. Rev. 24:*181 (1939).
331. E. K. Bauman, L. H. Goodson, and J. R. Thomsom, *Anal. Biochem. 19:*587 (1967).
332. T. M. S. Chang, L. J. Johnson, and O. J. Ransome, *Can. J. Physiol. Pharmacol. 45:*705 (1967).
333. H. J. Trurnit and W. M. Lawson, *Arch. Biochem. Biophys. 47:*251 (1953).
334. H. J. Trurnit and W. M. Lawson, *Arch. Biochem. Biophys. 51:*176 (1954).
335. M. Mandels, J. Kostick, and R. Parizek, *J. Polym. Sci., Part C, 36:*445 (1971).
336. I Langmuir and V. J. Schaefer, *J. Am. Chem. Soc. 60:*1351 (1938).
337. Anon., *Chem. Eng. News 49*(48):27 (1971).
338. E. G. Griffin and J. M. Nelson, *J. Am. Chem. Soc. 38:*722 (1916).
339. A. K. Sharp, G. Kay, and M. D. Lilly, *Biotechnol. Bioeng. 11:*363 (1969).
340. S. J. Updike and G. P. Hicks, *Science 158:*270 (1967).
341. R. Goldman and H. M. Lenhoff, *Biochim. Biophys. Acta 242:*514 (1971).

342. E. A. Tveritinova, E. Kirai, E. S. Chukrai, and O. M. Poltorak, *Vestn. Mosk. Univ. Khim. 24:*16 (1970).
343. G. Fletcher and S. Okada, *Radiat. Res. 11:*291 (1959).
344. Yu. A. Zhirkov, E. S. Chukrai, and O. M. Poltorak, *Vestn. Mosk. Univ. Khim. 12:*405 (1971).
345. E. S. Vorobeva and O. M. Poltorak, *Zh. Fiz. Khim. 40:*2596 (1966).
346. M. N. Thang, M. Graffe, and M. Grunberg-Manago, *Biochem. Biophys. Res. Comm. 31:*1 (1968).
347. R. Koelsch, J. Lasch, and H. Hanson, *Acta Biol. Med. Ger. 24:*833 (1970).
348. E. F. Gale and H. M. R. Epps, *Biochem. J. 38:*232 (1944).
349. E. A. Tveritinova, N. I. Loboda, E. S. Chukrai, and O. M. Poltorak, *Vestn. Mosk. Univ. Khim. 12:*526 (1971).
350. M. van der Ploeg and P. van Duijn, *J. Roy. Microscop. Soc. 83:*415 (1964).
351. P. Bernfeld, R. E. Bieber, and D. M. Watson, *Biochim. Biophys. Acta 191:*570 (1969).
352. J. Gryszkiewicz, E. Dziembor, and W. Ostrowski, *Bull. Acad. Sci. Ser. Sci. Biol. 18:*439 (1970).
353. W. Becker and E. Pfeil, *J. Am. Chem. Soc. 88:*4299 (1966).
354. N. A. Mkrtumova and G. A. Deborin, *Dokl. Akad. Nauk. SSSR 146:*1434 (1962).
355. K. Mosbach and P.-O. Larsson, *Biotechnol. Bioeng. 12:*19 (1970).
356. P. Johnson and T. L. Whateley, *J. Colloid Interface Sci. 37:*557 (1971).
357. G. G. Guilbault and J. G. Montalvo, Jr., *Anal. Lett. 2:*283 (1969).
358. G. G. Guilbault and J. G. Montalvo, Jr., *J. Am. Chem. Soc. 91:*2164 (1969).
359. J. G. Montalvo, Jr. and G. G. Guilbault, *Anal. Chem. 41:*1897 (1969).
360. G. G. Guilbault and E. Hrabankova, *Anal. Chim. Acta 52:*287 (1970).
361. Anon., *Chem. Eng. News 48*(50): 39 (1970).
362. P. V. Sundaram and E. M. Crook. *Can. J. Biochem. 49:*1388 (1971).
363. P. V. Sundaram and E. M. Crook, *7th Int. Congr. Biochem. Tokyo, Int. Union Biochem. 4:* (Abstr. F-212) 801 (1967).
364. T. M. S. Chang and M. J. Poznansky, *J. Biomed. Mater. Res. 2:*187 (1968).
365. T. M. S. Chang and N. Malave, *Trans. Am. Soc. Artif. Intern. Organs 16:*141 (1970).
366. S.-T. Chung, M. Hamano, K. Aida, and T. Uemura, *Agric. Biol. Chem. 32:*1287 (1968).
367. H. Filippusson and W. E. Hornby, *Biochem. J. 120:*215 (1970).
368. B. H. J. Hofstee and N. F. Otillio, *Biochem. Biophys. Res. Comm. 53:*1137 (1973).
369. T. Fukumura, Japanese Patent 74-15795 (1974).
370. T. Tosa, T. Mori, T. Watanabe, M. Fujimura, M. Ono, and I. Chibata, *Abstr. Annu. Meet. Agric. Chem. Soc. Japan,* 1977 p. 296.
371. T. Mori, T. Watanabe, T. Tosa, and I. Chibata, *Abstr. Annu. Meet. Agric. Chem. Soc. Japan,* 1977, p. 296.
372. P. S. Bunting and K. J. Laidler, *Biochemistry 11:*4477 (1972).
373. T. Tosa, T. Mori, N. Fuse, and I. Chibata, *Enzymologia 31:*225 (1966).

374. S. Kasuga, K. Oike, T. Kobayashi, and S. Suzuki, *Hakko Kyokaishi 28:*299 (1970).
375. I. Chibata, T. Tosa, T. Sato, T. Mori, and Y. Matsuo in *Proc. IV I.F.S. Fermentation Technology Today* (G. Terui, ed.), Society of Fermentation Technology, Kyoto, 1972, p. 383.
376. T. Barth and H. Maskova, *Collect. Czech. Chem. Commun. 36:*2398 (1971).
377. T. Tosa, T. Mori, and I. Chibata, *Agric. Biol. Chem. 33:*1053 (1969).
378. T. Tosa, T. Mori, N. Fuse, and I. Chibata, *Biotechnol. Bioeng. 9:*603 (1967).
379. E. K. Bauman, L. H. Goodson, G. G. Guilbault, and D. N. Kramer, *Anal. Chem. 37:*1378 (1965).
380. T. Tosa, T. Mori, N. Fuse, and I. Chibata, *Agric. Biol. Chem. 33:*1047 (1969).
381. G. G. Guilbault and D. N. Kramer, *Anal. Chem. 37:*1675 (1965).
382. I. Steenhoek and P. Kooiman, *Enzymologia 35:*335 (1968).
383. T. Tominaga, T. Niimi, and H. Sugihara, Japanese Patent, 68-23560 (1968).
384. J. A. Rothfus and S. J. Kennel, *Cereal Chem. 47:*140 (1970).
385. A. Ahmad, S. Bishayee, and B. K. Bacchawat, *Biochem. Biophys. Res. Comm. 53:*730 (1973).
386. L. H. Goodson, W. B. Jacobs, and A. W. Davis, *Anal. Biochem. 51:*362 (1973).
387. A. Y. Nikolaev, *Biokhimiya* (English trans.) *27:*843 (1962).
388. B. P. Surinov and S. E. Manoilov, *Biokhimiya* (English trans.) *31:*337 (1966).
389. T. Tosa, T. Mori, T. Sato, K. Yamamoto, I. Takata, Y. Nishida, and I. Chibata, *Biotechnol. Bioeng. 21:*1697 (1979).
390. A. D. Traher and J. R. Kittrell, *Biotechnol. Bioeng. 16:*413 (1974).
391. H. Brandenberger, *Rev. Ferment. Ind. Aliment 11:*237 (1956).
392. B. J. Abbott and D. S. Fukuda, *Antimicrob. Agents Chemother.* 8:282 (1975).
393. M. A. Mitz and R. J. Schlueter, *J. Am. Chem. Soc. 81:*4024 (1959).
394. S. Ogino, *Agric. Biol. Chem. 34:*1268 (1970).
395. F. X. Hasselberger, B. Allen, E. K. Paruchuri, M. Charles, and R. W. Coughlin, *Biochem. Biophys. Res. Comm. 57:*1054 (1974).
396. E. J. Vandamme, *Enzyme Microb. Technol. 5:*403 (1983).
397. H. Suzuki, Y. Ozawa, and H. Maeda, *Agr. Biol. Chem. 30:*807 (1966).
398. S. Usami, T. Yamada, and A. Kimura, *Hakko Kyokaishi 25:*513 (1967).
399. T. Tominaga, T. Niimi, and H. Sugihara, Japanese Patent, 69-1360 (1969).
400. A. Kimura, H. Shirasaki, and S. Usami, *Kogyo Kagaku Zasshi 72:*489–492 (1969).
401. S. Usami, M. Matsubara, and J. Noda, *Hakko Kyokaishi 29:*195 (1971).
402. T. Wieland, H. Determann, and Z. Bünnig, *Z. Naturforsch. b 21:*1003 (1966).
403. B. Solomon and Y. Levin, *Biotechnol. Bioeng. 17:*1323 (1975).
404. Y. K. Park and D. Caretti de Lima, *Colet. Inst. Technol. Aliment.* (Port.) *4:*147 (1971; publ. 1972).
405. S. Usami and N. Taketomi, *Hakko Kyokaishi 23:*267 (1965).

406. P. A. Srere, B. Mattiasson, and K. Mosbach, *Proc. Nat. Acad. Sci.* (Washington) 70: 2534 (1973).
407. K. L. Smiley, *Biotechnol. Bioeng. 13:*309 (1971).
408. B. Solomon and Y. Levin, *Biotechnol. Bioeng. 16:*1161 (1974).
409. G. P. Hicks and S. J. Updike, *Anal. Chem. 38:*726 (1966).
410. S. J. Updike and G. P. Hicks, *Nature 214:*986 (1967).
411. K. Miyamoto, T. Fujii, N. Tamaoki, M. Okazaki, and Y. Miura, *J. Ferment. Technol. 51:*566 (1973).
412. R. A. Messing, *Enzymologia 39:*12 (1970).
413. R. A. Messing, *Biotechnol. Bioeng. 16:*897 (1974).
414. N. Tsumura and M. Ishikawa, *Shokuhin Kogyo Gakkaishi 14:*539 (1967)
415. H. Maeda, H. Suzuki, and A. Sakimae, *Biotechnol. Bioeng. 15:*403 (1973).
416. L. B. Wingard, Jr., C. C. Liu, and N. L. Nagda, *Biotechnol. Bioeng. 13:*629 (1971).
417. P. Monsan and G. Durand, *FEBS Lett. 16:*39 (1971).
418. J. Boudrant and C. Cheftel, *Biotechnol. Bioeng. 17:*827 (1975).
419. H. Suzuki, Y. Ozawa, H. Maeda, and O. Tanabe, *Kogyu Gijutsuin. Hakko Kenkyusho Kenkyu Hokuku 31:*11 (1967).
420. S. Usami, J. Noda, and K. Goto, *J. Ferm. Tech. 49:*598 (1971).
421. G. Gestrelius, B. Mattiasson, and K. Mosbach, *Eur. J. Biochem. 36:* 89 (1973).
422. C. Schwabe, *Biochemistry 8:*795 (1969).
423. K. Kawashima and K. Umeda, *Biotechnol. Bioeng. 16:*609 (1974).
424. H. L. Brockman, J. H. Law, and F. J. Kézdy, *J. Biol. Chem. 248:*4965 (1973).
425. J. Visser and M. Strating, *FEBS Lett. 57:*183 (1975).
426. A. D. McLaren and E. F. Estermann, *Arch. Biochem. Biophys. 61:*158 (1956).
427. A. Traub, E. Kaufmann, and Y. Teitz, *Anal. Biochem. 28:*469 (1969).
428. G. G. Guilbault and E. Hrabankova, *Anal. Chem. 42:*1779 (1970).
429. R. A. Messing, *Enzymologia 38:*39 (1970).
430. H. D. Brown, A. B. Patel, and S. K. Chattopadhyay, *J. Chromatog. 35:*103 (1968).
431. D. L. Marshall and J. L. Walter, *Carbohydrate Res. 25:*489 (1972).
432. M. P. Coughlan, *Biotechnol. Genet. Eng. Revs. 3:*39 (1985).
433. J. P. Hummel and B. S. Anderson, *Arch. Biochem. Biophys. 112:*443 (1965).
434. R. A. Messing, *Enzymologia 38:*370 (1970).
435. T. Toganehara and K. Watanabe, Japanese Patent, 65-27319 (1965).
436. P. Bernfeld and J. Wan, *Science 142:*678 (1963).
437. Y. Degani and T. Miron, *Biochim. Biophys. Acta 212:*362 (1970).
438. T. T. Ngo, P. S. Bunting, and K. J. Laidler, *Can. J. Biochem. 53:*11 (1975).
439. H. M. Walton and J. E. Eastman, *Biotechnol. Bioeng. 15:*951 (1973).
440. C. Gruesbeck and H. F. Rase, *Ind. Eng. Chem. Prod. Res. Develop. 11:*74 (1972).
441. H. Maeda, H. Suzuki, and A. Yamauchi, *Biotechnol. Bioeng. 15:*607 (1973).
442. H. Maeda, A. Yamauchi, A. and H. Suzuki, *Biochim. Biophys. Acta 315:*18 (1973).

443. R. A. Llenado and G. A. Rechnitz, *Anal. Chem. 43:*1457 (1971).
444. Anon., *Enzyme Microb. Technol. 5:*304 (1983).
445. C. O. L. Boyce, *J. Am. Oil Chem. Soc. 61:*1750 (1984).
446. M. Gronow, *Trends Biochem. Sci. 9:* 336 (1984).
447. M. P. Coughlan, M. P. J. Kierstan, P. M. Border, and A. P. F. Turner, *J. Microb. Meth., 8:*1 (1988).

3

Covalent and Coordination Immobilization of Proteins

Joaquim M. S. Cabral

*Universidade Técnica de Lisboa
Lisbon, Portugal*

John F. Kennedy

*University of Birmingham
Birmingham, England*

I. INTRODUCTION

The immobilization of proteins on solid supports by covalent coupling and metal coordination usually leads to very stable preparations with extended active life when compared with immobilized protein preparations obtained with other coupling methods, namely, physical adsorption and ionic binding.

The purpose of this chapter is to review the chemistry involved in protein immobilization by covalent and metal coordination binding. The protein structure and properties which influence the molecular binding on the supports are described, namely, the various types of interaction through side groups, C and N termini with other molecules. The morphological and chemical classification of solid supports for protein immobilization are covered, leading into the surface functional groups and general hydrophilic and hydrophobic properties of supports.

The last section describes the major types of reactions between supports and proteins involving covalent and coordinative immobilization. Applications in terms of protein immobilization by chemical binding are also presented.

The covalent binding method is based on the covalent attachment of proteins to water-insoluble matrices. This method has been the most widespread and one of the most thoroughly investigated approaches to protein (enzyme) immobilization.

The selection of conditions for immobilization by covalent binding is more difficult than in the other carrier binding methods. The reaction conditions required are relatively complicated and not usually mild.

Covalent binding is strong, so that stable immobilized protein preparations have been obtained that do not lose protein into the solution, even in the presence of high ionic strength solutions. The immobilization of an enzyme by covalent attachment to a support matrix should involve only functional groups of the enzyme that are not essential for its catalytic action, and thus the active center of the enzyme must be unaffected by the various reagents that are used.

To achieve higher activities in the resulting immobilized enzyme preparations by preventing inactivation reactions with the essential amino acid residues of the active site, several attempts (1) have been made: (1) covalent attachment of the enzyme in the presence of a competitive inhibitor or substrate, (2) a reversible covalently linked enzyme–inhibitor complex, (3) a chemically modified soluble enzyme whose covalent linkage to matrix is achieved by newly incorporated residues, and (4) a Zymogen precursor.

The wide variety of binding reactions and of matrices with functional groups capable of covalent coupling, or susceptible to being activated to give such groups, makes this a generally applicable method of immobilization. Nevertheless, the compositional and structural complexity of proteins has not allowed, except in a very limited number of cases, the application of general rules by means of which the method best suited for a specific task could be predicted. It would be most satisfactory if the enzyme tertiary, primary, and active site structures were to be known so that a potential linkage point least likely to be involved in activity could be selected for a specific application study.

Three main factors have to be taken into account for covalent immobilization of proteins by a specific method: (1) the functional group of proteins suitable for covalent binding under mild conditions, (2) the coupling reactions between the proteins and the support, and (3) the functionalized supports suitable for protein immobilization. These factors are discussed in the following sections.

II. PROTEIN FUNCTIONAL GROUPS

Proteins are heteropolymers built up from more than 20 types of monomer units of amino acid residues. Table 1 summarizes the residues of amino acids that have functional groups in the side chain suitable for linking to a support matrix.

From this list, the amino acids with amide groups (L-glutamine and L-asparagine), those with hydrocarbon side chains (L-alanine, L-leucine, L-

Table 1 Reactive Residues of Proteins

$-NH_2$	*t*-Amino of L-lysine (L-Lys) and N-terminus amino group
$-SH$	Thiol of L-cysteine (L-Cys)
$-COOH$	Carboxyl of L-aspartate (L-Asp) and L-glutamate (L-Glu) and C-terminus carboxyl group
(phenyl)–OH	Phenolic of L-tyrosine (L-Tyr)
$-N(H)-C(=NH)NH_2$	Guanidino of L-arginine (L-Arg)
(imidazole ring: N, NH)	Imidazole of L-histidine (L-His)
$-S-S-$	Disulfide of L-cystine
(indole ring: N–H)	Indole of L-tryptophan (L-Trp)
CH_3-S-	Thioether of L-methionine (L-Met)
$-CH_2OH$	Hydroxyl of L-serine (L-Ser) and L-threonine (L-Thr)

isoleucine, L-valine, L-phenylalanine, and L-proline), and L-glycine are absent owing to their relatively low concentration on the exposed protein surfaces and, even more important, owing to their nonreactive hydrophobic natures (Table 1). Increased hydrophobicity tends to increase the changes of a residue being buried inside the protein. Table 2 shows the average amino acid composition of a number of proteins, from the point of view of the above-mentioned relative reactivity of residues. Means and Feeney (2) made an interesting comparison that summarizes the number of reactions in which each amino acid side chain can participate: L-Cys-31, L-Lys-27, L-Tyr-16, L-His-13, L-Met-7, L-Trp-7, L-Arg-6, glu-4, L-Asp-4, L-Ser-0, L-Thr-0.

Most of the reactions described below, which are the coupling reactions involving the active side chains, arc classificd as carbonyl-typc rcactions with the nucleophilic groups of the protein, $-NH_2$, -SH, and -OH. In terms of nucleophilic reactivity, the sulfur type of anion is one or two orders of magnitude larger than that of most nitrogen and oxygen compounds of comparable basicity. However, the thioesters formed are much less stable

Table 2 Average Composition of Proteins (Reactive Residues Only)

Residue	Percent
Ser	7.8
Lys	7.0
Thr	6.5
Asp	4.8
Glu	4.8
Arg	3.8
Tyr	3.4
Cys	3.4
His	2.2
Met	1.6
Trp	1.2

Table 3 Number of Reactions in Which Each Amino Acid Participates

Residue	Number of reactions
Cys	31
Lys	27
Tyr	16
His	13
Met	7
Trp	7
Arg	6
Glu	4
Asp	4
Ser	0[a]
Thr	0[a]

[a]Normally considered too unreactive to be of use.
Source: Ref. 2.

than the esters, and these in turn are less stable than the substituted amines that are formed.

Comparing the data in Tables 2 and 3, and taking into account these considerations, it appears that the most convenient residues for immobilization are the lysyl residues, followed by L-cysteine, L-tyrosine, L-histidine, L-aspartic acid, L-glutamic acid, L-arginine, L-tryptophan, L-serine, L-threonine, and L-methionine.

Table 4 Characteristics for a Protein Support

Large surface area
Permeability
Hydrophilic character
Insolubility
Chemical, mechanical, and thermal stability
High rigidity
Suitable shape and particle size
Resistance to microbial attack
Regenerability

III. SUPPORTS FOR PROTEIN COVALENT COUPLING

The major components of an immobilized protein are the protein, the support, and the mode of interaction of the protein with the carrier. One of the most important contributing components to the performance of an immobilized protein system is the carrier. A carrier judiciously chosen can enhance the immobilized protein preparation. Although there is no universal carrier, a number of desirable characteristics (see Table 4) should be common to any material considered for immobilizing proteins.

A variety of support properties, e.g., surface properties, morphology, configuration, composition, and modification, should be assessed with respect to the specific application prior to the selection for evaluation. Solid supports can be classified according to their composition as organics and inorganics, however, this description is not entirely adequate for the full characterization of very pertinent support parameters (3) such as surface area and pore diameter, both of which in turn affect the loading of the protein. To take account of these parameters the carriers can be reclassified based on their morphology as nonporous and porous matrices. These two classifications will be considered in the following sections.

A. Morphological Classification

1. Nonporous Supports

Nonporous supports usually present a major disadvantage for protein immobilization. The surface area of nonporous carriers is extremely low, and therefore the available area of protein coupling is very limited. In order to increase the protein loading, very fine particles or fibers can be used. However, in this last case additional problems arise, as these fine configurations are difficult both to remove from the reaction mixture and to employ in continuous systems since they lead to high pressure drops in packed bed columns and limited flow rates if they are used in fluidized bed mode.

However, nonporous carriers can present some advantages due to their morphological characteristics. The diffusion constraints with respect to the soluble compounds can be eliminated or certainly reduced by decreasing the particle size or increasing the fluid linear velocity, since the protein is immobilized on the external surface of the carrier and is in immediate contact with the surrounding environment.

2. Porous Supports

Porous carriers have a high surface area per unit weight, which allows a high protein loading and thus makes them ideal for large industrial processes. A major disadvantage of porous carriers, however, is that most of the surface available for protein coupling is internal surface. Porous supports must have an internal morphology that allows not only the protein coupling but also an easy accessibility to soluble molecules in order to minimize diffusional effects. On the other hand, proteins bound on the internal surface are protected from the turbulent and harsh external environment.

Porous carriers can have a broad or controlled pore distribution. With a broad pore carrier, only a limited number of the pores will be large enough to accommodate both protein and solutes, and thus only a small portion of the total surface is effectively usable. To overcome this problem, controlled pore supports with a wide range of pore size are available. One major factor in choosing a porous carrier is the inverse relationship between the pore size and the surface area (Table 5), which makes it possible to optimize the pore diameter for a particular protein–solute system.

Table 5 Relationship Between the Pore Size and the Surface Area

Pore size (nm)	Surface area[a] (m^2/g)
7.5	356
12.5	214
17.5	153
2	111
37	72
70	38
125	21
200	13

[a]Pore volume: 1.0 cm^3/g.

Table 6 Comparison of Protein Supports

Morphological classification	Advantages	Disadvantages
Nonporous	Low diffusional limitations	Low surface area Low enzyme loading Use of fine particles: Difficult to remove Difficult to use continuously High pressure drops Limited flow rates
Porous	High surface area High enzyme loading Protection from external environment	High internal surface area High diffusional limitations High cost (controlled pore) High pressure drops (gels) Low surface area available (broad pore)

A variety of porous matrices may be classified as gels. Due to their elastic character, these gels do not have truly stable pores and are readily compacted leading to high pressure drops in plug-flow columns.

Use of the gels as immobilization matrices leads to large protein (enzyme) loadings since a relatively high surface, mainly internal, is available. Although the enzymes immobilized on gels are very useful for transformation of small solutes, they are extremely limited insofar as large molecular weight materials or synthetic reactions are concerned.

The main advantage of a gel is that due to its elastic character it may readily be formed into the required or desirable conformation, e.g., the membrane, special particles, etc. Also the elastic carriers can be used in stirred vessels with a lower deformation than a rigid structure (e.g., inorganic carriers). Table 6 summarizes and compares the protein carriers based on the morphological classification by showing the advantages gained by one type of matrix.

B. Chemical Classification

Carriers can be classified according to their chemical composition as organic and inorganic supports, and the former can be further classified into natural and synthetic matrices (Table 7). The next sections describe these different types of carriers and the types of reactions involved with proteins. A more detailed discussion of commercially available supports for protein immobilization is presented in Chapter 4 of this text.

Table 7 Chemical Classification of Protein Supports

Organic			Inorganic
Natural Polymers			Minerals
	Polysaccharides	Cellulose	Attapulgite clays
		Starch	Bentonite
		Dextran	Kieselgur
		Agar and agarose	Pumic stone
		Alginate	
		Carrageenan	
		Chitin and chitosan	
	Proteins	Collagen	
		Gelatin	
		Albumin	
		Silk	
	Carbon materials		
Synthetic Polymers			Fabricated Materials
	Polystyrene		Nonporous glass
	Polyacrylate types	Polyacrylates and polymethacrylates	Controlled pore glass
		Polyacrylamide	Controlled pore metal oxides
		Hydroxylalkyl methacrylates	Metals
		Glycidyl methacrylates	
		Maleic anhydride polymers	
		Vinyl and allyl polymers	
		Polyamides	

1. Organic Supports

Although it has been suggested that in industrial processing inorganic carriers have many advantages over their organic counterparts (due to undesirable physical properties of the organic supports that lead to poor stability of protein preparations against physical, chemical, thermal, and microbial degradation), most of the commercially available immobilized enzymes and proteins are obtained with organic matrices. The reason for this is that there are a wide variety of functional reactive groups which can be put on organic supports. In contrast, the surface of the inorganic supports exhibit, in solution, hydroxyl groups which can only interact with the carboxyl or amino groups on the protein molecule by physical adsorption and/or ionic and metal-spliced coordination interactions. For covalent attachment there are limitations to the use of inorganic support as they are activated by only a few techniques.

Organic carriers are classified as either natural macromolecules or synthetic polymers, and as having hydrophilic or hydrophobic characteristics.

Natural Polymers.

POLYSACCHARIDES. The most widely used supports and those having a polysaccharide backbone, especially those constructed from algae and cellulose, have been used both for the entrapment (e.g., alginate, carrageenan) and coupling (e.g., agarose, dextran, cellulose derivatives) of enzymes and proteins.

The hydroxyl groups of polysaccharides can be activated directly by introduction of an electrophilic group, reactive towards enzyme, into the matrix. However, the nucleophilic character of polysaccharide supports is so weak that pendant functional groups, such as aliphatic or aromatic amino groups, carboxyl or thiol groups, have to be introduced as the activation for coupling or before activation (indirect coupling).

The different active derivatives that have been used for coupling of proteins are listed in Table 8. The major advantage of the polysaccharide derivatives for immobilization of proteins is the existence of residual hydroxyl groups, which provide a hydrophilic character, protecting the attached protein.

CELLULOSE. The polysaccharide cellulose (Fig. 1) was one of the first materials to be used as a matrix for the covalent binding of protein. Chemically, it is a vegetable fiber composed of β-D-glucopyranosyl units linked by (1 → 4)-bonds and with additional intrachain interaction through hydrogen bonds, some of which form the so-called elementary fibers. Elementary fibers contain highly ordered crystalline regions and more accessible amorphous regions of a low degree of order. Although the hydroxyl groups of polysaccharides are not reactive enough to form covalent bonds between the protein and the support without previous activation, cellulose undergoes all the reactions associated with polyhydric alcohols so that a wide range of active cellulose can be prepared.

The permeability, the surface area available for protein attachment, and the reactivity of cellulose largely depend on the degree of crystallinity, the nature and the size of the compound to be bound, and the swelling-induced capacity of the activation medium.

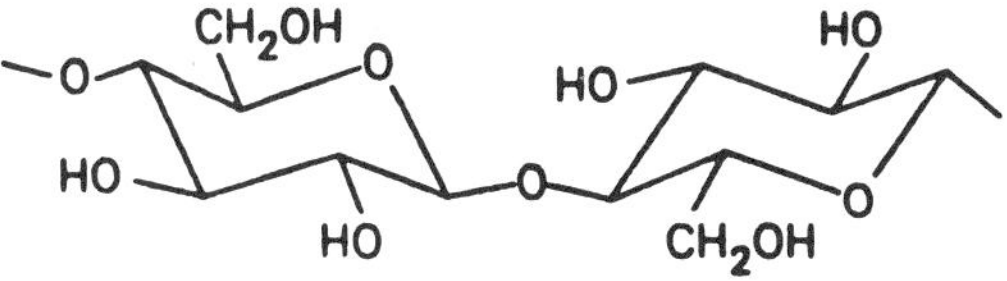

Figure 1 Structure of cellulose.

Table 8 Active Polysaccharide Derivatives for Protein Immobilization

Nucleophilic group of support	Activating reagent	Active derivative
Direct Coupling		
⊢OH, ⊢OH (vicinal)	CNBr	⊢O–, ⊢O– bridged to C=NH (cyclic imidocarbonate)
⊢OH, ⊢OH (vicinal)	$ClCO_2C_2H_5$	⊢O–, ⊢O– bridged to C=O (cyclic carbonate)
⊢OH	$ClCH_2-CH-CH_2$ (epoxide, O bridging CH and CH_2)	$⊢O-CH_2-CH-CH_2$ (epoxide, O bridging CH and CH_2)
⊢OH	$CH_2=CH-SO_2-CH=CH_2$	$⊢O-CH_2CH_2SO_2CH=CH_2$
⊢OH	p-benzoquinone (O=C₆H₄=O ring)	⊢O– substituted benzoquinone
⊢OH	$BrCOCH_2Br$	$⊢OOCOCH_2Br$
⊢OH	CH_2-CHCH_2Cl (epoxide, O bridging CH_2 and CH)	$⊢O-CH_2-CH(OH)-CH_2Cl$
⊢OH	Cl, Cl, X-substituted triazine ring (N, N, N)	O–, Cl, X-substituted triazine ring (N, N, N)
⊢OH, ⊢OH	IO_4^-	⊢CHO, ⊢CHO
Indirect Coupling		
$⊢NH_2$	Cl_2CS	⊢NCS

Table 8 (Continued)

Nucleophilic group of support	Activating reagent	Active derivative
$\vdash NH_2$	$OHC-(CH_2)_3-CHO$	$\vdash N=CH-(CH_2)_3-CHO$
$\vdash C_6H_4-NH_2$	Cl_2CS	$\vdash C_6H_4-NCS$
$\vdash C_6H_4-NH_2$	Cl_2CO	$\vdash C_6H_4-NCO$
$\vdash C_6H_4-NH_2$	$NaNO_2HCl$	$\vdash C_6H_4-N_2^+Cl^-$
$\vdash CONHNH_2$	$NaNO_2$	$\vdash CON_3$
$\vdash COOH$	$R-N=C=N-R', H^+$	$\vdash COO-C(=\overset{+}{N}HR')-NHR$
$\vdash COOH$	CH_3OH, NH_2NH_2, HNO_2	$\vdash CON_3$
	$C_5H_4N-S-S-C_5H_4N$	$\vdash S-S-C_6H_5$

Table 9 Commercially Available Modified Celluloses

4-Aminobenzyl-
Aminoethyl- (AE-)
Diethylaminoethyl- (DEAE-)
Carboxymethyl- (CM-)
Epichlorohydrin triethanolamine- (ECTEOLA-)
Oxy-
Phospho-
Sulfoethyl-
Triethylaminoethyl- (TEAE-)

Several types of chemically modified celluloses are commercially available (Table 9) and originally were used as ion-exchangers. These modified celluloses were used to directly immobilize proteins by ionic binding and also have permitted a wide range of covalent binding methods in which the protein binds mainly through amino groups. Some of the common cellulose derivatives used for immobilization of proteins are listed in Table 10.

Table 10 Cellulose Derivatives for Covalent Coupling of Enzymes

Triazinyl-cellulose
Bromacetyl-cellulose
Cellulose *trans*-2,3-carbonate
Cellulose imidocarbonate
Cellulose azide
Cellulose carbonyl
Diazo-cellulose
Isocyanat-cellulose

Cellulose is currently employed as a carrier to a lesser extent than other polysaccharides, e.g., dextran and agarose. The reasons are its microbial degradation and the forms (fibers of various sizes) in which it is commercially available. Macroporous cellulose in the form of beads was also reported (4) with good permeation, mechanical properties, and higher capacities of binding. Regeneration of cellulose supports is in most instances impossible, but this is not very relevant since it is a relatively cheap material due to its natural abundance.

STARCH. Starch (Fig. 2) is the least suitable common polysaccharide for protein immobilization owing to its ease of microbial degradability, it is rarely employed as a carrier compared with other polysaccharides. Starch was mainly used in the derivatized form of dialdehyde, which is obtained by oxidation of its chain with periodic acid or its salts (5).

DEXTRAN. Dextran (Fig. 3) is a basically linear, water-soluble polysaccharide composed of (1→6)-linked α-D-glucopyranosyl units (although dextrans are frequently branched). Insoluble and porous supports based on dextran (cross-linked) and possessing molecular sieving properties were originally developed as support for gel filtration chromatography and are commercially available (Sephadex) in grades characterized by their water regain and molecular exclusion limits. Because of these properties they have gained wide acceptance in enzyme and protein immobilization.

The commercially available dextran gels are prepared by cross-linking the water-soluble polymer with epichlorohydrin (6). By control of linear dextran and the degree of cross-linking, a range of well-defined water region and molecular sieving properties has been obtained (7).

Cross-linked dextrans have to be activated for use as support of immobilized proteins. One of the most commonly and extensively used procedures is the cyanogen bromide activation method in which inter alia a cyclic *trans*-2,3-imidocarbonate derivative is obtained (8). Some other covalent coupling methods of immobilization of proteins that are used involve the use of

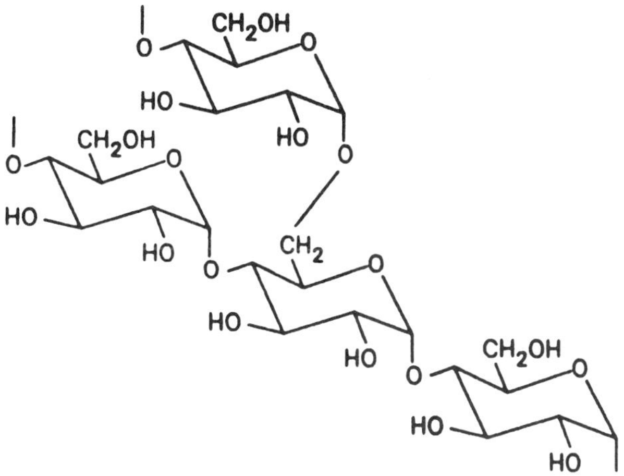

Figure 2 Structure of starch.

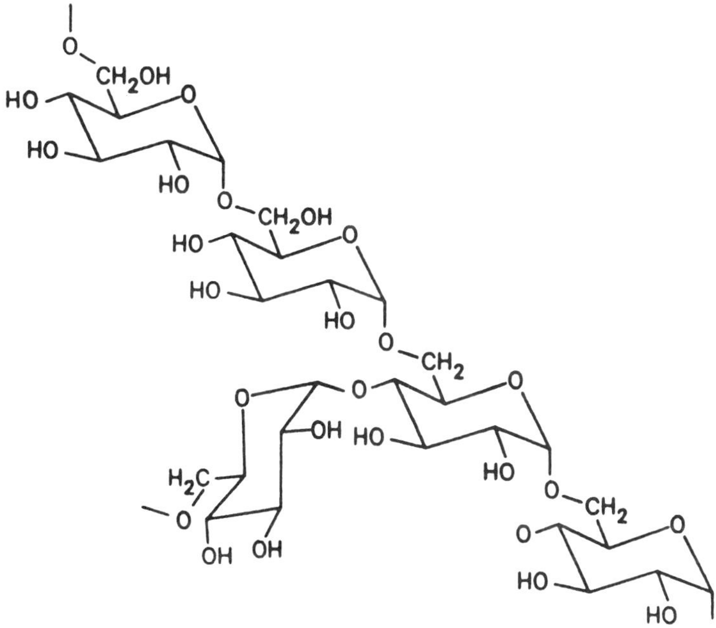

Figure 3 Structure of dextran.

cyanuric chloride, benzoquinone, and epichlorohydrin for activation of hydroxyl groups (direct coupling), carbodimides for carboxyl derivatives, and 2-pyridine disulfide for thiol derivatives. Ionic derivatives of cross-linked dextran are also commercially available and used for protein coupling by ionic interactions.

ALGAL POLYSACCHARIDES. Algal polysaccharides are the most widely used supports for immobilization of biocatalysts. Figure 4 shows the sources of algae extracts, agar, alginate, and carrageenan. Agar and agarose (fractionated agar of low charge density) are D-galactans isolated from certain red seaweeds that form rigid gels when their solutions are cooled to temperatures below about 45°C. These materials in bead form have been used (9) as support media in zone electrophoresis and as molecular sieves. Somewhat later, agarose began to be used as a support for immobilized proteins.

Chemically, agarose is a purified linear-galactan hydrocolloid isolated from agar with a repeating unit, agarobiose (Fig. 5), possessing an alternating (1→3)-linked β-D-galactopyranose and (1→4)-linked, 3,6-anhydro-α-L-galactopyranose structure.

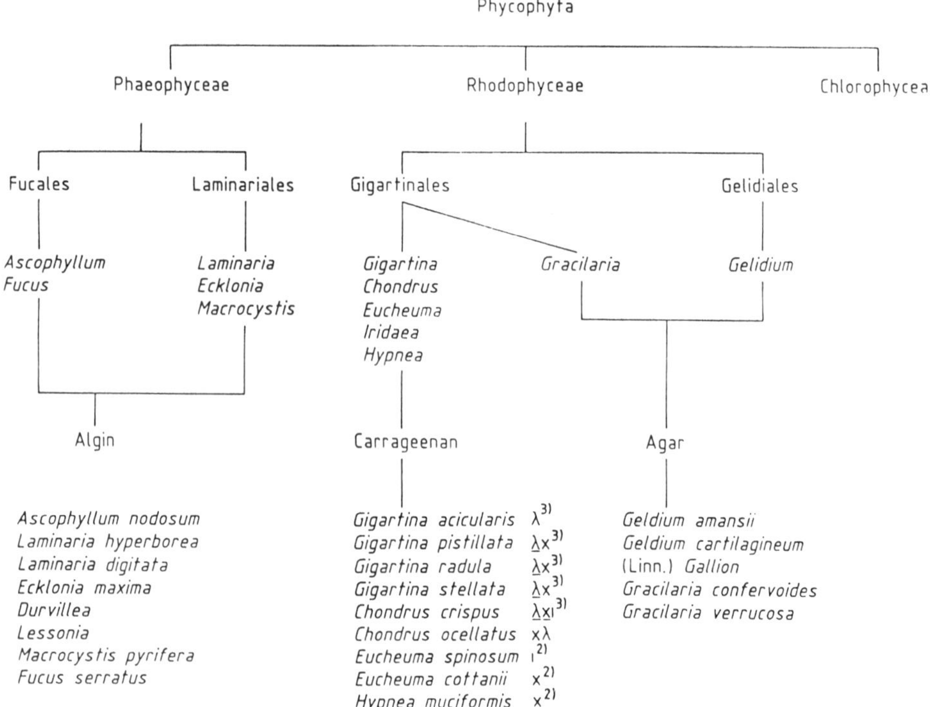

Figure 4 Sources of algae extracts.

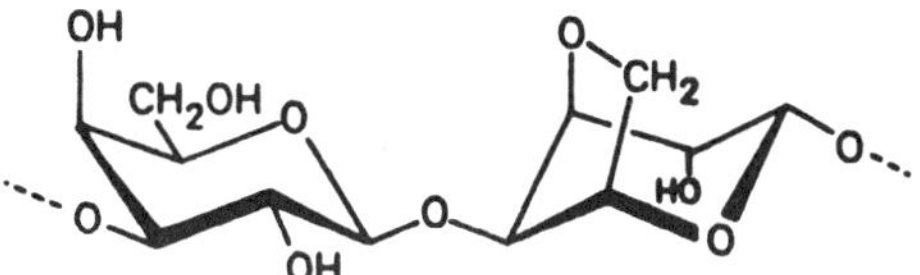

Figure 5 Structure of agarose.

Many procedures for the isolation of agarose exist, some have been developed into commercial processes for preparing beaded agarose. In such procedures, agarose is isolated from natural agar by dispergation of a 6–10% hot water solution of agar in an organic solvent in the presence of a suitable emulsifier; the particles that result from cooling below 45°C are separated.

Agarose gels are mechanically more stable and have greater pore size than other gels. They are resistant to microbial degradation, as agar-degrading enzymes (agarases) have been found only in certain microorganisms living on seaweeds. Despite superior properties as macroporous hydrophilic and nonadsorbing supports, however, they present several disadvantages: Agarose gels cannot be heat-sterilized; they disintegrate in strong alkaline solutions and in organic solvents; even at neutral pH values there is a possibility of solubilization; and they must be stored in wet form since they shrink irreversibly on drying. However, improvement of the mechanical properties and chemical resistance may be achieved by cross-linking with bifunctional reagents (e.g., epichlorohydrin, 2,3-dibromopropanol, or divinylsulfone), which eliminates most of these deficiencies.

Despite its superior properties, macroporous cross-linked agarose is only used for analytical purposes because it is a very expensive matrix and unsuitable economically for industrialization. Comparatively inexpensive crude agar, which contains negatively charged sulfate and carboxyl groups that may undesirably affect its applicability, can be treated with alkali to remove these groups, and thus the material may be rendered suitable for industrial application as a matrix for immobilized protein.

Although agar and agarose may potentially be used for entrapment methods of proteins, cross-linked agarose (Sepharose^{R}), which has been activated for use as a support of immobilized proteins by procedure similar to the cross-linked dextran, would appear to be more satisfactory.

Natural acidic polysaccharides can be used as supports for protein immobilization. One example of a suitable acidic polysaccharide is alginic acid (or alginate). Alginate is a glycuronan extracted from brown algae and chemically is an unbranched co-polymer consisting of residues of D-mannuronic acid and L-guluronic acid. Three structural elements are conventionally considered to occur (Fig. 6): (1→4)-α-L-guluronan (G) blocks, (1→4)-β-D-mannuron (M) blocks, and a polyuronid consisting of alternating L-guluronic and D-mannuronic acid residues arranged in a blockwise fashion.

Alginate gels can be formed by ionic-network formation in the presence of calcium ions or other multivalent counterions. Enzymes and proteins can be immobilized by entrapment in these gels or by covalent coupling on their surface either by reaction with the polysaccharide carboxyl groups or by reaction with the alginate gel after activation of their carboxyl groups with carbodiimides to form an O-acylurea derivative (10) or with hydrazine followed by a treatment with nitrous acid to form an azide derivative (11).

Alginates are chemically very stable between pH 5 and 10. High acid concentrations as well as high temperatures cause decarboxylation of alginates. These compounds are also degraded by direct-oxidation agents such as halogens or periodate and by redox systems such as polyphenols or

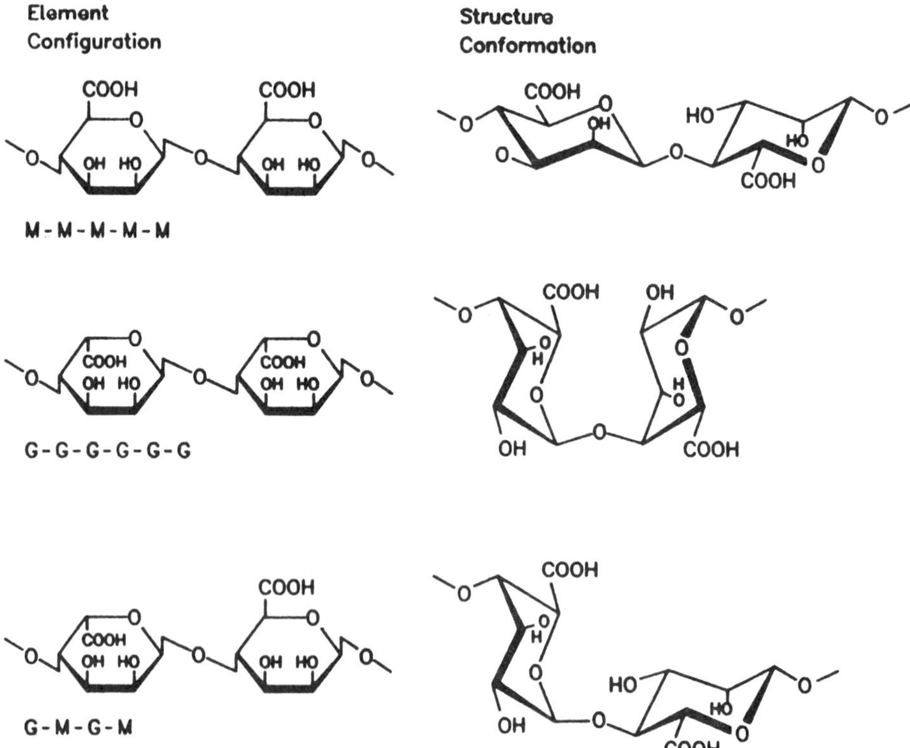

Figure 6 Structural elements in alginate.

thiols. Alginate gels are mechanically stable except when high concentrations of K^+, Mg^{2+}, phosphate, or chelating agents are present. Besides this disadvantage, leakage of protein and solute mass transfer effects limit the applicability of this type of support. However, immobilization in alginate gels is a safe, fast, mild, simple, cheap, and versatile technique, which may be applied to a wide range of biocatalysts and proteins.

Carrageenan is the general name for the galactans extracted from red algae, which consist of galactose units which are partly in the anhydro form and partly esterified with sulfuric acid (Fig. 7).

Carrageenan has properties very similar to alginate and agarose, which render it a suitable matrix for immobilization of biocatalysts. Of the three known types of carrageenan—gamma, kappa, and iota—only the latter two are suitable as a support for immobilization, the κ-form being an excellent matrix (12). Potassium and calcium salts form gels of carrageenan at ambient temperatures as a result of the formation of double helixes. The gel strength increases with ion concentration. However, these gels are unstable in the presence of sodium ions, and at low pH because of the weak (1→4)-glycosidic bond connecting the galactose residues. On the other hand, above pH 4.5, the carrageenan gels are very stable even under heat-sterilization conditions.

CHITIN AND CHITOSAN. Chitin and chitosan are polysaccharides containing amino groups which have been used as inexpensive supports for protein

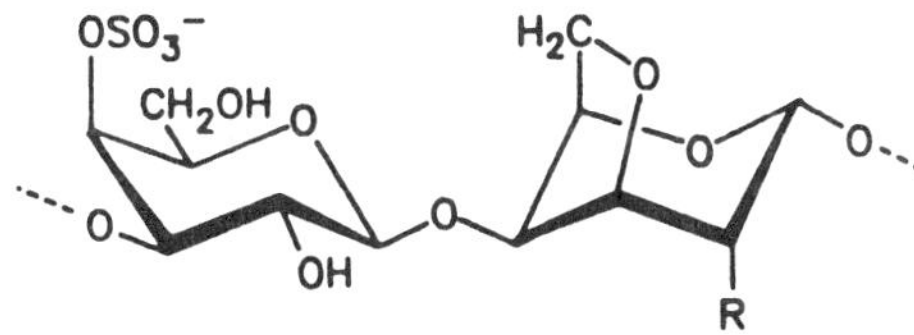

R = OH : κ - Carrageenan

R = OSO_3^- : ι - Carrageenan

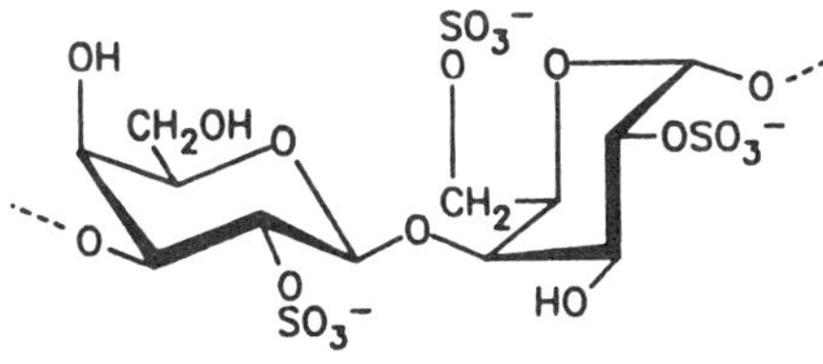

λ - Carrageenan

Figure 7 Structures of the three main forms of carrageenan: kappa-, iota- and lambda-carrageenan.

immobilization. Chitin is an abundant by-product of the fishing (crab, shrimp, and prawn) and the fermentation (citric acid and pharmaceuticals) industries. Chemically, chitin (Fig. 8) is a polysaccharide composed of (1→4)-linked 2-acetamido-2-deoxy-β-D-glucopyranosyl residues. About one of every six residues is not acetylated, whereas in chitosan (Fig. 9) essentially all the residues are not acetylated (2-amino-2-deoxy-D-glucose). Water-soluble chitosan can be obtained from chitin by deacetylation in concentrated sodium hydroxide solutions.

Enzymes and proteins can be bound to chitin by adsorption but this is usually followed by crosslinking with glutaraldehyde (13). Covalent linkage onto carbonyl derivatives, obtained by a previous treatment with glutaraldehyde, is also possible (14).

Chitosan, in a soluble form, can be mixed with a protein solution, and gel can be formed either by adding a multifunctional cross-linking agent, usually glutaraldehyde (15), or by ionotropic gelation (16) with several multivalent anionic conterions, e.g., $Fe(CN)_6^{4-}$, polyphosphates, etc. In other techniques (17) reprecipitated chitosan obtained from chitosan acetate or epichlorohydrin-cross-linked chitosan is first treated with glutaraldehyde to obtain a carbonyl derivative, and this links the protein covalently.

Chitin and chitosan derivatives have been reported to have favorable characteristics and have been qualified as attractive supports for enzyme immobilization. However, their uses have been limited and glutaraldehyde has been practically the only reagent used in their activation, the use of other reagents and techniques on these polysaccharides being as yet unexplored.

Proteins. Proteins that are now known to be enzymes have been used in several methods of immobilization, such as complexation of enzymes with collagen (18), entrapment of enzymes within a gelatin matrix (19), and cross-linking of enzymes together with inactive proteins (albumin, gelatin)

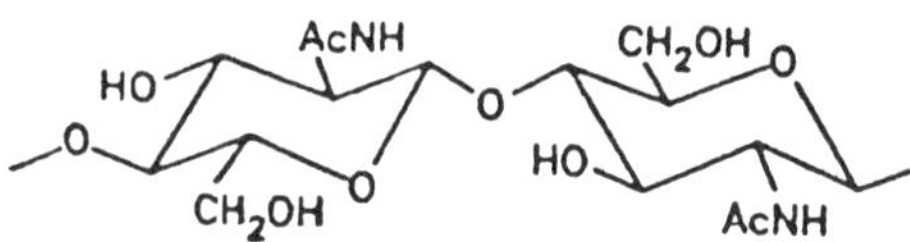

Figure 8 Structure of chitin.

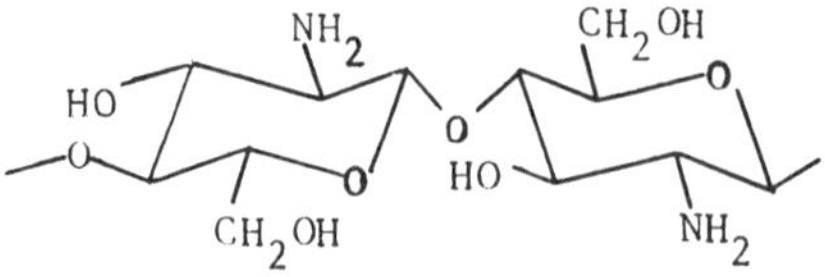

Figure 9 Structure of chitosan.

with or without a preexisting support. Proteinaceous supports have several advantages, due to their hydrophilic nature and easy handling, for enzyme immobilization, however, their chemical, mechanical, and thermal stabilities depend on the nature of the protein. Also, proteinaceous supports are subject to microbial attack, and their regeneration is impossible. The most used proteinaceous supports are described in the following sections.

COLLAGEN. Collagen is the most abundant protein constituent of higher vertebrates. It is easily isolated from many biological sources and can be reconstituted into various forms without losing its native structure. Thus, its availability from a large number of biological species, from fish to cattle, makes it an useful carrier.

Collagen has several advantages as an enzyme carrier. Its hydrophilicity facilitates the accessibility of enzymes to the binding sites. Its insoluble proteinaceous nature makes possible a strong physical adsorption of enzymes. Its open internal structure provides a high concentration of binding sites. Its fibrous structure and high swellability in aqueous solutions also contribute to its use as an enzyme support matrix.

Enzymes can be immobilized on collagen by adsorption, complexation, entrapment, and covalent coupling. Coulet and co-workers (20) described a covalent binding technique in which an acidic derivative of collagen was used for the preparation of several collagen-enzyme conjugates. Due to its biological origin, collagen may be particularly useful in biomedial applications.

GELATIN. The protein gelatin has several properties that make it attractive as a support for the immobilization of enzymes. It is easily obtained in solution from collagen by boiling with water. Gelatin does not have the strength of collagen and to use it as a water-insoluble enzyme carrier it is necessary to cross-link the gel with multifunctional reagents (e.g., glutaraldehyde). However, the low gel strength of gelatin, due to its very hydrophilic nature, makes it an useful matrix allowing a homogeneous distribution of the enzyme molecules in contrast with collagen, as a solution of enzymes and gelatin can be obtained at relatively low temperatures (35–40°C). The gel, containing enzyme, is obtained by temperature decreasing, however, it has to be cross-linked in order to be used at operational conditions. The chemical, mechanical, and thermal stabilities largely depend on the degree of cross-linking as well as the rigidity and the porosity of the gelatin support.

Enzymes can be immobilized on gelatin by gel entrapment followed by cross-linking (21) or by covalent coupling on the residual active groups of the multifunctional reagent (e.g., carbonyl groups of glutaraldehyde) at the surface of the gel (21).

ALBUMIN. Albumin is another protein useful for enzyme immobilization by miscellaneous methods, similar to those used for gelatin. Broun et al.

(22) used in their studies an enzymically inert L-lysine-rich protein (bovine serum albumin) which was co-cross-linked with the enzyme molecules by means of glutaraldehyde. By this method insoluble proteinaceous matrices, containing active enzymes, were obtained in the form of particles on membranes. The same method has been used by several others, but with a modification in which the bovine serum albumin was replaced by a cheaper, albumin-rich matrix: crude egg white. There is a not a direct comparison between albumin and gelatin as enzyme carriers, thus they can be used indistinctly as their physical and chemical properties and stabilities depend very much on their cross-linking conditions.

SILK. Raw silk is composed of two proteins, the water-insoluble fibroin and the soluble sericin, and it has been used as an enzyme support in the form of industrial woven silk. Grasset et al. (23) reported some physical and chemical properties of silk that make it useful as a support, namely, thermal stability, chemical stability from pH 3 to 8, microbial resistance, availability in a useful woven form that is easy to manipulate, low compaction, and resistance to abrasion.

The use of silk as enzyme support requires its surface activation. Although several processes of activation, based on the availability of functional groups, are possible, one choice made was based on the amino acid composition of the fibroin and involved enzyme fixation on the carboxyl groups after activation through the azide derivative. This derivative was obtained by acidic methylation, in a first step, producing methylated silk (which itself serves as support for adsorption of enzymes). The methylated silk is successively reacted with hydrazine and then nitrous acid to form an active azide silk.

Carbon Materials. Carbon materials are attractive as immobilized protein support because of their reasonable price and mechanical strength, and because they are obtainable in several forms, including porous structures. Carbon materials are already used in numerous medical, food, and fine chemical processing operations, which are also the potential fields of application of immobilized proteins.

One of the most used carbon materials for immobilization of proteins is activated carbon, which is a highly porous carbonaceous material with a large internal pore surface (24). It is prepared by dehydration and carbonization, followed by activation, of organic substances of mainly vegetable origin. Its pore structure has been reported (25) as "tridispersed"—containing micropores (0–0.2 nm radii), transitional pores (0.2–50 nm), and macropores (50–2000 nm). Activation of carbon is performed with O_2, chemicals, CO_2, or steam at high temperatures, carboxylic, vinyl, phenolic groups, and other oxides being obtained on the carbon surface, the matrix of these functional groups and its hydrophilicity depending on the type of oxidation process.

Almost all of the research on protein immobilization on activated carbon has dealt with adsorption, but covalent coupling methods for enzyme immobilization are also reported (26) based on the possible modification (amino silanized, carbodiimide derivative) of a specific activated-carbon preparation.

Activated carbon possesses a mechanical strength comparable to process-glass materials, but it suffers from abrasion. This support is also thermally stable for the operational conditions of immobilized proteins and it is completely inert to microbial degradation. Regeneration can easily be carried out in various ways (27).

Activated carbon can be produced in such a way that it has many advantageous properties for use as support in the immobilization of proteins.

Synthetic Polymers. Synthetic carriers are the largest number of supports available for protein immobilization. This is due to their physical and chemical characteristics and the ease of preparing various polymers for a particular protein or enzyme and application. Some advantages of this type of carriers are their inertness to microbial attack, the degree of porosity and their chemical composition which can be achieved by either copolymerization of very different available monomers or by chemical modification of preformed polymers. The most important synthetic carriers for protein immobilization are described in the following sections.

POLYSTYRENE. This polymer was the first synthetic polymer to be used for the immobilization of proteins (28). The main derivative is the polyaminostyrene, obtained from polystyrene by nitration and reduction. Due to its relative inertness for chemical coupling of protein molecules, this derivative has been activated by diazotization (28), carbonylation (29), and by reaction with thiophosgene (30).

Despite the rather high concentration of reactive groups on this type of carriers, the bound protein and coupling yields are usually poor due to the inherent hydrophobicity of the polymer matrix. This shortcoming has been bypassed by copolymerization with hydrophilic monomers, acrylic and methacrylic acids. The major advantage of polystyrene as a protein carrier is its low cost and availability.

POLYACRYLATE TYPES. Acrylic polymers are among the most acid synthetic polymers in the field of protein immobilization. They have been used in enzyme coupling methods as well as in entrapment. Some of them, such as polyacrylates, poly(hydroxy alkyl methacrylates), and derivatives, are available from commercial sources, which led to their broad use in protein immobilization. The most important polyacrylates used in the immobilization of proteins are listed in Table 11.

Table 11 Polyacrylate Derivatives for Protein and Enzyme Immobilization

Functional group of original support	Activating reagent	Active derivative
⊢CO_2H Acrylic acid methacrylic acid	R–N=C=N–R′, H+	⊢COO–C(NHR)(=$\overset{+}{N}$HR′)
⊢CO–O–CO⊣ (cyclic) Methacrylic acid anhydride	—	—
⊢$CONH_2$ Acrylamide	$OHC(CH_2)_3CHO$	⊢$CON{=}CH(CH_2)_3CHO$
⊢$CONH_2$ Acrylamide	H_2NNH_2: HNO_2	⊢CON_3
⊢$CONH_2$ Acrylamide	$H_2N(CH_2)_2NH_2$: $OHC(CH_2)_3CHO$	⊢$CONH(CH_2)_2N{=}CH(CH_2)_3Cl$
⊢$CONH_2$ Acrylamide	—	—
⊢$CONH_2$ Acrylamide	—	⊢$CONHNHCOCH_2CO_2H$
⊢$CONH_2$ Acrylamide	$H_2N(CH_2)_2NH_2$;O_2N–C_6H_4–CON_3 $Na_2S_2O_4$: $NaNO_2$ + HCl	⊢$CONH(CH_2)_2NHCO$–C_6H_4–$N_2^+Cl^-$
⊢OH ⊢OH Hydroxyalkyl methacrylate	CNBr	⊢O–C(=NH)–O⊣ (cyclic)
⊢CH–CH–CH_2 (epoxide, O) Glycidyl methacrylate	—	—

Polyacrylates and polymethacrylates are obtained by polymerization of acrylic and methacrylic acids, respectively. They have been used mainly in copolymers with other organic compounds to form a more hydrophobic matrix or to prepare negatively charged matrices. These types of carriers are usually activated with a soluble carbodiimide for enzyme coupling.

Polyacrylamides and their derivatives are among the synthetic matrices most often employed for the immobilization of proteins due their chemical structure, with hydrophilic character.

Because of the solubility of the linear polymers in water, these polymers have to be insolubilized by cross-linking with bifunctional compounds. One of the most popular methods is identical to that employed for the preparation of gel used for electrophoresis, based on the free-radical polymerization of acrylamide in an aqueous solution containing a crosslinking agent, usually, *N,N'*-methylene bisacrylamide.

Polyacrylamide can also be obtained by irradiating a frozen monomer solution using as irradiation sources ^{60}Co or ^{137}Cs.

Polyacrylamide matrices have been largely used as an entrapment technique for enzymes. However, they can be activated by several methods for chemical coupling of enzymes and proteins. Polyacrylamide and some of its chemically modified derivatives are commercially available (Bio-Gel^R, Bio-Rad Laboratories Ltd.). In addition to direct chemical modifications of performed polyacrylamides other chemically modified polyacrylamides are obtained by copolymerization of acrylamide and other monomers, which are also cross-linking agents.

Some of these copolymers are also commercially available (Enzacryls^R, Koch-Light Laboratories Ltd.; Trisacryl, IBF Reactifs) (Table 12). Ultragel is a mixture of polyacrylamide and agarose gelled together, which provides the mechanical stability that agarose lacks.

Hydroxyalkyl methacrylates are hydrophilic organic carriers that correspond, on account of their content of hydroxyl groups, to polysaccharide carriers with improved mechanical properties and biological resistance. Macroporous carriers are prepared in the form of beads by radical suspension copolymerization of hydroxyalkyl methacrylate and a cross-linking agent. Most widely used are the copolymers of 2-hydroxyethyl methacrylate and ethylene dimethacrylate (Spheron^R, Lachema). Like other organic polymeric carriers, chemically modified derivatives with hydroxyl, carboxyl, sulfonyl, and aminoacryl functions may be obtained by (1) direct chemical modification of preformed polymer or (2) copolymerization with monomers containing reactive groups or precursors of functional groups.

Glycidyl methacrylates are carriers containing oxirane (epoxide) groups, enabling both the direct coupling of proteins and also chemical modifications. The oxirane group is neutral and does not give rise to charge on

Table 12 Polyacrylamide and Derivatives

Name	Functional groups of original matrices
Acrylamide (Bio-Gel[R] P)	$-CH(CONH_2)-CH_2-CH(CONH_2)-$
Enzacryl[R] AA	$-CH(CONH_2)-CH_2-CH(CONH-C_6H_4-NH_2)-$
Enzacryl[R] AH	$-CH(CONH_2)-CH_2-CH(CONHNH_2)-$
Enzacryl[R] Polythiol	$-CH(CONH_2)-CH_2-CH(CONHCH(COOH)-CH_2SH)-$
Enzacryl[R] Polythiolactone	$-CH(CONH_2)-CH_2-CH(CONHCH-CH_2)-$, ring closed by $CO-S$
Enzacryl[R] Polyacetal	$-CH(CO-NH-CH_2-CH-(OCH_3)_2)-CH_2-CH(CO-NH-CH_2-CH-(OCH_3)_2)-$
Acryalmide/2-hydroxyethyl methacrylate	$-CH(CONH_2)-CH_2-C(CH_3)(COOCH_2CH_2OH)-CH_2-$
Acrylamide/acrylic acid	$-CH(CONH_2)-CH_2-CH(COOH)-$

Table 12 (Continued)

Name	Functional groups of original matrices
Acrylamide/methacrylate	$-CH-CH_2-C(CH_3)(COOCH_3)-CH_2-$, with $CONH_2$ on the first CH
4-Acrylamidosalicylic acid	$-CH-CH_2-CH-$, with $CONH$ on each CH; each CONH bears a benzene ring with OH and COOH
Acryloyl-*N,N′*-bis(2,2′-dimethoxyethyl)amine	$-CH-CH_2-CH-$, with $CON[CH_2CH_2(OCH_3)_2]_2$ on each CH
Acryloyl-*N,N′*-bis(2,2′-dimethoxyethyl)amine/acryloyl morpholine	$-CH-CH_2-CH-$, with $CON[CH_2CH_2(OCH_3)_2]_2$ on the first CH and $C{=}O$ on the second, bonded to the N of a morpholine ring (H_2C, CH_2, H_2C, CH_2, O)

reactions with proteins, but the resultant enzyme-carrier bond is very stable. This type of carrier is commercialized (Eupergit[R], Rohm Pharma) for covalent immobilization.

Maleic anhydride–based polymers, namely, copolymers of maleic anhydride and ethylene, are common supports for the immobilization of enzymes. They allow the direct coupling of proteins by adsorption due to their highly negatively charged groups. Neutralization of the negative charge can be accomplished by the addition of diamines during the immobilization procedure, which also acts as a crosslinking agent to produce a highly water-insoluble enzyme preparation.

Vinyl and allyl polymers are carriers with neutral and hydrophilic characteristics that can be obtained by chemical modification of polyvinyl alcohol, polyallyl alcohol, or vinyl ether copolymers. Reactive carriers based on polyvinyl alcohol are obtained by crosslinking the soluble polyvinyl alcohol with terephthaldehyde and by reaction of this crosslinked polymer with 2-3-amino-phenyl-1,3-dioxolane followed by diazotization or with 2,4,6-trichloro-*sym*-triazine.

Polyallyl alcohol can be chemically modified to yield reactive forms with isocyanate compounds, which serve as cross-linking agent. The insoluble polymers can couple proteins directly to the isocyanato groups.

Polyamides, known as nylons, are a family of condensation polymers of α,ω-dicarboxylic acids and α,ω-diamines.

Several types of nylon, differing only in the number of methylene groups in the repeating alkane segments, are available in a variety of physical forms, such as fibers, membranes, powders, and tubes. These types of supports have several advantages, such as mechanical strength, biological resistance, and relative hydrophilicity. For covalent immobilization of proteins, however, the chemical inertness of the polyamide backbone leaves only the terminal carboxyl and amino groups as possible reactive groups. Thus, to increase the binding capacity of nylon, it is necessary to treat chemically the support, namely, by a mild acid depolymerization. Carboxyl and amino pairs are generated on the surface of the nylon structure, and condensing reactions involving an aldehyde and an isocyanide in conjunction with those groups may occur, yielding the polyisonitrile nylon (Scheme 1).

2. Inorganic Carriers

Some of the earliest work on protein (enzyme) immobilization was with inorganic matrices (31), however, due to the high reactivity of the organic supports, these have been used more widely in the subsequent stages of immobilized enzyme development. However, inorganic carriers, due to their physical properties, are suitable for industrial use and offer several advantages over their organic counterparts: high mechanical strength, thermal stability, resistance to organic solvents and microbial attack, easy handling, excellent shelf life, and easy regenerability by a simple pyrolysis process. Moreover, inorganic materials do not change in structure over wide ranges of pH, pressure, and temperature.

A great variety of inorganic supports have been used as matrices for the immobilization of proteins. Materials ranging from fabricated particles with specifically tailored properties to inexpensive minerals have been examined as inorganic supports. The scarcity of natural macroporous materials has made it desirable to fabricate porous inorganic supports for the immobilization of enzymes. One of the most important prefabricated inorganic solid supports is controlled-pore glass (CPG). CPG is prepared by heating certain borosilicate glasses to 500–700°C for prolonged periods of time followed by acid leaching of the borate component, leaving a porous structure of very high silica content. Careful control of the physical and chemical treatments allows the production of glass particles of various diameters with narrow pore-size distributions in the range from 4.5 to 400 nm.

CPG has a wide solvent compatibility, but is unstable in alkaline environments ($pH > 8$) during prolonged dynamic operation, due to the solubility of silica in alkali. This solubility of CPG in alkali severely limits its application as a protein support. To increase the durability of this support, coating with metal oxides, less soluble in alkali, has been employed. This coating is produced by vacuum-impregnating the porous glass with a solution of metal salt, particularly of zirconium, and subsequent calcination to give an oxide layer on the CPG surface.

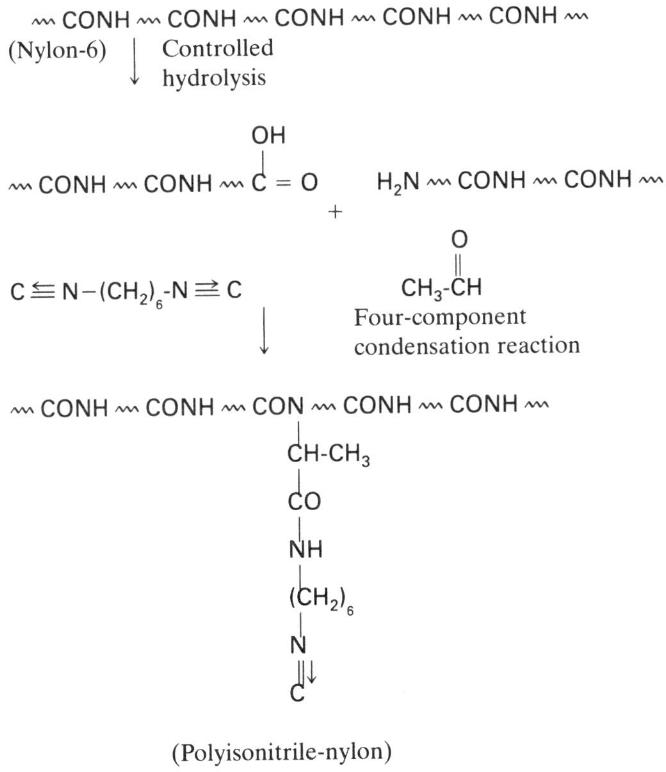

~ CONH ~ CONH ~ CONH ~ CONH ~ CONH ~
(Nylon-6) ↓ Controlled hydrolysis

~ CONH ~ CONH ~ C(OH) = O + H_2N ~ CONH ~ CONH ~

$C \equiv N-(CH_2)_6-N \equiv C$ + CH_3-CH=O
↓ Four-component condensation reaction

~ CONH ~ CONH ~ CON ~ CONH ~ CONH ~
CH-CH_3
CO
NH
$(CH_2)_6$
N
⇊
C

(Polyisonitrile-nylon)

Scheme 1

In spite of this, CPG is unsuitable for industrial purposes owing to its high price. Other carriers, controlled pore ceramics, silica, aluminum, and titanium, have been developed much more cheaply. The choice of one of those carriers is dependent on their chemical durability at several pH solutions. Thus alumina and titanium are suitable as carriers at alkaline solutions and silica is better suited to acidic solutions.

Naturally occurring porous minerals, such as kieselgur, attapulgite clays, pumice stone, and bentonite, can also be used as enzymic carriers, however, they have a broad pore distribution.

Nonporous carriers, such as pure metals or metal oxides with ferromagnetic properties such as magnetic iron oxide, nickel, and stainless steel, are advantageous carriers as they can readily be separated from the reaction mixture and, in conjunction with their destiny, could be useful in fluidized bed columns. Inorganic materials that have been used as carriers for protein (enzyme) immobilization are listed in Table 13.

The surface of most inorganic carriers is mainly composed of oxide and hydroxyl groups, such as silanol groups in glass, which provide a mildly reactive surface for activation and protein binding.

The relative inertness of the inorganic supports makes their derivatization difficult. However, this can be accomplished by an inorganic and/or a cross-linked organic coating. One of the most popular derivatization methods of inorganic materials is the silanization method, which involves the use of trialkoxy silane derivatives containing an organic functional group (Table 14) (32). Coupling of these reagents to the carrier takes place presumably by displacement of the alkoxy residues on the silane, by hydroxyl groups or the oxidized surface of the inorganic support to form a metal-O-Si linkage (Fig. 10).

In some cases the silanized supports can be reacted directly with proteins, but in most cases the organic groups are modified to produce activated intermediates, which in turn are reacted with enzymes. The most popular and versatile silane compound is the 3-aminopropyltriethoxy silane, which produces an alkylamine derivative. The preparation of its reactive intermediates has been discussed in the literature (32) (see Table 15). The pioneering work on silanization was done by Weetall and his coworkers at Corning Glass Works.

A method that permits the covalent binding of enzymes to porous glass has been developed by Cabral et al. (33). In this method the inorganic support is coated with the oxychloride derivative of transition metal salts, namely titanium and zirconium chlorides, which are subsequently reacted with a suitable diamine reagent in an organic hydrophobic solvent. This yields an alkylamine derivative similar to that obtained with the silanization process.

Table 13 Inorganic Supports for Protein and Enzyme Immobilization

Minerals	Fabricated materials
Bentonite	Controlled-pore glass
Attapulgite	Porous titania
Pumice stone	Porous alumina
Hornblende	Porous silica
Sand	Alumina catalyst
Diatomaceous earth	Silochrome
	Aluminosilicate
	Iron oxide
	Nickel/nickel oxide
	Stainless steel

Table 14 Organic Functional Groups of Silicane Coupling Agents for Protein and Enzyme Immobilization

Amine (aliphatic)	$-NH_2$
Amine (aromatic)	$-C_6H_4-NH_2$
Halide	$-I; -Br; -Cl$
Aldehyde	$-CHO$
Acetal	$-CH(OC_2H_5)_2$

```
|                                  |   |
O                                  O   O
|                                  |   |
M-OH                               M-O-Si-(CH2)nX
|                                  |   |
O    + (RO)3Si(CH2)nX  ---->       O   O
|                                  |   |
M-OH                               M-O-Si-(CH2)nX
|                                  |   |
O                                  O   O
|                                  |   |
```

Figure 10 Silanization method.

Table 15 Silane Derivatives of Inorganic Carriers for Enzyme Immobilization

Original derivatives	Activating reagents	Reactive derivative	Enzyme coupling method
$\vdash OSi(CH_2)_3NH_2$	$NO_2-C_6H_4-COCl$;	$\vdash OSi(CH_2)_3NHC(=O)-C_6H_4-N_2^+Cl^-$	Diazotization
	$Na_2S_2O_4$: HNO_2		
	$OHC(CH_2)_3CHO$	$\vdash OSi(CH_2)_3N{=}CH(CH_2)_3CHO$	Schiff's base formation
	Cl_2CSCl	$\vdash OSi(CH_2)_3NCS$	Peptide bond formation (isothiocyanato derivative)
	H_2NNH_2, HNO_2	$\vdash OSi(CH_2)_3N_3^+$	Peptide bond formation (azide derivative)
	$F-C_6H_3(NO_2)-N_3^+$	$\vdash OSi(CH_2)_3-C_6H_3(NO_2)-N_3^+$	Peptide bond formation (azide derivative)

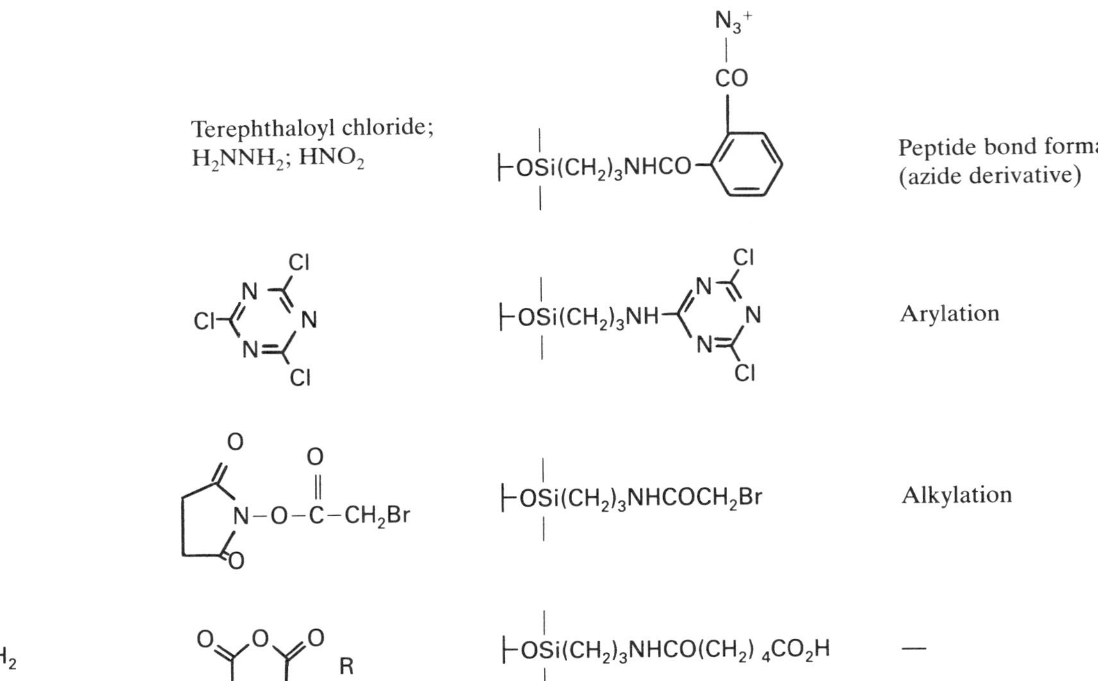
Terephthaloyl chloride; H_2NNH_2; HNO_2
$\vdash OSi(CH_2)_3NHCO-C_6H_4-CO-N_3^+$
Peptide bond formation (azide derivative)
Arylation
$\vdash OSi(CH_2)_3NHCOCH_2Br$
Alkylation
$\vdash OSi(CH_2)_3NH_2$
R
$\vdash OSi(CH_2)_3NHCO(CH_2)_4CO_2H$
—

Table 15 (Continued)

Original derivatives	Activating reagents	Reactive derivative	Enzyme coupling method
$\vdash OSi(CH_2)_3NHCO(CH_2)_4CO_2H$	$R-N=C=N=R'$	$\vdash OSi(CH_2)_3NHCO(CH_2)_4COOC(-NH-R)(=N-R')$	Peptide bond formation (condensing agents)
	$SOCl_2$	$\vdash OSi(CH_2)_3NHCO(CH_2)_4COCl$	Peptide bond formation (acyl chloride derivative)
	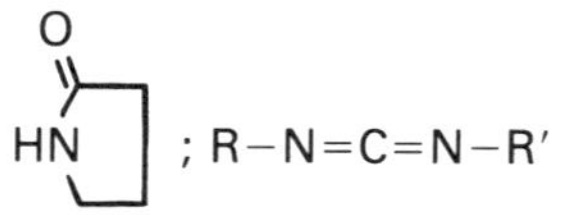 $\mathrm{O{=}C1CCCN1}$; $R-N=C=N-R'$	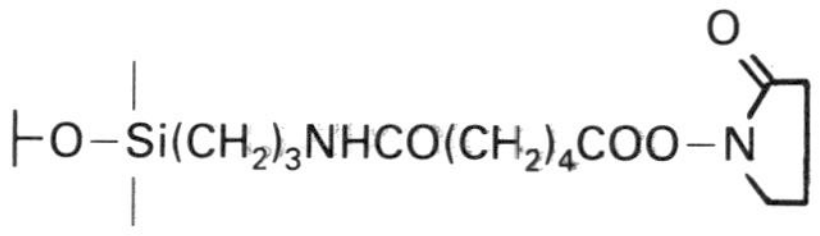 $\vdash O-Si(CH_2)_3NHCO(CH_2)_4COO-N1C(=O)CCC1$	Peptide bond formation

IV. COUPLING REACTIONS

A. Covalent Methods

The coupling of protein molecules to solid supports involves mild reactions between amino acid residues of the protein and several groups of functionalized carriers. The functional reaction groups of carriers for direct coupling with proteins, the major reacting groups of proteins, and the coupling reactions are listed in Table 16.

The simplest procedure for producing water-insoluble, surface-bonded, protein-support conjugates consists of contacting an enzyme solution with a preformed reactive carrier. However, only a few supports contain these reactive groups for direct coupling of proteins. Examples of such preformed reactive carrier synthetic polymers are maleic anhydride–based copolymers, methacrylic acid anhydride–based copolymers, nitrated fluorocryl methacrylic copolymers, and iodoalkylmethacrylates.

The support materials most commonly used, however, do not possess these reactive groups but rather hydroxy, amino, amide, and carboxy groups, which have to be activated for immobilization of proteins. The various methods of activation of these supports are described below.

As can be seen from Table 16, several types of coupling reactions can be used, and most of the common covalent coupling reactions involve coupling through protein amino groups, thiol groups, carboxy groups, or aromatic rings of L-tyrosine and L-histidine. The major classes of coupling reactions used for the immobilization of proteins are:

1. Diazotization
2. Amide (peptide) bond formation
3. Alkylation and arylation
4. Schiff's base formation
5. Ugi reaction
6. Amidination reactions
7. Thiol–disulfide interchange reactions
8. Mercury–enzyme interactions
9. γ-Irradiation induced coupling

1. Diazotization

The diazo coupling method is one of the most commonly used for coupling enzyme to carrier. Historically, it is among the oldest described in the literature (28,34). It is based on the diazo linkage between proteins and aryldiazonium electrophilic groups of the carrier. These diazonium carriers are prepared by treating the support containing aromatic amino groups

Table 16 Immobilization Methods of Proteins by Covalent Binding

Reactive group of support	Reacting group of protein	Coupling reaction
$-C_6H_4-N_2^+Cl^-$ Diazonium salt	$-NH_2$ $-SH$ $-C_6H_4-OH$	Diazotization
$-C(=O)-O-C(=O)-$ Acid anhydride	$-NH_2$	Peptide bond formation
$-CH_2CON_3$ Acyl azide	$-NH_2$ $-SH$ $-C_6H_4-OH$	Peptide bond formation
$(-O)_2C=NH$ Imidocarbonate	$-NH_2$	Peptide bond formation
$-R-NCS$ Isothiocyanate	$-NH_2$	Peptide bond formation
$-C_6H_4-NCO$ Isocyanate	$-NH_2$	Peptide bond formation
$-CH_2COCl$ Acyl chloride	$-NH_2$	Peptide bond formation
$(-O)_2C=O$ Cyclic carbonate	$-NH_2$	Peptide bond formation

Table 16 (Continued)

Reactive group of support	Reacting group of protein	Coupling reaction
$-COO-C(-NH-R^1)=^{+}NH-R^2$ *O*-Acylisourea	$-NH_2$	Peptide bond formation
$-COO-C(=CHCONHC_2H_5)-C_6H_4-SO_3^+$ Woodward's reagent K derivative	$-NH_2$	Peptide bond formation
F, NO_2, NO_2 (substituted benzene ring) δ-Fluoramdinitroanilide	$-NH_2$	Arylation
$-O-$ triazine ring with two Cl Triazinyl	$-NH_2$	Arylation
$-O-CH-CH_2$ (ring closed through X)	$-NH_2$	Alkylation
X = >NH, >O, >S, e.g., Oxirane	$-OH$ $-SH$	Alkylation

Table 16 (Continued)

Reactive group of support	Reacting group of protein	Coupling reaction
$-O-CH_2-CH_2-SO_2-CH=CH_2$ Vinylsulfonyl	$-NH_2$ $-SH$ $-OH$	Conjugation
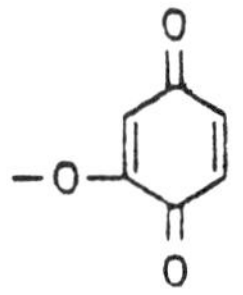 $-O-$ (benzoquinone ring with two C=O groups) Vinyl keto	$-NH_2$ $-SH$ $-OH$	Arylation
$-CHO$ Aldehyde	$-NH_2$	Schiff's base formation
$-\overset{+}{N}H=C(R^1)(R^2)$ Imine	$-CO_2H$ $-NH_2$	Ugi reaction
$-C(=NH)-OC_2H_5$ Imidoester	$-NH_2$	Amidination
$-CN$ Cyanide	$-NH_2$	Amidination
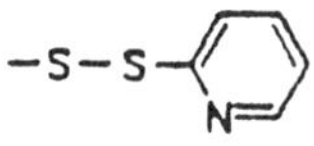 $-S-S-$(2-pyridyl) Disulfide residue	$-SH$	Thioldisulfide interchange
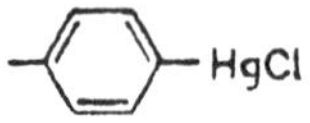 $-C_6H_4-HgCl$ Mercury derivative	$-SH$	Mercury–enzyme interaction
M′ Matrix radical	E′ Enzyme radical	γ-irradiated induced coupling
$-NH_2$ Amine	$-NH_2$ $-CO_2H$	Peptide bond formation (in presence of condensing agents)
$-CONHNH_2$ Acylhydrazide	$-NH_2$ $-CO_2H$	"

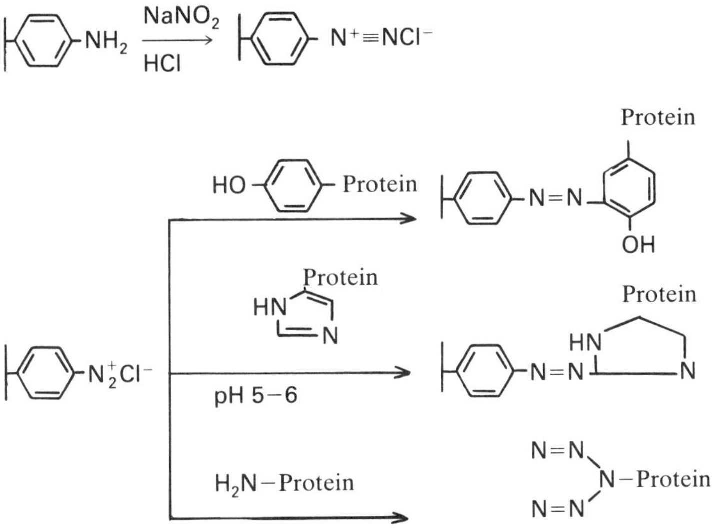

NH_2 $\xrightarrow[HCl]{NaNO_2}$ $N^+{\equiv}NCl^-$ (I)

Protein

HO— —Protein —N=N— OH (II)

Protein

HN N Protein HN —N=N— N (III)

$N_2^+Cl^-$ pH 5–6

N=N N–Protein N=N (IV)

H_2N–Protein

Scheme 2

with nitrite in an acidic medium (derivative I, Scheme 2). The functional groups on the protein participating in diazo coupling are mainly activated aromatic rings such as phenol (L-tyrosine) and imidazole (L-histidine), to form the corresponding azo derivatives.

In addition to these groups, several other amino acids in proteins react under similar conditions with the diazonium salt to give the azo derivatives II and III or the disubstituted bisazo derivative of primary amines (IV), as shown in Scheme 2. The guanido group (L-arginine) and indole (L-tryptophan) can also be attacked by the electrophilic aryldiazonium ion.

Examples of immobilized enzymes that have been prepared covalently by the diazo coupling method are listed in Table 17.

2. Amide Bond Formation

The amide binding method is based on the formation of amide (peptide) bonds, or similar bonds, between the protein and a water-insoluble carrier by the application of the peptide synthesis technique. Several procedures are available depending on the different derivatized solid matrices (see Table 16).

The common mechanistic feature in the peptide bond formation method is the attack of the nucleophilic groups (amino, hydroxy, thiol) of protein at an activated functional group on the carrier. These nucleophiles are most effective in their unprotonated forms ($-NH_2$, $-C_6H_4-O^-$, $-S^-$) at pH values above their pK_a values. Owing to the irreversible denaturation of the pro-

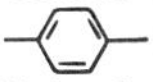

Table 17 Examples of Immobilized Enzymes Prepared by Diazotization Coupling

Carrier	Enzyme	Ref.
Organic supports		
Cellulose	β-D-Galactosidase	35
	D-Glucose oxidase	35
	Papain	35
	Pepsin	35
	Trypsin	35,36
Enzoacryl AA	β-D-Xylosidase	37
	Dextransucrase	38
	Tyrosinase	39
Styrene/maleic anhydride	Chymotrypsin	40
Glycidyl methacylate/ethylene dimethyacrylate	Penicillin amidase	41
Polyacrylamide/nylon	Urease	42
Polythene terephtalate	Trypsin	43
Inorganic supports		
Controlled pore glass	Aldehyde dehydrogenase	44
	Phospholipase A_2	45
Controlled pore silica	Hydrogenase	46

tein at high pH, however, these reactions are commonly carried out at intermediate pH values (7.5–8.5) and at low temperature (4°C).

Table 18 lists some immobilized proteins that have been prepared by the amide bond formation method.

Acid Anhydride Derivatives. Acid anhydrides react with amino groups of proteins as shown in Scheme 3. As can be seen in this scheme, carboxylic acid groups are formed (in addition to the amide linkages) by cleavage of anhydride rings; these do not react with functional groups of protein but are ionized in the aqueous solutions spontaneously by OH^- ions, that is, at higher pH values, generating free carboxy groups with a highly negative charge. The negative charge gives undesirable ionogenic properties to the preparation and may unfavorably affect the protein structure. To overcome

$$\text{cyclic anhydride } (-C(=O)-O-C(=O)-) \xrightarrow{H_2N-\text{Protein}} -C(=O)-NH-\text{Protein} \;+\; -C(=O)-OH$$

Scheme 3

Table 18 Examples of Immbolized Enzymes Prepared by Amide Bond Formation

Carrier	Enzyme	Ref.
Acid Anhydride Derivatives		
Ethylene/maleic anhydride	Alkaline phosphatase	47
Butanediol divinylether/anhydride	Lactate dehydrogenase	48
	Trypsin	48
	Chymotrypsin	48
	Papain	48
	Bromelain	58
	Subtlisin	48
Methylvinylether/maleic anhydride	Alkaline phosphatase	47
Acyl Azide Derivatives		
Enzacryl AH	β-D-Xylosidase	37
	Tyrosinase	49
Glycidyl methacrylate/ethylene dimethacrylate	Penicillin amidase	41
Poly(vinyl alcohol)	β-D-Glucosidase	49
Nylon/polyacrylamide	β-D-Galactosidase	50
	Papain	50
Collagen	D-Glucose oxidase	51,52
	Aspartate aminotransferase	20,53
	Luciferase	54
Controlled pore silica	Parathion hydrolase	55
Cyclic Imidocarbonate Derivatives		
Agarose	Tyrosinase	39
	β-D-Galactosidase	56
	Hydroxysteroid dehydrogenase	57
	Xanthine oxidase	58,59
	Superoxide dismutase	58
	Catalase	58
	Trypsin	60
	Bilirubin oxidase	61
	Cellulase	62
	Aldehyde oxidase	63
	Multienzyme enniatin synthetase	64
Cellulose	Xanthine oxidase	59
Hornblende	Xanthine oxidase	59
Glass beads	Chymotrypsin	65
Isocyanate and Isothiocyanate Derivatives		
Cellulose	β-Amylase	66
Polyurethane	β-D-Fructofuranosidase	67

Table 18 (Continued)

Carrier	Enzyme	Ref.
Polyethylene terephtalate	Trypsin	43
Glycidyl methacrylate/ethylene dimethacrylate	Penicillin amidase	41
Activated carbon	D-Glucose oxidase	26
Polyethylene glycol	Cellulase	68
Acyl Chloride Derivatives		
Amberlite[R] IRC-50	Triacylgylcerol lipase	69
Bentonite	Dextranase	70
Cyclic Carbonate Derivatives		
Cellulose trans-2,3-carbonate	Dextranase	71
	β-D-Glucosidase	72,73
	Trypsin	74
Carbodiimide Mediated Derivatives		
Akrilex[R]	Hexokinase	75
	3-Phosphoglycerate kinase	76
	Glucoamylase	77
	Glucose-phosphate isomerase	78
Bone	Invertase	79
	Pepsin	79
	Pectinase	79
	Lactase	79
	Catalase	79
Agarose	β-D-Xylosidase	37
Activated Carbon	D-Glucose oxidase	80
	Glucoamylase	80
	D-Gluconolactonase	80
Glass	Trypsin	81

this disadvantage it is usual to neutralize some of the negatively charged species by the addition of diamines during the immobilization procedure.

Acyl Azide Derivatives. The azide method is one of the most frequently employed procedures in the field of peptide synthesis and is one of the oldest methods of immobilizing proteins (enzymes). The azide carriers are prepared from hydroxy or carboxy supports, as indicated in Scheme 4.

The acyl azide derivative reacts with protein involving predominantly the primary amino groups of protein to form the amide linkage.

Side reactions with aliphatic or aromatic hydroxy groups or thiol groups occur under the same conditions.

Cyclic Imidocarbonate Derivatives. The imidocarbonate method is probably one of the most commonly used for laboratory preparations of immo-

$$\vdash OH \xrightarrow[NaOH]{ClCH_2CO_2H} \vdash OCH_2CO_2H$$

$$\vdash OCH_2CO_2H \xrightarrow[HCl]{CH_3OH} \vdash OCH_2CO_2CH_3$$

$$\vdash OCH_2CO_2CH_3 \xrightarrow{H_2NNH_2} \vdash OCH_2CONHNH_2$$

$$\vdash OCH_2CONHNH_2 \xrightarrow[HCl]{NaNO_2} \vdash OCH_2CON_3$$

$$\vdash OCH_2CON_3 + H_2N\text{–Protein} \rightarrow \vdash OCH_2CONH\text{–Protein}$$

Scheme 4

bilized protein derivatives as well as insoluble adsorbents for affinity chromatography. It was first developed by Axen et al. (8).

The method consists in the reaction of cyanogen halide (mostly CNBr) with hydroxy groups of the carrier at high pH (10–11.5) to give an inert carbamate (II) and reactive imidocarbonate (III) through the intermediate cyanide (I), which is very labile, as indicated in Scheme 5.

Immobilization is then carried out in a basic medium (pH 9–10) and results in the formation of the different types of structures: N-substituted isoureas (IV), N-substituted imido carbonates (V), and N-substituted carbamates (VI), as shown in Scheme 5. However, it seems from additional evidence accumulated in the past few years that major reaction products of the imidocarbonates with amines are probably the substituted isourea structures.

Although this immobilization procedure is simple, there are some disadvantages such as the high toxicity of cyanogen bromide, the high pH of the derivatizing and coupling reactions, and also the presence in the products of charged groups, which cause interfering and unreliable nonspecific adsorptions.

An analogous active product is obtained by reaction of hydroxy groups of polysaccharides with alkyl or aryl cyanates (Scheme 6). The results obtained are similar to those from the cyanogen activation, but the cyanate method has not become widely used, possibly because the preparation of the alcohol cyanate involves the reaction of cyanogen halide with alcohol or phenol and thus is not only one reaction step longer but requires continued handling of toxic compounds. Furthermore, cyanogen bromide activated polysaccharide matrices are available commercially.

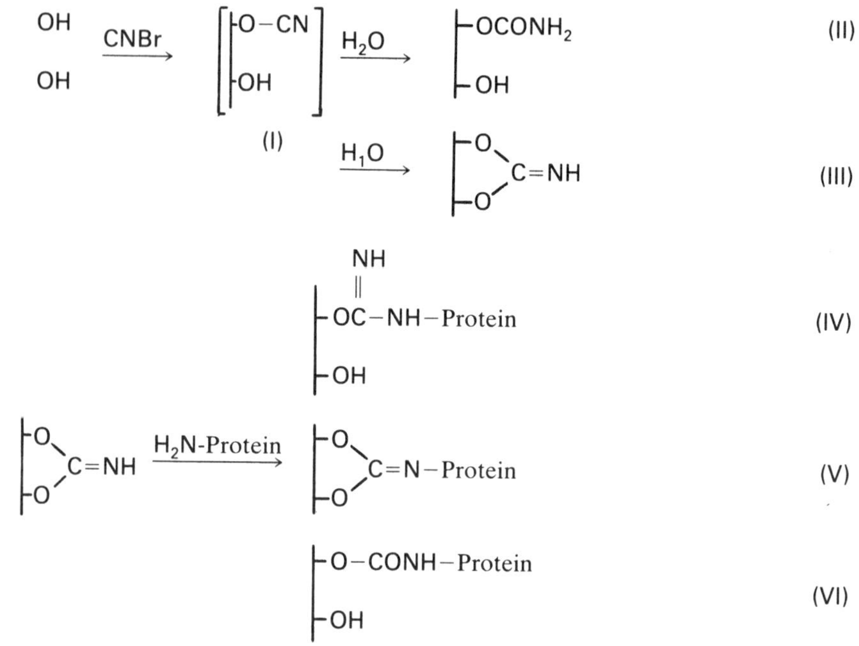

$$\begin{matrix}\text{OH} \\ \text{OH}\end{matrix} \xrightarrow{\text{CNBr}} \left[\begin{matrix}\vdash\text{O}-\text{CN} \\ \vdash\text{OH}\end{matrix}\right] \xrightarrow{H_2O} \begin{matrix}\vdash\text{OCONH}_2 \\ \vdash\text{OH}\end{matrix} \qquad \text{(II)}$$

(I)

$$\xrightarrow{H_1O} \begin{matrix}\vdash\text{O} \\ \vdash\text{O}\end{matrix}\!\!>\text{C}=\text{NH} \qquad \text{(III)}$$

$$\begin{matrix}\vdash\text{O}\overset{\text{NH}}{\overset{\|}{\text{C}}}-\text{NH}-\text{Protein} \\ \vdash\text{OH}\end{matrix} \qquad \text{(IV)}$$

$$\begin{matrix}\vdash\text{O} \\ \vdash\text{O}\end{matrix}\!\!>\text{C}=\text{NH} \xrightarrow{H_2N\text{-Protein}} \begin{matrix}\vdash\text{O} \\ \vdash\text{O}\end{matrix}\!\!>\text{C}=\text{N}-\text{Protein} \qquad \text{(V)}$$

$$\begin{matrix}\vdash\text{O}-\text{CONH}-\text{Protein} \\ \vdash\text{OH}\end{matrix} \qquad \text{(VI)}$$

Scheme 5

$$\begin{matrix}\vdash\text{OH} \\ \vdash\text{OH}\end{matrix} \xrightarrow{\text{ROCN}} \begin{matrix}\vdash\text{O}-\overset{\text{NH}}{\overset{\|}{\text{C}}}\text{OR} \\ \vdash\text{OH}\end{matrix} \longrightarrow \begin{matrix}\text{O} \\ \text{O}\end{matrix} \quad \text{C}=\text{NH}$$

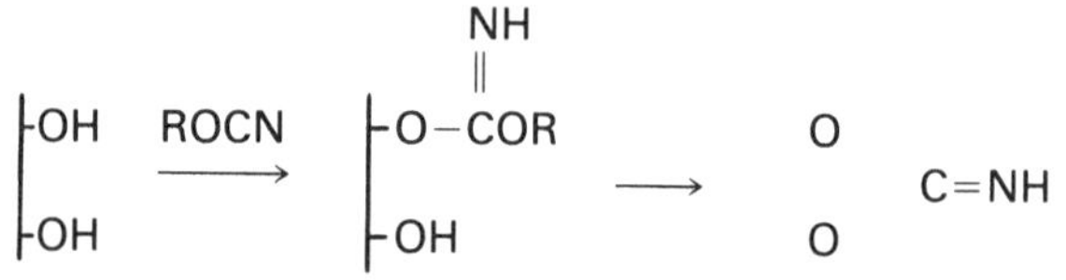

Scheme 6

Isocyanate and Isothiocyanate Derivatives. Carriers containing aromatic amino acid and acyl azide groups can be modified at alkaline pH to the corresponding isocyanate or isothiocyanate derivatives, by reaction, respectively, with phosgene or thiophosgene. These isocyanate or isothiocyanate functional groups have been used as electrophiles. They react with free amino groups of the protein to form an amide or thioamide linkage of the substituted urea or thiourea (Scheme 7).

The reaction with other nucleophiles of the proteins, such as thiol, imidazole, or aromatic hydroxy and carboxy groups, leads to relatively unstable derivatives that decompose at mildly alkaline pH values or in the presence of nucleophiles such as hydrazine (2).

$$\vdash NH_2 \xrightarrow{CXCl_2} \vdash N{=}C{=}X \xrightarrow{H_2N-\text{Enzyme}} \vdash NH\overset{X}{\overset{\|}{C}}-NH-\text{Enzyme}$$

$X = -O$ or $-S$

$$\vdash CON_3 \xrightarrow{HCl} \vdash N{=}C{=}O \xrightarrow{H_2N-\text{Enzyme}} \vdash NHCONH-\text{Enzyme}$$

Scheme 7

The isothiocyanate derivative has a higher stability and for this reason it has been used more frequently than the isocyanate derivative. The isothiocyanate derivative can also be obtained from aliphatic amino-type carriers.

Acyl Chloride Derivatives. Carboxy functional supports can be activated by refluxing in a solution containing thionyl chloride: An acyl chloride derivative is obtained (derivative I, Scheme 8). The acyl chloride derivative can be reacted with proteins at low temperatures for immobilization according to Scheme 8.

This method of immobilization was used by Brandenberger (69) in 1956 for the immobilization of triacylglycerol lipase (EC 3.1.1.3) on the carboxychloride derivative of a cation exchanger, Amberlite IRC-50, and this report is considered to be the first on the amide binding methods.

Cyclic Carbonate Derivatives. A simple method to attach to the carrier under very mild conditions is described by Barker et al. (74). The activated support is obtained by the reaction of the hydroxy groups of the support with ethylchloroformate in a dimethylsulfoxide medium with the addition of dioxane and triethylamine. It is completed within a few minutes (derivative I, Scheme 9).

Immobilized protein preparations can be obtained by reaction of nucleophilic groups of the protein at pH 7–8 with various modes of binding, obtaining N-substituted imidocarbonate (II) and N-substituted carbonate (III) (Scheme 9), as with the cyclic imidocarbonate method.

By comparison with the cyclic imidocarbonate reaction method using cyanogen bromide, the work with ethyl chloroformate or phosgene is much

$$\vdash CO_2H \xrightarrow{SOCl_2} \vdash COCl \qquad \text{(I)}$$

$$\vdash COCl + H_2N-\text{Protein} \rightarrow \vdash CONH-\text{Protein} \qquad \text{(II)}$$

Scheme 8

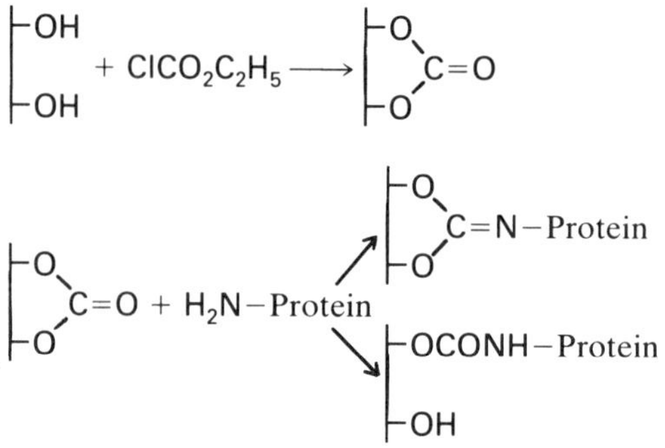

$$\left.\begin{matrix}\vdash OH \\ \vdash OH\end{matrix}\right. + ClCO_2C_2H_5 \longrightarrow \left.\begin{matrix}\vdash O \\ \vdash O\end{matrix}\right\rangle C{=}O \qquad (I)$$

$$\left.\begin{matrix}\vdash O \\ \vdash O\end{matrix}\right\rangle C{=}O + H_2N{-}\text{Protein} \nearrow \left.\begin{matrix}\vdash O \\ \vdash O\end{matrix}\right\rangle C{=}N{-}\text{Protein} \qquad (II)$$

$$\searrow \begin{matrix}\vdash OCONH{-}\text{Protein} \\ \vdash OH\end{matrix} \qquad (III)$$

Scheme 9

cheaper, simpler, and, in the case of ethyl chloroformate, less toxic. However, it has not been used much.

Condensing Reagents. The most general methods for activation of supports containing carboxy groups involve the use of carbodiimides and similar reagents. As shown in Scheme 10, carbodiimides react with carboxy groups at slightly acid pH values to give O-acylisourea derivatives (I), and the activated carboxy groups subsequently can rearrange to an acyl urea (II) or react with amino groups of the protein to yield the corresponding amides (III). The intermediate can also react with other nucleophiles, e.g., hydroxy and thiol groups, to give different carboxylate derivatives but at much lower reaction rates.

Similar carbodiimide reagents, such as Woodward's reagent K, can be used in the same way, as shown in Scheme 11.

These types of condensing reagents also have been used to activate carriers containing amino or hydrazide groups to bind proteins through their carboxy groups. The main difference from the former process is that the carrier and the condensing reagent are added simultaneously to the

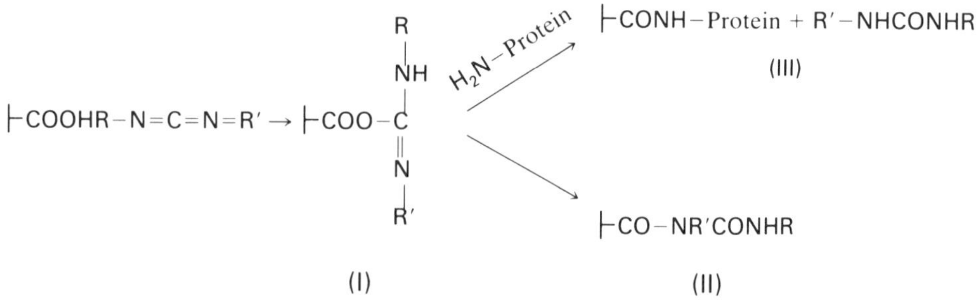

$$\vdash COOH R{-}N{=}C{=}N{=}R' \rightarrow \vdash COO{-}\underset{\underset{\underset{R'}{|}}{N}}{\overset{\overset{\overset{R}{|}}{NH}}{\underset{\|}{\overset{|}{C}}}} \;(I) \xrightarrow{H_2N-\text{Protein}} \vdash CONH{-}\text{Protein} + R'{-}NHCONHR \;(III)$$

$$\longrightarrow \vdash CO{-}NR'CONHR \;(II)$$

Scheme 10

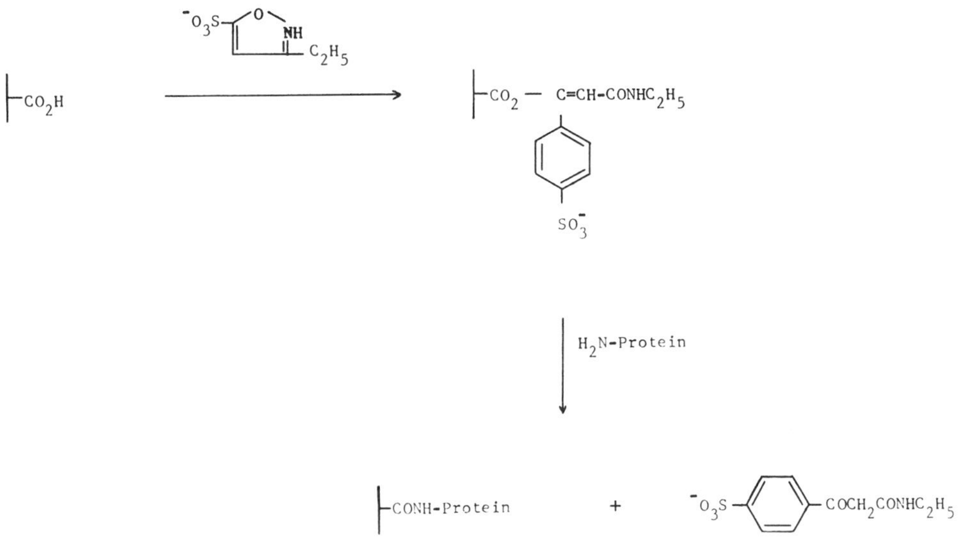

$^{-}O_3S$ (isoxazolium, NH, C_2H_5)

$\vdash CO_2H$ → $\vdash CO_2 - C{=}CH{-}CONHC_2H_5$ (p-SO_3^--phenyl)

↓ H_2N-Protein

$\vdash CONH$-Protein + $^{-}O_3S-C_6H_4-COCH_2CONHC_2H_5$

Scheme 11

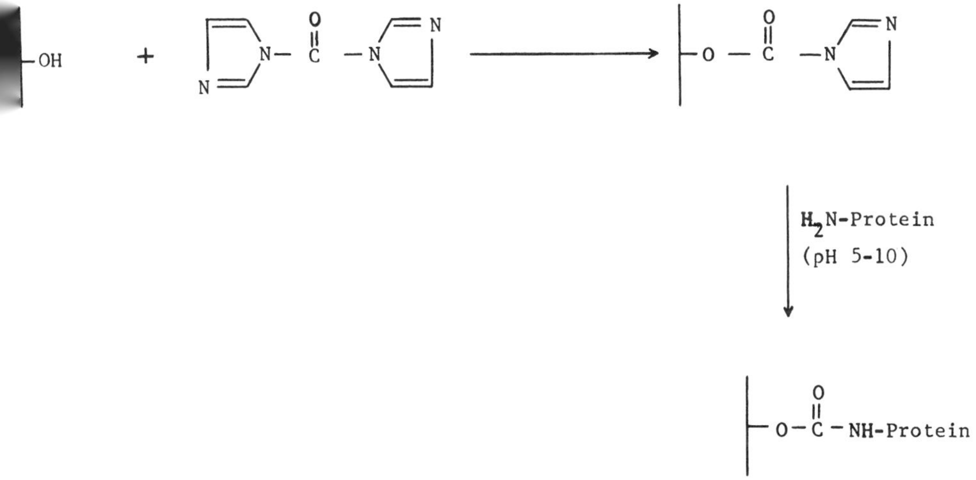

$\vdash OH$ + (Im)N–C(=O)–N(Im) → $\vdash O-C(=O)-N$(Im)

↓ H_2N-Protein (pH 5-10)

$\vdash O-C(=O)-NH$-Protein

Scheme 12

protein solution and stirred. Amide bonds are formed between amino groups of the carrier and carbonyl groups of the protein thereby effecting an immobilization.

The free hydroxyl groups on polysaccharides gels, glycerol propyl-silica, and other hydroxyl copolymers supports can be converted by carbonyl-diimidazole (Scheme 12) into imidazolyl-carbonates (I) which, on reaction with *N*-nucleophiles, result in the formation of immobilized protein preparation.

This method has been used for both immobilization of enzymes and affinity ligands for chromatography of proteins (82).

3. Alkylation and Arylation

The alkylation and arylation methods based on the alkylation of amino, phenolic, or thiol groups of a protein with reactive supports containing

Table 19 Examples of Immobilized Proteins Prepared by Alkylation and Arylation

Carrier	Enzyme	Ref.
Halogeno Acetyl Derivatives		
Bromoacetyl cellulose	Aminoacylase	83
	Glucoamylase	84
Chloroacetyl cellulose	Aminoacylase	83
Iodoacetyl cellulose	Aminoacylase	83
	Glucoamylase	84
Triazinyl Derivatives		
Cellulose	Subtilisin	85
	Glucoamylase	86
	Trypsin	36
Bentonite	Dextranase	70
Agarose	Chymotrypsin	87
	Trypsin	87
	Lactate dehydrogenase	87
Oxirane Derivatives		
Alginic acid/Poly(glycidyl methacrylate)	Trypsin	88
Eupergit C	Carboxypeptidase	89
	Endopolygalacturonate	90
	β-D-Galactosidase	91
	Glucose oxidase	92
Tresylate Derivatives		
Agarose	α-Chymotrypsin	93,95
	T4DNA ligase	93
	Alcohol dehydrogenase	93
	β-Galactosidase	93
	Concanavalin A	93
	Protein A	93
	12-α-Hydroxysteroid dehydrogenase	94
Silica	Alcohol dehydrogenase	93
	Soybean trypsin inhibitor	93

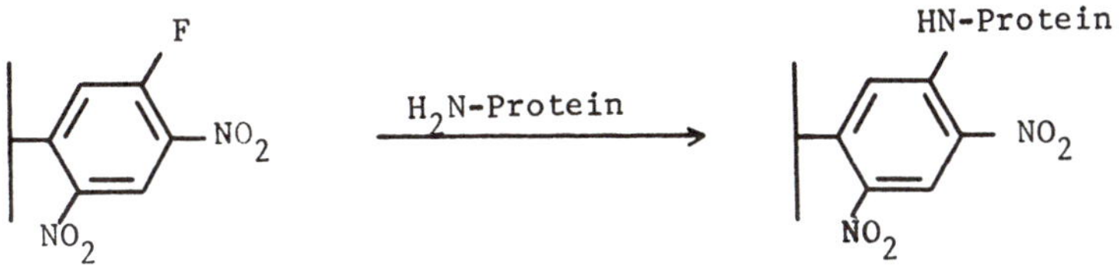

F

NO_2

NO_2

H_2N-Protein

HN-Protein

NO_2

NO_2

Scheme 13

active halides, oxirane, vinylsulfonyl, or vinylketo groups. Proteins that have been immobilized by this method are listed in Table 19.

Supports with halogen-substituted aromatic rings containing nitro groups, such as copolymers of methacrylate and methacrylate-5-fluoro-2,4-dinitroanilide, have been used to couple the amino groups of protein via arylation, as indicated in Scheme 13. This support is synthesized from monomeric units that contain reaction groups, and no activation is required.

However, this method can be used with less reactive supports, in which a halogen-substituted heterocyclic ring is incorporated by reaction with cyanuric chloride or its derivatives (Scheme 14).

Jagendorf et al. (96) introduced a method using carriers with monohalogenoacetyl functional groups which for immobilization of protein involved alkylation of the amino groups on the protein (Scheme 15). Although other nucleophiles on the protein can be used, such as thiol, this method has been applied in a limited number of cases.

Among these three halides, iodide confers the best reactivity stability properties to the carrier. The use of chlorides is disadvantageous; although the activation is performed readily in organic solvents, the transfer into aqueous solution when the protein immobilization proceeds is accompanied by a competitive hydrolysis reaction of the chloride, which decreases the activity of the carrier and introduces a large negatively charged carboxy group. A substantial disadvantage of this technique overall is the formation between the carrier and the halogenoacetyl group of ester bonds that are sometimes insufficiently stable.

A more stable bond can be introduced by the reaction of a support matrix with dihalogenoketone or epichlorohydrin (Scheme 16).

Another possibility for activation of carriers containing hydroxy groups involves the introduction of oxiranes, highly strained three-membered electrophilic cycles (Scheme 17), into the matrix. However, epithio groups are relatively stable and not directly suited for bonding enzymes.

The most frequently used are epoxide rings, particularly for their easy accessibility. The activation of the carrier with epoxide compounds is shown in Scheme 18.

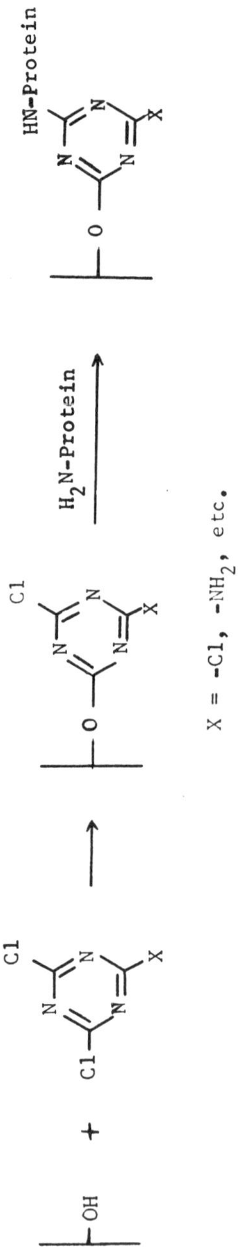

—OH + Cl, Cl, X (triazine) → —O— (triazine, Cl, X) —H_2N-Protein→ —O— (triazine, HN-Protein, X)

X = -Cl, -NH_2, etc.

Scheme 14

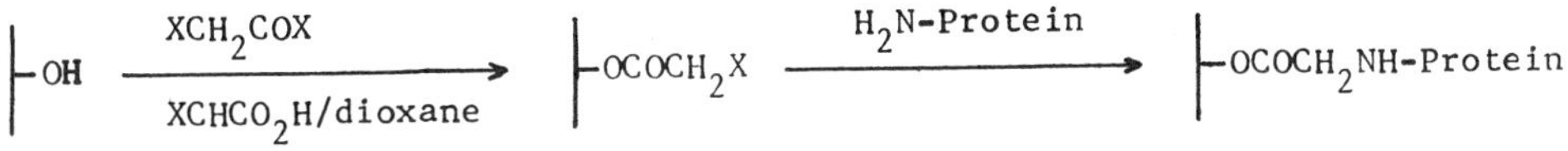

$$\vdash OH \xrightarrow[XCHCO_2H/dioxane]{XCH_2COX} \vdash OCOCH_2X \xrightarrow{H_2N\text{-}Protein} \vdash OCOCH_2NH\text{-}Protein$$

X= -Cl, -Br, -I

Scheme 15

$$\vdash OH + \underset{\diagdown O \diagup}{CH_2 - CHCH_2Cl} \longrightarrow \vdash O\text{-}CH_2\text{-}\underset{OH}{CH}\text{-}CH_2Cl$$

Scheme 16

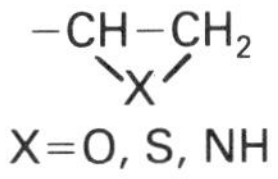

$$-\underset{\diagdown X \diagup}{CH-CH_2}$$

X=O, S, NH

Scheme 17

$$\vdash OH + \underset{\diagdown O \diagup}{CH_2-CH}\sim\underset{\diagdown O \diagup}{CH-CH_2} \rightarrow \vdash OCH_2\underset{OH}{CH}\sim\underset{\diagdown O \diagup}{CH-CH_2}$$

$$\vdash \underset{\diagdown O \diagup}{CH-CH_2} + H_2N-\text{Protein} \rightarrow \vdash \underset{OH}{CH}-CH_2NH-\text{Protein}$$

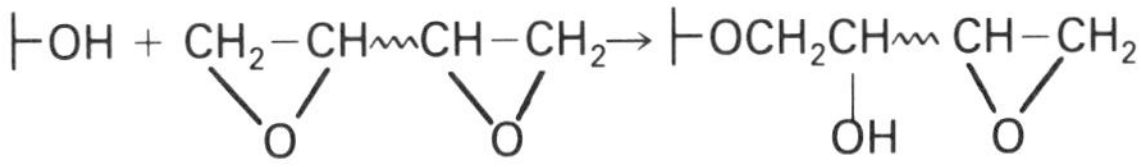

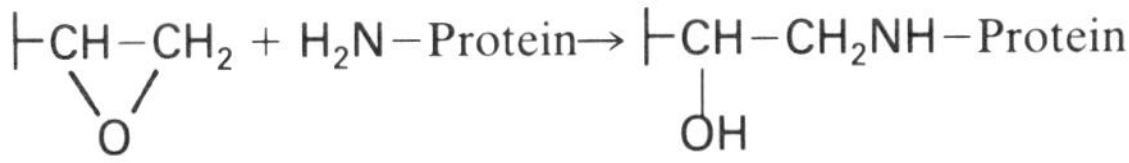

Scheme 18

The reactivity of this epoxide derivative towards nucleophilic groups of the proteins has the known order of thiol > amino > hydroxy. Nucleophilic attack on aliphatic hydroxy groups requires a strongly basic medium (pH ~11), whereas the amino groups react at a lower pH and thiol groups at a still lower pH value (pH 7). The reaction is depicted in Scheme 18. Aromatic hydroxy (L-tyrosine), guanidino (L-arginine), and imidazole (L-histidine) may also react.

The linkages of type C-N, C-O, and C-S formed between the support and protein are extremely stable, and no charged groups arise during their realization, thus reducing or avoiding the nonspecific adsorptions (97). The reaction with proteins is slow, however, and the reaction time may have to be extended to several days or even weeks.

Alkyl sulfonate esters, $R'SO_2OR''$ have been used for affinity ligand and protein immobilization onto hydroxyl group carrying supports, such as

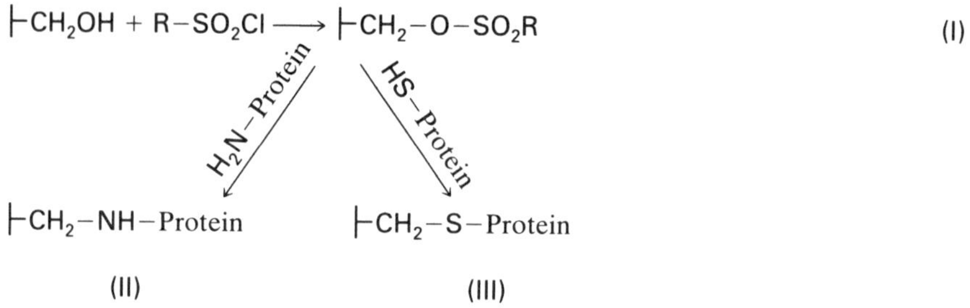

$$\vdash CH_2OH + R-SO_2Cl \longrightarrow \vdash CH_2-O-SO_2R \quad (I)$$

H_2N–Protein ↙ ↘ HS–Protein

$\vdash CH_2-NH-$Protein (II) $\vdash CH_2-S-$Protein (III)

R = CH_2CF_2 (tresylates); $C_6H_4CH_3$ (tosylates)

Scheme 19

agarose, cellulose, glycophase glass, and glycerolpropyl-silica (93). The activation and coupling of the protein to the support are thought to involve the steps shown in Scheme 19.

Vinylsulfonyl groups can be introduced into hydroxy-containing polymers by treatment of the latter with divinylsulfone in alkali (98) (derivative I, Scheme 20). The unavoidable simultaneous crosslinking improves the properties of original carrier (98).

The nucleophilic attacks of functional groups of enzyme follow in the same order as for oxirane: thiol > amino > hydroxy. They proceed at somewhat lower pH values, but also at somewhat higher reaction velocity. The main disadvantage is a slow release of the low molecular weight ligand at pH > 8. The interaction of the activated support with amino groups of proteins can be seen in Scheme 20.

The hydroxy groups of the carrier can also be more readily attacked by quinones (99) (Scheme 21). Proteins are bound analogously to Scheme 20, that is, through their nucleophilic groups with high coupling yields. Actu-

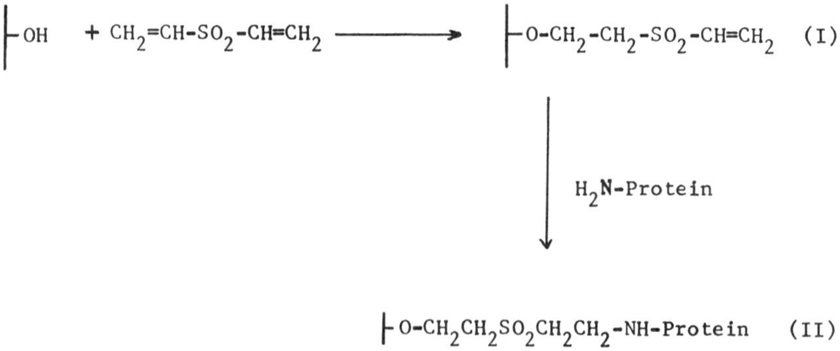

$$\vdash OH + CH_2{=}CH-SO_2-CH{=}CH_2 \longrightarrow \vdash O-CH_2-CH_2-SO_2-CH{=}CH_2 \quad (I)$$

↓ H_2N-Protein

$$\vdash O-CH_2CH_2SO_2CH_2CH_2-NH-\text{Protein} \quad (II)$$

Scheme 20

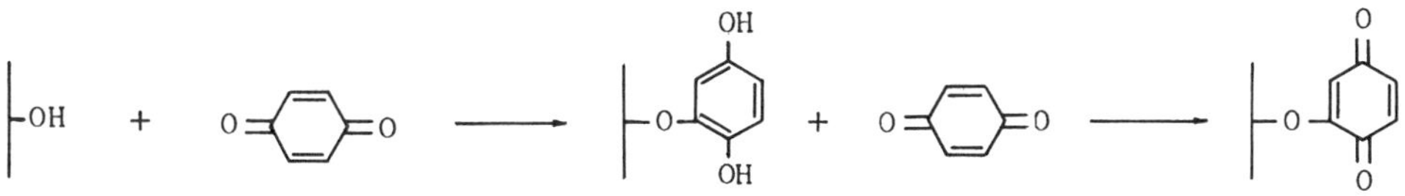

Scheme 21

ally, the coupling reaction can occur in the broad region of pH 3–10. However, undesirable side reactions are obtained, indicated by the color of the products.

4. Schiff's Base Formation

This method is based on the formation of a Schiff's base (aldimine) linkage between carbonyl groups of the activated support and free amino groups on the protein (Table 20).

Supports containing primary amino groups have been activated with glutaraldehyde, initially introduced as a crosslinking reagent, to yield an active carbonyl derivative by formation of a Schiff's base (derivatives I, Scheme 22).

Aldehyde derivatives react with amino groups on the protein to form a Schiff's base linkage (Scheme 22). Sulfhydryl and imidazole groups may undergo similar reactions. One disadvantage of this method is the reversibility of Schiff's base formation in aqueous media, particularly at low pH values. This can be avoided, however, by hydrogenation of the imine bonds with $NaBH_4$ to yield stable alkylamino groups.

Although the activation of amino carriers with glutaraldehyde is based on the idea that one aldehyde group reacts with the solid matrix forming a Schiff's base and the other remains for the further reaction with the protein, the nature of the reaction with protein and amino carriers is not fully understood. Proteins are bound irreversibly to the glutaraldehyde-treated carrier by a reaction presumably analogous to that occurring during in-

$$\vdash NH_2 + OHC(CH_2)_3CHO \longrightarrow \vdash N{=}CH(CH_2)_3CHO \quad (I)$$

$$\vdash CHO + H_2N\text{-Protein} \longrightarrow \vdash CH{=}N\text{-Protein} \quad (II)$$

Scheme 22

Table 20 Examples of Immobilized Proteins Prepared by Schiff's Base Formation

Carrier	Enzyme	Ref.
Porous silica	Cellobiase	100
	Glutamate decarboxylase	101
Corn stover	Glucoamylase	102,103
Bentonite	Dextranase	70
Glass	Cellulase	104
	D-Xylanase	104
Controlled pore glass	NAD$^+$-synthetase	105
	Cellulase	106
	D-Glucose oxidase	106
	Peroxidase	106
	Pyruvate kinase	107
	Lactate dehydrogenase	107
	Naringinase	108,109
	Acetate kinase	110
	Phosphotransacetylase	110
	Glucoamylase	111
Polystyrene	Anti-S-methylcytosine antibodies	112
Separon^R H1000	Dipeptide Gly-DL-Phe	113
Sorba ide	Endopolygalacturonase	114
Bone	β-D-Fructofuranosidase	79
	Pepsin	79
	Pectinase	79
	Lactase	79
	Catalase	79
Cellulose	Penicillinase	115
	β-D-Galactosidase	35
	D-Glucose oxidase	35
	Trypsin	35
	Papain	35
	Pepsin A	35

termolecular cross-linking with the same reagents. The intermolecular crosslinking of a protein with glutaraldehyde is shown in Scheme 23. The linkages formed between the enzyme and glutaraldehyde are irreversible and survive extremes of pH and temperature. However, the straightforward participation of the reagent's aldehyde groups to form an aldimine (Schiff's base) bond with protein amino groups has been questioned. It has been suggested that the glutaraldehyde reaction most probably involves conjugate addition of enzyme amino groups to ethylene double bonds of

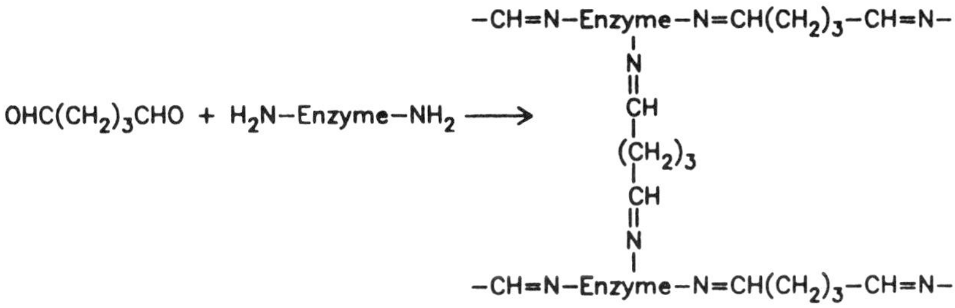

$$OHC(CH_2)_3CHO + H_2N\text{–Enzyme–}NH_2 \longrightarrow$$

$$\begin{array}{c} -CH{=}N\text{–Enzyme–}N{=}CH(CH_2)_3\text{–}CH{=}N- \\ | \\ N \\ \| \\ CH \\ | \\ (CH_2)_3 \\ | \\ CH \\ \| \\ N \\ | \\ -CH{=}N\text{–Enzyme–}N{=}CH(CH_2)_3\text{–}CH{=}N- \end{array}$$

Scheme 23

$$OHC(CH_2)_3CHO$$

$$\Updownarrow$$

$$-CH{=}\underset{}{\overset{CHO}{C}}\text{–}(CH_2)_2\text{–}CH{=}\overset{CHO}{C}\text{–}(CH_2)_2\text{–}CH{=}\overset{CHO}{C}\text{–}(CH_2)_2-$$

$$\Big\downarrow \text{NH–Enzyme}$$

$$-\underset{\text{NH–Enzyme}}{CH}\text{–}\overset{CHO}{CH}\text{–}(CH_2)_2\text{–}CH{=}\overset{CHO}{C}\text{–}(CH_2)_2\text{–}\underset{\text{NH–Enzyme}}{CH}\text{–}\overset{CHO}{CH}\text{–}(CH_2)_2-$$

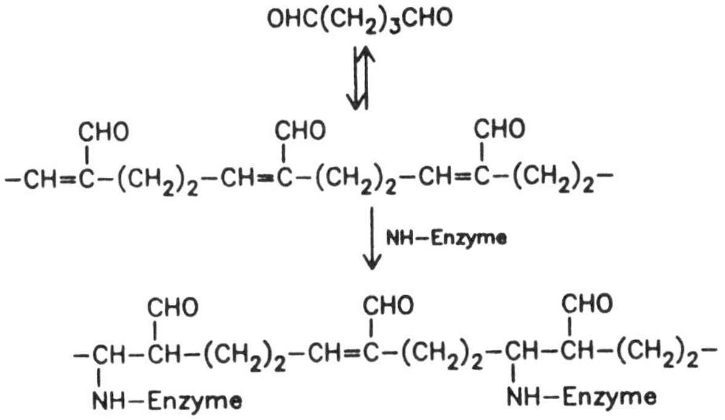

Scheme 24

α,β-unsaturated oligomers, contained in the commercial aqueous glutaraldehyde solutions usually used (Scheme 24).

According to Richards and Knowles (119) this mechanism explains the stability of the bond, which cannot be due to a simple Schiff's base formation, and it also explains the lower reactivity on proteins, including enzymes, of solutions of freshly distilled glutaraldehyde.

5. Ugi Reaction

In 1962, Ugi (116) described a reaction involving four functional groups simultaneously—carboxylate, amine, aldehyde, and isocyanide—resulting in the formation of an N-substituted amide (Scheme 25). In the first step, a protonated Schiff's base (immonium ion) (I) is produced by the reaction between the amino and the carbonyl compounds. This ion complexes with isocyanide compound and with a carboxylate ion; the three-component complex then undergoes an intramolecular rearrangement to give the *N*-substituted amide (II). The reaction proceeds in acidic medium and is mild enough to be used for protein immobilization.

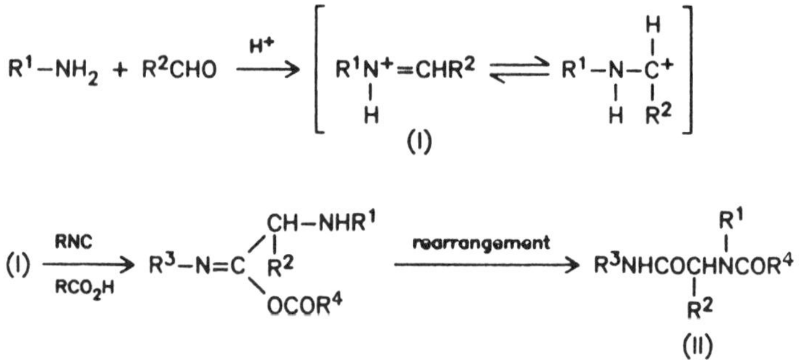

$$R^1{-}NH_2 + R^2CHO \xrightarrow{H^+} \left[R^1\underset{\displaystyle H}{\underset{|}{N^+}}{=}CHR^2 \rightleftharpoons R^1{-}\underset{\displaystyle H}{\underset{|}{N}}{-}\overset{\displaystyle H}{\overset{|}{\underset{\displaystyle R^2}{\underset{|}{C^+}}}} \right] \quad (I)$$

$$(I) \xrightarrow[RCO_2H]{RNC} R^3{-}N{=}C\begin{matrix} CH{-}NHR^1 \\ | \\ R^2 \\ OCOR^4 \end{matrix} \xrightarrow{\text{rearrangement}} R^3NHCOCHNCOR^4 \;(\text{with } R^2 \text{ on CH and } R^1 \text{ on N}) \quad (II)$$

Scheme 25

This method was optimized for proteins by Axen et al. (117). It is versatile with respect to the location of the participating functional groups. Any of the reacting groups (-CO_2H, -NH_2, -CHO, -NC) may be present in the original matrix. Direct coupling of proteins can be achieved toward either amino or carboxy groups on the protein using carboxy and amino functional group carriers, respectively. Also, proteins can be insolubilized in an isocyanide-containing carrier through their amino or carbonyl group in the presence of a water-soluble aldehyde.

The major disadvantage of this process is that the final product in the reaction contains four different functional groups. However, it is precisely these reactive groups that provide versatility to the reaction and allow one to attach proteins by different ways. This process, therefore, can be used advantageously when a protein cannot be immobilized to an active aldehyde carrier.

6. Amidination Reactions

Supports containing imidoester functional groups have been employed for immobilization of proteins. Such carriers can be prepared by treating polyacrylonitrite with absolute ethanol and bubbling hydrogen chloride through the mixture to produce the corresponding imidoester derivatives (Scheme 26). Imidoesters are readily attacked by nucleophiles and react selectively with amino groups of protein at basic pH to yield amidines (II), which are stable in neutral or acidic solution, but are hydrolyzed slowly at high pH values (2).

In another procedure, carriers containing amino groups can be activated with CNB and reacted with amino protein groups, obtaining guanidine linkages between the amino groups of the proteins and carriers (118) (Scheme 27).

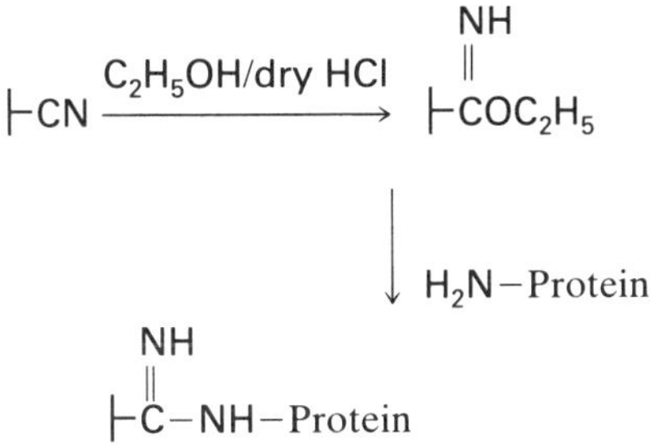

$$\vdash CN \xrightarrow{C_2H_5OH/\text{dry } HCl} \vdash \overset{NH}{\overset{\|}{C}}OC_2H_5 \qquad (I)$$

$$\downarrow H_2N-\text{Protein}$$

$$\vdash \overset{NH}{\overset{\|}{C}}-NH-\text{Protein} \qquad (II)$$

Scheme 26

$$\vdash NH_2 \xrightarrow{CNBr} \vdash NH-CN \xrightarrow{H_2N-\text{Protein}} \vdash NH-\overset{NH}{\overset{\|}{C}}-NH-\text{Protein}$$

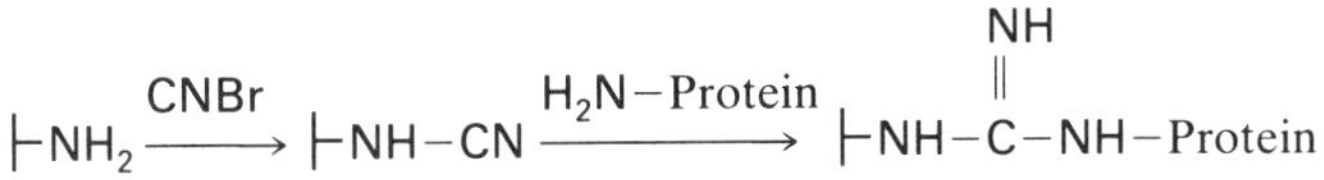

Scheme 27

7. Thiol–Disulfide Interchange Reactions

This method is used for protein bonding via thiol groups of both carrier and protein, which works in an acidic medium (120). A first step involves the preparation of a mixed disulfide derivative of a thiol-containing carrier by treating the thiolated carrier with 2,2′-dipyridyldisulfide according to Scheme 28.

The coupling of proteins through its thiol groups is accomplished with the liberation of 2-thiopyridone (II) (Scheme 28). However, the linkage between the protein and the support via -S-S- bonds is only stable under nonreducing conditions, and the coupling can be reversed with low molecular weight reagents (dithiothreitol or 2-mercaptoethanol), allowing subsequently the reactivation of the support. An alternative procedure

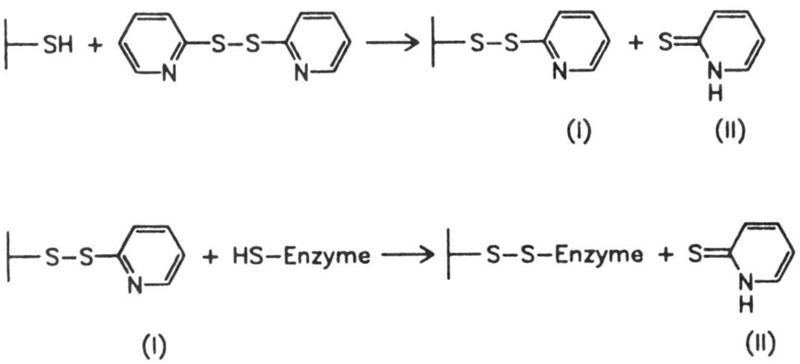

$\vdash$SH + (2-pyridyl)–S–S–(2-pyridyl) → $\vdash$S–S–(2-pyridyl) (I) + S=(2-pyridone, N–H) (II)

$\vdash$S–S–(2-pyridyl) (I) + HS–Enzyme → $\vdash$S–S–Enzyme + S=(2-pyridone, N–H) (II)

Scheme 28

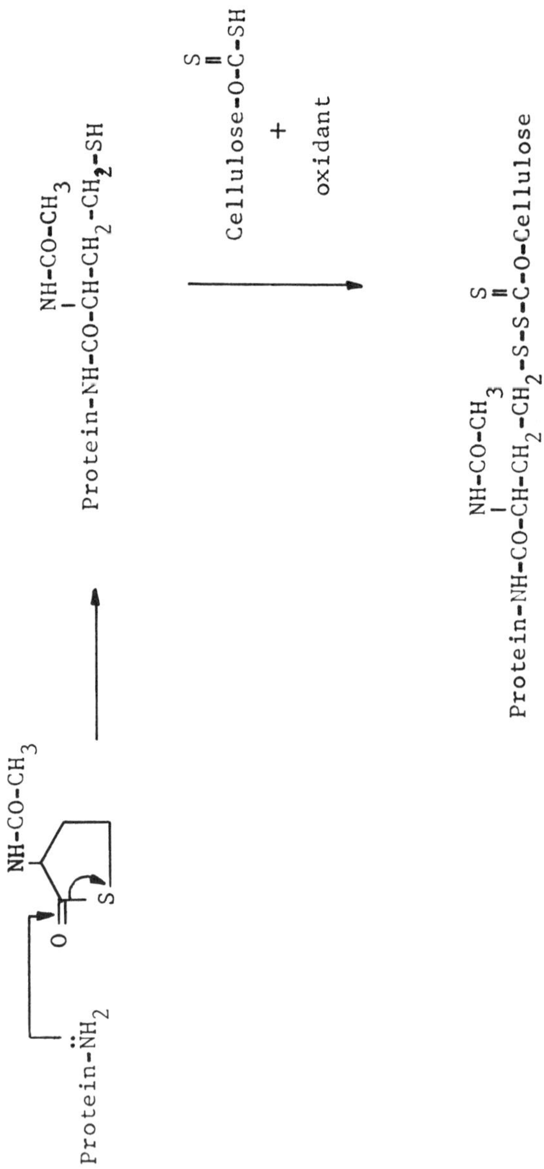

Protein-$\ddot{N}H_2$ + (N-acetylhomocysteine thiolactone: O=C, S ring, NH-CO-CH_3) → Protein-NH-CO-CH(NH-CO-CH_3)-CH_2-CH_2-SH

Protein-NH-CO-CH(NH-CO-CH_3)-CH_2-CH_2-SH + Cellulose-O-C(=S)-SH + oxidant → Protein-NH-CO-CH(NH-CO-CH_3)-CH_2-CH_2-S-S-C(=S)-O-Cellulose

Scheme 29

with probably more versatility is the reaction of the protein with *N*-acetylhomocysteine-thiolactone (Scheme 29), which provides a spacer arm (121) and utilizes the more readily available amino group(s) of the protein. Again, reversibility is possible via oxidation-reduction to a third type of matrix, which may be as above or cellulose xanthate (121) (Scheme 29).

8. Mercury–Enzyme Interactions

This method is based on the interaction of mercury derivatives carriers and the thiol groups of the protein (122) (Scheme 30).

The linkage between carrier and protein is formed at pH 4–8, but it is reversed by low molecular weight thiol reagents. This method can also be classified as physical adsorption, since the complex bond formed does not have a purely covalent character.

9. γ-Irradiation-Induced Coupling

This method has been used by Brandt and Anderson (123) to couple proteins to agarose and dextran by γ irradiation of the protein in the presence of the carrier. The radicals formed on both components then combine, giving rise to the covalent bond (Scheme 31).

This method is nonspecific, however, the immobilization yields are low and there is a high loss of activity arising from radiation damage. The advantage of this method is its independence of temperature and pH.

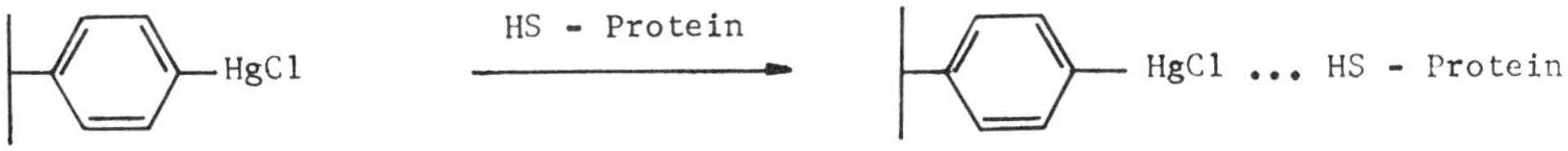

$$\text{Matrix}-C_6H_4-HgCl \xrightarrow{\text{HS - Protein}} \text{Matrix}-C_6H_4-HgCl \cdots \text{HS - Protein}$$

Scheme 30

$$MH \xrightarrow{\gamma\text{-rays}} M^{\bullet} + H^{\bullet}$$

$$PH \xrightarrow{\gamma\text{-rays}} P^{\bullet} + H^{\bullet}$$

$$M^{\bullet} + P^{\bullet} \longrightarrow M\text{-}\mathbf{P}$$

M = Matrix, P = Protein

Scheme 31

B. Metal-Linking Method

Immobilized biomolecules have been prepared using transition metal compounds, mainly titanium and zirconium metal salts, for the activation of the surface of organic carriers or using the correspondent hydrous metal oxides, precipitated from a metal salt, as carriers, Table 21. This method is based on the chelation properties of the transition metals, which can be employed to couple proteins.

The interaction of the transition metal compounds with biopolymers is illustrated in Figure 11, from the viewpoint of a simple system of titanium(IV) chloride and a biopolymer.

In the titanium(IV) chloride solution, the titanium ions are coordinated with molecules or ionic species that are essentially the ligands of the complex ion (see Figure 12 for the example of complex ions). Nucleophilic (hydroxy, amino, thiol, etc.) groups are effective ligands for the transition metal ions, and therefore it is to be expected that transition metal ions may complex both supports and enzymes. Supports, such as cellulose and silica, have hydroxyl groups which act as new ligands replacing others. For example, cellulose contains vicinal diol groups not involved in the glycosidic linkages between residues, which therefore are amenable to chelation by transition metal ions, the chelate being produced by replacement of two of the titanium ion ligands by polysaccharide hydroxyl groups.

It should be noted that, for steric reasons, it is impossible for all the water or chloride ligands of the titanium ion to become replaced by other

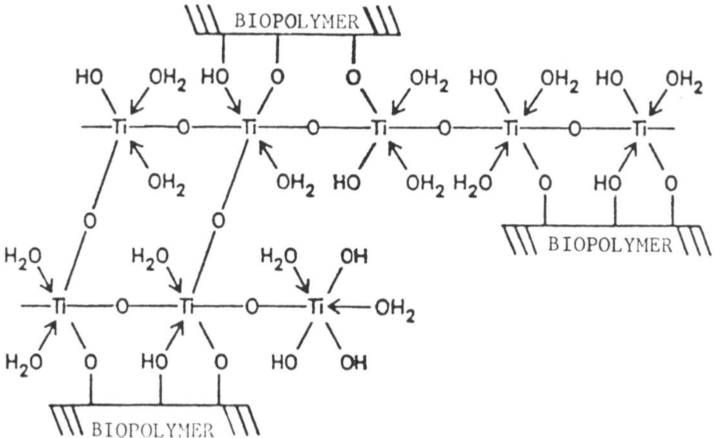

Figure 11 Representative structure of hydrous titanium (IV) oxide-biopolymer chelate.

Table 21 Examples of Immobilized Enzymes Prepared by Coordination Binding

Carrier	Enzyme	Ref.
Cellulose	α-Amylase	124
	Glucoamylase	124,125
	D-Glucose oxidase	124
	β-D-Fructofuranosidase	124
	Urease	125
Glass	Glucoamylase	125
	Papain	126
Silica gel	Nuclease P_1	127
Porous glass	Glucoamylase	128
	Nuclease P_1	127
	β-D-Fructofuranosidase	129
Pumice stone	Nuclease P_1	127
Spheron	Glucoamylase	130
	Trypsin	130
Hornblende		129
Nylon	Urease	125
Alumina	Glucoamylase	131
Silica	Glucoamylase	131
Nickel/alumina	Glucoamylase	131
Polystyrene sulfonate	Adenosine (phosphate) deaminase	132
Cation exchangers	Nuclease P_1	132
Gelatin	Glucoamylase	21
	β-D-Fructofuranosidase	21
Hydrous iron (III) oxide	Chymotrypsin	132
	Dextranase	133
	Glucoamylase	133
	D-Glucose oxidase	133
	β-D-Glucosidase	133
	Peroxidase	133
	Trypsin	133
Hydrous zirconium (IV) oxide	Chymotrypsin	133
	Dextranase	133
	Glucoamylase	133
	D-Glucose oxidase	133
	β-D-Glucosidase	133
	Peroxidase	133
	Trypsin	133
Hydrous vanadium (III) oxide	Chymotrypsin	133
	Dextranase	133
	Glucoamylase	133
	D-Glucose oxidase	133
	β-D-Glucosidase	133
	Peroxidase	133
	Trypsin	133

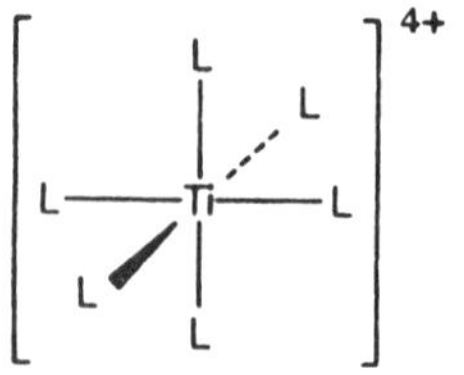

Figure 12 Titanium (IV) ions in $TiCl_4$-HCl solution.

hydroxyl groups of the one polysaccharide chain. Furthermore, these remaining ligands will not be extensively replaced by the hydroxyl groups of adjacent cellulose chains on account of the insolubility and lack of mobility of the cellulose molecules. Chelation may be achieved for any molecule-containing groups appropriate to replace the ligands of the titanium bound to cellulose. Thus proteins have groups that can act as ligands such as the free carboxyl groups from the C-terminus of acidic amino acids, the phenolic hydroxyl groups of tyrosyl residues, the alcoholic hydroxyl groups of seryl and threonyl residues, the free sulfydryl groups from any cysteinyl residues, and the amino groups from the N-terminus of ϵ-amino groups of lysyl residues. The production of such a bond between the titanium-treated polymer and the protein would be expected to depend on: (1) the availability in the protein molecule of residual groups which can act as ligands; (2) steric factors which allow such groups to come in contact with the titanium atoms; (3) the noninvolvement of such groups in the region of the active site(s) of the enzyme; and (4) the close proximity of enzyme molecules already bound in the polymer matrix.

The metal-linking method for immobilization of enzymes was initially developed by Novais (134). In this method cellulose is steeped in a solution of titanium (IV) chloride, an oxychloride derivative being obtained, followed by drying at moderate temperatures (45°C). The solid is washed with the buffer to remove excess salt solution, and the enzyme coupling is achieved by contacting the enzyme with the activated support, usually in a buffer solution at the optimum pH of the enzymes. The transition metal salts used in this activation process have been the chlorides and sulfates of titanium, iron, zirconium, vanadium, and tin.

The same technique of drying a mixture of the carrier and the metal salt solution has been used to activate not only organic supports, such as cellulose, sawdust, chitin, or alginic acid, but also inorganic supports such as silica-based carriers, celite, glass wool, or hornblende.

On the inorganic supports, however, some variable results were obtained with the former technique, this is mainly due to the availability of free silanol groups as ligands on the surface of the silica-based supports (136). Silica gel is relatively easy to activate, but nonporous glass is not likely to be activated by this procedure. Thus, several procedures have been developed to achieve better results. These techniques are based on the precipitation of a hydrous metal oxide layer on the support, by heating (135) or by neutralization (136).

In most cases of immobilization via transition metal chelation using metal chlorides, pure or in an acidic solution, the supports employed have been aerogel-type (glass, silica, etc.) or organic crosslinked polysaccharides aerogel-xerogel hybrids (e.g., cellulose derivatives). In such cases activation with low pH solutions can be carried out easily without disruption of the support matrix. In the case of more sensitive supports, namely those of proteinaceous nature (e.g., gelatin), the activation conditions should be as mild as possible so that the maximum level of active surface bound enzyme is obtained under conditions of minimum deterioration of the support.

Several mild techniques in which the transition metal (e.g., titanium(IV)) remains in a soluble and reactive form at higher pH values and therefore gives use to less disruption of proteins have been developed (21). Titanium(IV) chloride complexes with acrylamide, citric acid, and lactose were prepared, which can interact with crosslinked gelatin. Cation-exchangers, sulfonated polystyrene-based, have been equilibrated with several transition metal salts (Ti^{4+}, Zr^{4+}, VV^{5+}, etc.) and linked to immobilize enzymes via chelation (133).

Kennedy and co-workers (137) developed a method of protein immobilization based on the chemistry of the hydrous transition metal oxides per se. With this technique the only requirement, one reagent besides the protein is commercially available and of adequate stability. A one-step immobilization process such as this would be more desirable from an economic point of view. This technique uses the hydrous metal oxides as the only supports (internal) for protein immobilization, as these hydrous oxides can be produced by purification after the hydrolysis of the corresponding chlorides, the proteins being immobilized by chelation.

The metal-linking method leads to relatively high specific activity retention (30–80%). The operational stability of the immobilized enzymes prepared by this method is also variable, low stabilities being obtained with high molecular weight substrates (e.g., starch) (128). The metal activator, e.g., titanium, also seems to influence the stability of the immobilized enzyme preparation, probably due to inhibition of the enyzme by the metal activator (135,138).

REFERENCES

1. O. R. Zaborsky, *Immobilized Enzymes,* CRC Press, Cleveland, 1973.
2. G. Means and R. E. Feeney, *Chemical Modification of Proteins,* Holden-Day, San Francisco, 1971.
3. R. A. Messing, *Adv. Biochem. Eng. 10:*51 (1978).
4. L. F. Chen and G. T. Tsao, *Biotechnol. Bioeng. 18:*1507 (1976).
5. L. Goldstein, M. Pecht, S. Blumberg, D. Atlas, and Y. Levin, *Biochem. 9:*2322 (1970).
6. J. Porath and P. Flodin, *Nature 18:*1657 (1959).
7. J. Porath, *Biochim. Biophys. Acta 39:*193 (1960).
8. R. Axen, J. Porath, and S. Ernbaeck, *Nature 214:*1302 (1967).
9. A. Polson, *Biochim. Biophys. Acta 50:*565 (1961).
10. P. J. Hill, M. A. Gresswell, and J. G. Feinberg, U.S. Patent 4003792 (1977).
11. P. Vretblad and R. Axen, *FEBS Lett. 18:*254 (1971).
12. I. Takata, T. Tosa, and I. Chibata, *J. Solid-Phase Biochem 2:*225 (1977).
13. W. L. Stanley, G. G. Watters, S. H. Kelly, and A. C. Olson, *Biotechnol. Bioeng. 20:*135 (1978).
14. L. Iyengar and A. V. S. P. Rao, *Biotechnol. Bioeng. 21:*1333 (1979).
15. O. H. Friedrich, M. Chun, and M. Sernetz, *Biotechnol. Bioeng. 22:*157 (1980).
16. J. Klein and K. D. Vorlop, *Biotechnol. Lett. 3:*9 (1981).
17. J. F. Leuba and F. Widmer, *J. Solid-Phase Biochem. 2:*257 (1979).
18. W. R. Vieth and K. Venkatsubramanian, *Methods. Enzymol. 44:*243 (1976).
19. A. G. van Velzen, U.S. Patent 383007 (1974).
20. P. R. Coulet, C. Godinot, and D. C. Gautheron, *Biochim. Biophys. Acta 391:*272 (1975).
21. J. F. Kennedy, B. Kalogerakis, and J. M. S. Cabral, *Enzyme Microb. Technol. 6:*68 (1984).
22. G. B. Broun, *Methods Enzymol. 44:*263 (1976).
23. L. Grasset, D. Cordier, and A. Ville, *Process Biochim. 14:*2 (1979).
24. W. A. Beverloo and S. Bruin, in *New Processes of Waste Water Treatment and Recovery* (G. Mattock and W. J. Eilbeck, eds.), Ellis Harwood, Chichester, U.K., 1978, p. 307.
25. M. M. Derbinin, *Ukr. Khim. 24:*3 (1955).
26. Y. K. Cho and J. E. Bailey, *Biotechnol. Bioeng. 21:*461 (1979).
27. J. S. Mattson and H. B. Mark, *Activated Carbon: Surface Chemistry and Adsorption from Solution,* Marcel Dekker, New York, 1971.
28. N. Grubhofer and L. Schleith, *Naturwissenschaften 45:*440 (1953).
29. J. Beitz, A. Schellenberger, J. Lasch, and J. Fischer, *Biochim. Biophys. Acta 612:*451 (1980).
30. H. Brandenberger, *J. Polymer Sci. 20:*215 (1956).
31. J. M. Nelson and E. G. Griffith, *J. Am. Chem. Soc. 38:*1109 (1916).
32. H. H. Weetall, *Methods Enzymol. 44:*134 (1976).
33. J. M. S. Cabral, J. M. Novais, and J. P. Cardoso, *Biotechnol. Bioeng. 23:*2083 (1981).

34. D. H. Campbell, F. Leuscher, and L. S. Lerman, *Proc. Natl. Acad. Sci. US 37:*575 (1951).
35. C. G. Beddows, R. A. Minauer, and J. T. Guthrie, *Biotechnol. Bioeng. 22:*311 (1980).
36. P. Gemeiner, C. Polaik, A. Brier, and L. Petrus, *Enzyme Microb. Technol. 8:*109 (1986).
37. G. B. Ogumtimein and P. J. Reilly, *Biotechnol. Bioeng. 22:*1127 (1980).
38. H. Kaboli and P. J. Reilly, *Biotechnol. Bioeng. 22:*1055 (1980).
39. J. L. Iborra, A. Manjon, and J. A. Lozano, *J. Solid-Phase Biochem. 2:*85 (1977).
40. T.-S. Lai and P.-S. Cheng, *Biotechnol. Bioeng. 20:*773 (1978).
41. J. Drobnik, V. Saudek, F. Svec, J. Kalal, V. Vojtisek, and M. Barta, *Biotechnol. Bioeng. 21:*1317 (1979).
42. L. Shemer, R. Granot, A. Freeman, M. Sokolovsky, and L. Goldstein, *Biotechnol. Bioeng. 21:*1607 (1979).
43. D. Blassberger, A. Freeman, and L. Goldstein, *Biotechnol. Bioeng. 20:*309 (1978).
44. C. Y. Lee, *J. Solid-Phase Biochem. 3:*71 (1978).
45. M. Adamich, H. F. Voss, and E. A. Dennis, *Arch. Biochem. Biophys. 189:*417 (1978).
46. E. C. Hatchikian and P. Monsan, *Biochem. Biophys. Res. Commun. 92:*1091 (1980).
47. R. A. Zingaro and M. Uziel, *Biochim. Biophys. Acta. 213:*371 (1970).
48. W. Brummer, N. Heinrich, M. Klockow, H. Lang, and H. D. Orth, *Eur. J. Biochem. 25:*129 (1972).
49. S. Miyairi, *Biochim. Biophys. Acta 571:*374 (1979).
50. C. G. Beddows, R. A. Mirauer, J. T. Guthrie, F. I. Abdel-Hay, and C. E. J. Mouish, *Polym. Bull.* (Berlin) *1:*749 (1979).
51. P. R. Coulet, R. Steinberg, and D. R. Thevenot, *Biochim. Biophys. Acta 612:*317 (1980).
52. G. Bardeletti, B. Maisterrena, and P. R. Coulet, *Enzyme Microb. Technol. 8:*365 (1986).
53. N. Arrio-Dupont and P. R. Coulet, *Biochem. Biophys. Res. Commun. 89:*345 (1975).
54. L. J. Blum and P. R. Coulet, *Biotechnol. Bioeng. 28:*1154 (1986).
55. D. M. Munnecke, *Biotechnol. Bioeng. 21:*2247 (1979).
56. B. Davidson, B. Matthiasson, R. Karlsson, and F. Winquist, *Biotechnol. Bioeng. 21:*1749 (1979).
57. G. Carrea, F. Colombi, G. Mazzola, P. Cremonesi, and E. Antonini, *Biotechnol. Bioeng. 21:*39 (1979).
58. J. Tramper, F. Miller, and H. C. van der Plas, *Biotechnol. Bioeng. 20:*1507 (1978).
59. D. B. Johnson and M. P. Coughlan, *Biotechnol. Bioeng. 20:*1085 (1978).
60. K. Martinek, V. V. Mozhaev, and I. V. Berezin, *Biochim. Biophys. Acta 615:*425 (1980).

61. C. Sung, A. Lavin, A. Klibanov, and R. Langer, *Biotechnol. Bioeng. 28:*1531 (1986).
62. P. Chim-Anage, Y. Kashiwagi, Y. Magae, T. Ohta, and T. Sasaki, *Biotechnol. Bioeng. 28:*1876 (1986).
63. S. A. G. F. Angelino, F. Muller, and H. C. van der Plas, *Biotechnol. Bioeng. 27:*447 (1985).
64. N. Siegbahn, K. Mosbach, K. Grodzki, R. Zocher, and N. Madry, *Biotechnol. Lett. 7:*297 (1985).
65. V. Janasik, F. Barta, K. Krettschmer, and J. Lash, *J. Solid-Phase Biochem. 3:*291 (1978).
66. H. Maeda, G. T. Tsao, and L. F. Chen, *Biotechnol. Bioeng. 20:*383 (1978).
67. S. Fukushima, T. Nagai, K. Fujita, A. Tanaka, and S. Fukui, *Biotechnol. Bioeng. 20:*1465 (1978).
68. M. Kumakura and I. Kaetsu, *Biotechnol. Lett. 7:*773 (1985).
69. H. Brandenberger, *Rev. Ferment. Ind. Aliment. 11:*237 (1956).
70. Madhu and K. A. Prabhu, *Enzyme Microb. Technol. 7:*279 (1985).
71. N. W. H. Cheetham and G. N. Richards, *Carbohydr. Res. 30:*99 (1973).
72. J. F. Kennedy and A. Zamir, *Carbohydr. Res. 29:*497 (1973).
73. C. J. Gray and T. H. Yeo, *Carbohydr. Res. 27:*235 (1973).
74. S. A. Barker, S. H. Doss, C. J. Gray, J. F. Kennedy, M. Stacey, and T. H. Yeo, *Carbohydr. Res. 20:*1 (1971).
75. L. M. Simon, M. Nagi, M. Abraham, B. Szajani, and L. Boross, *Enzyme Microb. Technol. 7:*275 (1985).
76. L. M. Simon, J. Szelei, B. Szajani, and L. Boross, *Enzyme Microb. Technol. 7:*357 (1985).
77. B. Szajani, G. Klamar, and L. Ludvig, *Enzyme Microb. Technol. 7:*488 (1985).
78. L. M. Simon, M. Kotoman, B. Szajani, and L. Boross, *Enzyme Micro. Technol. 8:*222 (1986).
79. C. J. Findlay, K. L. Parkin, and R. Y. Yada, *Biotechnol. Lett. 8:*649 (1986).
80. Y. K. Cho and J. E. Bailey, *Biotechnol. Bioeng. 20:*1651 (1978).
81. A. Borchet and K. Bucholz, *Biotechnol. Lett. 1:*15 (1979).
82. M. T. Hearn, *Methods Enzymol. 135:*102 (1987).
83. T. Sato, T. Mori, T. Tosa, and I. Chibata, *Arch. Biochem. Biophys. 147:*788 (1972).
84. H. Maeda and H. Suzuki, *Agr. Biol. Chem. 36:*1581 (1972).
85. H. Schutt, G. Schmidt-Kastner, A. Arers, and M. Preiss, *Biotechnol. Bioeng. 27:*420 (1985).
86. M. Tomar and K. A. Prabhu, *Enzyme Microb. Technol. 7:*557 (1985).
87. T. H. Finlay, V. Troll, M. Levy, A. J. Johnson, and L. T. Hodgins, *Anal. Biochem. 87:*77 (1978).
88. C. R. Reddy, C. Rami Reddy, K. Raghunath, and T. Joseph, *Biotechnol. Bioeng. 28:*609 (1986).
89. B. Solomon, R. Koppel, G. Pines, and E. Katchalski-Katzir, *Biotechnol. Bioeng. 28:*1213 (1986).

90. E. Stratilova, M. Capka, and L. Rexova-Benkova, *Biotechnol. Lett. 9:*511 (1987).
91. K. Ajisaka, H. Nishida, and H. Fujimoto, *Biotechnol. Lett. 9:*387 (1987).
92. G. Wehnert, A. Sauerbrei, and K. Schugerl, *Biotechnol. Lett. 7:*827 (1985).
93. K. Nilsson and K. Mosbach, *Methods Enzymol. 135:*65 (1987).
94. G. Carrea, R. Bovara, R. Longhi, and S. Riva, *Enzyme Microb. Technol. 7:*597 (1985).
95. T. Mori, K. Nilsson, P.-O. Larsson, and K. Mosbach, *Biotechnol. Lett. 9:*455 (1987).
96. A. T. Jagendorf, A. Patchsinik, and M. Sela, *Biochim. Biophys. Acta 78:*516 (1963).
97. R. F. Murphy, J. M. Conlon, A. Iman, and G. J. C. Kelly, *J. Chromatogr. 135:*427 (1977).
98. J. Porath, *Methods Enzymol. 34:*13 (1974).
99. J. Brandt, L. O. Anderson, and J. Porath, *Biochim. Biophys. Acta 386:*196 (1975).
100. J.-P. Lenders, P. Germain, and R. R. Crichton, *Biotechnol. Bioeng. 27:*572 (1985).
101. R. H. D. Lambrecht, G. Stegens, G. Mannehs, and A. Claeys, *Enzyme Microb. Technol. 9:*221 (1987).
102. I. Vallat, P. Monsan, and J. P. Riba, *Biotechnol. Bioeng. 27:*1274 (1985).
103. I. Vallat, P. Monsan, and J. P. Riba, *Biotechnol. Bioeng. 28:*151 (1986).
104. J. Rogalski, J. Szizodrak, A. Damidowicz, Z. Ilazuk, and A. Leonowicz, *Enzyme Microb. Technol. 7:*395 (1985).
105. R. H. D. Lambrecht, G. Stegens, A. Claeys, and A. Vuye, *Enzyme Microb. Technol. 7:*493 (1985).
106. J. Lobarzewski and A. Paszczynski, *Enzyme Microb. Technol. 7:*564 (1985).
107. G. Stegens, S. De Laet, R. H. Lambrecht, and C. Block, *Enzyme Microb. Technol. 8:*153 (1986).
108. A. Bodalo, J. L. Gomez, and J. Bastida, *Enzyme Microb. Technol. 8:*433 (1986).
109. A. Manjon, J. Bastida, C. Romero, A. Jimeno, and J. L. Ibora, *Biotechnol. Lett. 7:*477 (1985).
110. G. Mannens, G. Slegens, R. Lambrecht, H. De Langhe, K. Puttemans, and C. Block, *Enzyme Microb. Technol. 9:*285 (1987).
111. F. Toldra, N. B. Jansen, and G. T. Tsao, *Biotechnol. Lett. 8:*785 (1986).
112. D. N. Deobagkar, V. Shankar, and D. D. Deobagkar, *Enzyme Microb. Technol. 8:*97 (1986).
113. R. Koelsch, M. Fusek, Z. Hostomska, J. Lasch, and J. Turkova, *Biotechnol. Lett. 8:*283 (1986).
114. J. Omelkova, I. Rexcva-Benkova, V. Kubanek, and B. Veruovic, *Biotechnol. Lett. 7:*99 (1985).
115. J. Klemes and N. Citri, *Biotechnol. Bioeng. 21:*897 (1979).
116. I. Ugi, *Augiw. Chem. 74:*8 (1962).
117. R. Axen, P. Vretblad, and J. Porath, *Acta. Chem. Scand. 25:*1129 (1971).

118. J. Schnapp and Y. Shalitin, *Biochem. Biophys. Res. Commun. 70:*8 (1976).
119. F. M. Richards and J. R. Knowles, *J. Mol. Biol. 37:*231 (1968).
120. J. Carlsson, R. Axen, K. Brocklehurst, and E. M. Crook, *Eur. J. Biochem. 44:*189 (1974).
121. J. F. Kennedy and A. Zamir, *Carbohydr. Res. 41:*227 (1975).
122. J. R. Shainoff, *J. Immunol. 100:*187 (1968).
123. J. Brandt and L. O. Anderson, *Acta. Chem. Scand. B30:*815 (1976).
124. S. A. Barker, A. N. Emery, and J. M. Novais, *Process Biochem. 6:*11 (1971).
125. A. N. Emery, J. S. Hough, J. M. Novais, and P. T. Lyons, *Chem. Eng.* (London) *259:*71 (1972).
126. J. F. Kennedy and V. W. Pike, *Enzyme Microb. Technol. 1:*31 (1978).
127. K. Rokugawa, T. Fujishima, A. Kuminaka, and H. Yoshino, *J. Ferment. Technol. 58:*509 (1980).
128. J. M. S. Cabral, J. P. Cardoso, and J. M. Novais, *Enzyme Microb. Technol. 3:*41 (1981).
129. J. F. Kennedy and P. M. Watts, *Carbohydr. Res. 32:*155 (1974).
130. C. J. Gray, C. M. Lee, and S. A. Barker, *Enzyme Microb. Technol. 4:*143 (1982).
131. M. Charles, R. W. Coughlin, E. K. Panuchuri, B. R. Allen, and F. X. Hasselberger, *Biotechnol. Bioeng. 17:*203 (1975).
132. K. Rokugawa, T. Fujishima, A. Kuminaka, and H. Yoshino, *J. Ferment. Technol. 58:*423 (1980).
133. J. F. Kennedy and J. M. S. Cabral, *Methods Enzymol. 135:*117 (1987).
134. J. M. Novais, Studies on the insolubilization and use of amylolytic enzymes. Ph.D. thesis, University of Birmingham, 1971.
135. J. P. Cardoso, M. F. Chaplin, A. N. Emery, J. F. Kennedy, and L. P. Revel-Chion, *J. Appl. Chem. Biotechnol. 28:*775 (1978).
136. J. F. Kennedy, S. A. Barker, and C. A. White, *Stärke 29:*240 (1977).
137. J. F. Kennedy, S. A. Barker, and J. D. Humphreys, *J. Chem. Soc. Perkin Trans. I:*962 (1986).
138. A. Flynn and D. B. Johnson, *Int. J. Biochem. 8:*243 (1977).

4

Commercially Available Supports for Protein Immobilization

Richard F. Taylor

Arthur D. Little, Inc.
Cambridge, Massachusetts

I. INTRODUCTION

Throughout this text, details are presented on basic reactions and methodologies to prepare protein and cell immobilization media and supports. Such preparation is critical to the development of new immobilization methods and applications and for studies on basic aspects of immobilization kinetics and reaction mechanisms.

Many users of immobilized proteins and workers just entering the field may, however, not need to develop and produce immobilization supports in their laboratories, or may not have the capabilities or resources to do so. Fortunately for such workers, there exist a variety of commercially available supports which can provide a near-immediate solution for a support need. While in most cases such commercial supports may appear expensive, their availability and, in most cases, their reliability can offset the cost of preparing supports in the laboratory. Commercially available supports are especially cost-effective in rapidly assessing the feasability of an applications-oriented immobilization effort.

The purpose of this chapter is to provide an introduction to commercially available supports which can be used for protein and (to a more limited extent) cell immobilization. The supports discussed are not meant to include all commercially available support products. Rather, representative examples of the major types of supports are provided. The main focus is on preactivated supports: supports which are ready for one-step protein (or cell) immobilization. Other supports which require simple activation prior to immobilization are also represented. Many chromatographic supports are not listed due to the large redundancy of support physical and chemical characteristics, and the fact that they would require treatment through multistep reactions before they would be ready for protein immobilization. Membrane and other filters such as those made of cellulose, cellulose acetate, nitrocellulose, nylon, polyvinyl chloride, glass fibers, etc., are also not considered. While many of these filters are used extensively for binding biological molecules including proteins, they are less appropriate for protein immobilization than other membranes and filters specifically activated for protein immobilization. These latter, preactivated filters will be considered here.

II. SUPPORT PRODUCT CATEGORIES

While there is a great deal of overlap among support products in their basic chemistry, there is also a large divergence in how they are marketed. Thus, appropriate protein immobilization supports may be marketed for this direct purpose or as chromatographic and purification media. Affinity chromatography supports are the best example of products suitable for protein immobilization which are marketed almost exclusively for separation purposes.

Affinity chromatography utilizes a specific immobilized ligand for the purification of a compound which binds reversibly to the ligand from a mixture of compounds (Fig. 1). After binding of the target compound and washing out of the nonbinding contaminating material, the target compound is desorbed from the immobilized ligand by adding a low concentration of free ligand (or ligand analog) to the elution buffer, by the use of salts (such as NaCl, NH_4Cl, or sodium thiocyanate), or by a change in pH (e.g., antibody–antigen complexes may be dissociated by exposure to pH 2–3 buffer). The eluted analyte can then be further purified and separated from the elution material by desalting on a gel permeation column. A number of excellent reviews on affinity chromatography are available (1–3).

Affinity supports containing a variety of surface chemistries are seldom sold as protein immobilization media even though they may be identified as antibody immobilization supports. Many of these products are, however, well suited for protein immobilization. The original agarose affinity chromatography supports from Pharmacia, for example, can be used for protein immobilization and have led to the development of more advanced supports with greater versitility and stability. In a similar manner, affinity filtration, the use of ligands bound on a filter for rapid purifica-

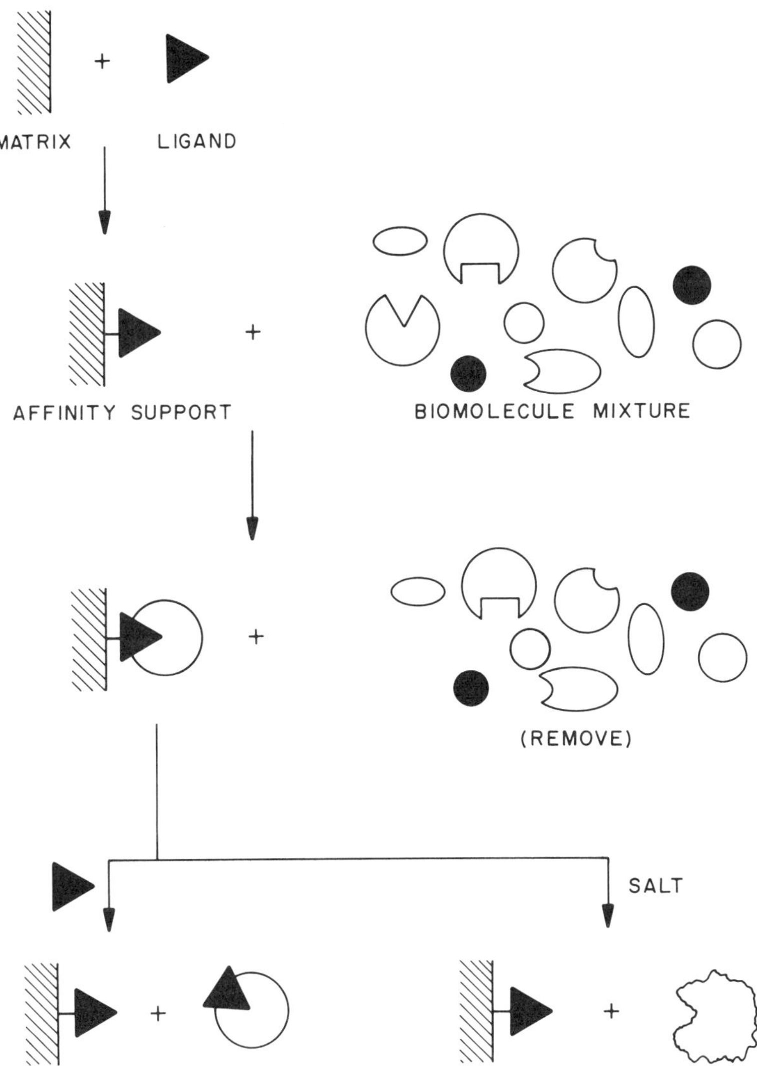

Figure 1 Basic principle of affinity chromatography. After adsorption of the molecule to be purified by the immobilized ligand, contaminating material is washed away. The target molecule is then desorbed using free ligand (or ligand analog) or salt.

tion, evolved from micro- and ultrafiltration and provide new supports for protein immobilization.

There are thus three categories of products which are useful for protein immobilization: identified protein immobilization supports, affinity chromatography supports, and affinity filters. These catagories are addressed here.

III. SUPPORT CHARACTERISTICS AND CHEMISTRIES

A. General Characteristics

The primary base supports available for protein immobilization include agarose, cellulose, dextrans, polymers (such as PVC, acrylates, nylons, and polystyrenes), and silicas (such as silica gel and glass beads). There are also hybrids of these base materials such as agarose-acrylamide and polymer-coated silica. The primary consideration for these materials as applied to biological molecules are low nonspecific binding, high surface interaction areas, storage and use stability (both mechanical and chemical), high porosity, and surface chemistry variety.

Table 1 lists a number of commercially available supports suitable for protein immobilization, and Table 2 lists the companies marketing these supports. In considering these products, a number of generalizations can be made.

1. Particle Size

All of the supports listed in Table 1, if particulate, have particle sizes in the range of 5 to 500 μm. In general, the most useful particle size for separations applications using immobilized proteins is from 25 to 75 μm. Process use of immobilized enzymes may require larger particle sizes due to back pressures in flow streams.

2. Porosity

Most of the supports listed in Table 1 have high porosity. In the case of macroporous particulate supports, pore sizes best suited for use with proteins are at least 250 Å (25 nm) in diameter. Smaller pore sizes can result in restructed mass transfer and pore penetration of the protein, thus effectively limiting protein interaction with the total (internal + external) surface area of the support particles. In the case of filter or membrane supports, typical pore sizes range from 0.2 to 5 μm.

3. Stability

The oldest protein immobilization supports, those based on cross-linked agarose, polyacrylamide, and cellulose, are the least stable to physical

Table 1 Examples of Commercially Available Supports for Protein and Cell Immobilization

Base support[a]	Supplier[b]
CNBr-Activated	
45–165 μm Agarose beads (Sepharose 4B)	Pharmacia LKB
30 μm, Macroporous (500Å) irregular silica particles	Serva
Glutaraldehyde-Activated	
25–50 μm Agarose beads (ActiBind ALD) with 5 to 12 atom spacer arms	ICN
45–165 μm 4% Cross-linked agarose beads (Amino Link)	Pierce
Cross-linked agarose beads (ACTIGEL A)	Sterogene
25–100 μm Cellulose beads with a 10 atom spacer arm (Matrex Cellufine)	Amicon
Cellulosic filter membranes (Nalgene U12 and U38)	Nalge
Cellulosic/acyllic polymer membrane (Zetaffinity)	CUNO
60–140 μm, 2% Polyacrylamide/2% agarose beads with (Act-Magnogel) or without (Act-Ultrogel) Fe_3O_4 particles	IBF
0.25–1 μm Monodisperse polyacrolein microspheres (Polybead)	Polysciences
5 μm Cellulosic-polymer membranes	Memtek
Hydrophilic, microporous (0.2–2 μm) PVC sheets, filters and cartridges impregnated with silica and treated with polyethyleneimine (Acti-Disk)	FMC
40 μm Macroporous (275 Å) silica beads (Bakerbond)	Baker
30 μm Macroporous (500 Å) silica irregular particles (HiPac)	ChromatoChem
50 μm Macroporous (200, 500, or 1000 Å) polymer-coated silica beads with 7 to 12 atom spacer arms (NuGel-AF)	Separation Industries
30 μm Macroporous (500 Å) irregular silica particles	Serva
100–1600 μm Macroporous (400–1500Å) polyethyleneimine impregnated alumina particles (Ketomax, Aldomax)	UOP
Epoxy-Activated	
45–165 μm Agarose beads with a 12 atom spacer arm (Sepharose 6B)	Pharmacia LKB
Cellulose cartridge (MEMSEP)	Domnick Hunter
15–150 μm and 150–250 μm Cross-linked, regenerated cellulose with a 6 atom spacer (Indion)	Phoenix
Microporous (0.1–3 μm) hydrophilic nylon sheets and membranes (Immunodyne)	Pall
30–250 μm Macroporous (>1000 Å) beads of copolymerized methacrylamide and N-methylenebismethacrylamide (Eupergit)	Rohm Pharma
60 μm Macroporous (350 Å) beads of copolymerized 2-hydroxyethylmethacrylate and ethylene dimethacrylate (Separon HEMA)	Tessek

Table 1 (Continued)

Base support[a]	Supplier[b]
50–200 μm Macroporous (300Å) vinyl acetate-divinylethylene urea beads (VA-Epoxy BIOSYNTH)	Crescent
12 μm Macroporous (300 Å) silica beads coated with a hydrophilic polymer (Hydropore-EP)	Rainin
7 μm Silica beads (Durasphere AS)	Alltech
40 μm Macroporous (275 Å) silica beads (Bakerbond)	Baker
10 μm Macroporous silica particles (Ultraffinity-EP)	Beckman
50 μm Macroporous (200, 500, or 1000 Å) polymer-coated silica beads with 7–12 atom spacer arms (NuGel-AF).	Separation Industries
30 μm Macroporous (500 Å) irregular silica particles	Serva
Hydrazine (and Azide)-Activated	
80–180 μm Agarose beads (Affi-Gel HZ)	BioRad
45–165 μm Agarose beads (adipic acid hydrazine) with a 9 atom spacer arm (Sepharose 4B)	Pharmacia LKB
45–165 μm Agarose beads (ImmunoPure)	Pierce
Proprietary polymeric beads (AvidGel F)	BioProbe
6 mm Nonporous polystyrene beads	Pierce
40 μm Macroporous (275 Å) silica beads activated with diazofluoroborate (Bakerbond)	Baker
30 μm Macroporous (500 Å) irregular silica particles	Serva
N-Hydroxysuccinimide Activated	
80–180 μm Agarose beads with 10 or 15 atom spacer arms (Affi-Gel 10 and 15)	BioRad
45–165 μm Agarose beads with a 6 atom spacer arm (CH Sepharose 4B)	Pharmacia LKB
Proprietary polymer macroporous beads with a 10 atom spacer arm (Affi-Prep 10)	BioRad
30 μm Macroporous (500 Å) irregular silica particles	Serva
Carbonyldiimidazole-Activated	
45–165 μm, 6% Cross-linked agarose beads (Reacti-Gel 6X)	Pierce
40–80 μm Beads of polymerized N-acryloyl-2-amino-hydroxymethyl-1,3-propane diol (Reacti-Gel GF-2000)	Pierce
32–63 μm Proprietary polymer beads (Reacti-Gel HW-65F)	Pierce
30 μm Macroporous (500 Å) irregular silica particles	Serva
Divinylsulfone-Activated	
40–80 μm Agarose beads (MINI LEAK)	Kem-En-Tec
Cellulose cartridge (Memsep)	Domnick Hunter
60 μm Macroporous (350 Å) beads of copolymerized 2-hydroxyethylmethacrylate and ethylene dimethylacrylate (Separon HEMA)	Tessek
Acrylate-coated microporous (0.65–5 μm) polyvinylidene difluoride discs and sheets (Immobilon)	Millipore

Table 1 (Continued)

Base support[a]	Supplier[b]
Tresyl-Activated	
45–165 μm Agarose beads (Sepharose 4B)	Pharmacia LKB
10 μm Macroporous (300 Å) silica beads (Selecti Sphere-10)	Pierce
Fluoromethylperidinium-Activated	
Coss-linked agarose beads (Avid Gel Ax)	BioProbe
Polyhydroxylalkylamidoacrylate beads (Avid Gel T)	BioProbe
Proprietary polymeric beads (Avid Gel F)	BioProbe
Biologically Activated	
Avidin-coated glass beads for immobilization of biotin-labeled glycoproteins (BioSpheres)	Kontes
Nonactivated, Functional Group Supports	
80–180 μm Agarose beads with 9–12 atom spacer arms and $-NH_2$, $-COOH$ and $-SH$ end groups (Affi-Gels 102, 202, and 401)	BioRad
45–165 μm Agarose beads derivitized with hexanoic acid and hexylamine (Sepharose 4B)	Pharmacia LKB
100–300 μm Macroporous kiselguhr (diatomite) silica/6% cross-linked agarose beads with available hydroxyl groups (Macrosorb KAX)	Sterling
25–100 μm Carboxy- or amino-terminal cellulose beads with 5–9 atom spacer arms (Matrex Cellufine Amino or Carboxyl)	Amicon
Aminophenylthioether- and aminobenzyloxymethyl-terminal cellulose paper for activation by diazotization (Transa-Bind)	S & S
40–300 μm Amino- and carboxyl-terminal polyacrylamide beads (Bio-Gel)	BioRad
250–1000 μm, 5% Cross-linked high porosity (90 to 97%) granules of polystyrene, polychloromethylstyrene or sulfonated polystyrene (POLYHIPE)	National Starch and Chemical
6 mm Nonporous polystyrene beads with available $-NH_2$ groups	Pierce
0.04–10 μm Monodisperse latex beads with available amino, carboxyl, hydroxyl, sulfate, and amidino groups, with/without spacer arms (Microspheres; CML particles; Polybead)	IDC; Bangs; Polysciences
Proprietary polymer beads with $-COOH$ and NH_2 terminal groups on 16–20 atom spacer arms (Avid Gel F)	BioProbe
Amino- and hydroxyl-terminal cellulose/acryllic polymer membranes with 5 or 18 atom spacer arms (Zetaffinity)	CUNO
Polymeric microporous hollow-fiber (300 μm diameter) bioreactors for immobilization of enzymes by entrapment	Sepracor
Aminopropyl-, succinylaminopropyl-, hexamethylenediamine-, and diol-terminal 40 μm macroporous (275 Å) silica beads (Bakerbond)	Baker
Glycerol-coated, 37–74 μm controlled pore glass beads with available hydroxyl groups (Glycophase)	Pierce

Table 1 (Continued)

Base support[a]	Supplier[b]
Aminopropyl- and alkylamine-terminal 125–177 μm silylated controlled-pore glass beads	Pierce
Amino-, carboxyl-, and hydroxyl-terminal 50 μm macroporous (200, 500, or 1000 Å) polymer-coated glass beads with 10 or 12 atom spacer arms (NuGel-AF)	Separation Industries
Aminohexyl-, carboxymethyl-, and mercaptopropyl-terminal 30 μm macroporous (500 Å) irregular silica particles	Serva
Microcarriers	
200–5000 μm, 2–4% Calcium alginate beads for cell entrapment and immobilization	Protan
500 μm Beads of collagen-glycosaminoglycan copolymer crosslinked with glutaraldehyde (Informatrix)	Biomat
Pepsin-solubilized collagen membranes (Cellagen)	ICN
90–150 μm and 150–210 μm Collagen-coated plastic beads	Whatman
Chitosan flakes with available amino groups	Protan
180 μm Dextran beads with available N,N-diethylaminoethyl groups throughout the matrix (Cytodex 1)	Pharmacia LKB
N,N,N-Trimethyl-2-hydroxyl-aminopropyl coated 155 μm dextran beads (Cytodex 2)	Pharmacia LKB
Collagen-coated 175 μm dextran beads (Cytodex 3)	Pharmacia LKB
Silica (diatomite celite) carriers for enzymes and cells (Celite)	Manvillle
90 to 150 μm and 150 to 210 μm Glass-coated plastic beads (Bio-Spheres)	Whatman
120 to 180 μm Glass and collagen-coated plastic beads	Bellco

[a]See Table 3 for ligand specificities and coupling reaction conditions.
[b]See Table 2 for full names and addresses.

deterioration and compression under constant use. Few of these supports can be used at pressures above 10 psi. Newer, more rigid polymeric supports such as hydroxymethacrylates and other copolymer based supports do not have this problem, and many can be used in high-flow streams with pressures in excess of 3000 psi.

For example, in a case where a high-flow affinity chromatography system was required for the purification of the enzyme acetylcholinesterase (AChE), a comparison was made between affinity supports prepared with a soft agarose gel and a rigid polymer (Fig. 2). In this example, a ligand which binds AChE, trimethyl(*p*-aminophenyl)ammonium chloride (TAPA), was immobilized onto the carboxyl forms of Sepharose and Separon HEMA using carbodiimide coupling (4). Because of the pressure stability of the

Table 2 Representative Companies Providing Supports for Protein and Cell Immobilization

Alltech Associates, Inc., Deerfield, IL 60015
Amicon Division, Danvers, MA 01923
J.T. Baker, Inc., Phillipsburg, NJ 08865
Bangs Laboratories, Inc., Carmel, IN 46032
Beckman Instruments, Inc., San Ramon, CA 94583
Bellco Glass, Inc., Vineland, NJ 08360
Biomat Corp., Belmont, MA 02178
BioProbe International, Inc., Tustin, CA 92680
Bio-Rad Laboratories, Richmond, CA 94804
ChromatoChem, Missoula, MT 59801
Crescent Chemical Co., Inc., Hauppauge, NY 11788
CUNO, Inc., Meriden, CT 06450
Domnick Hunter Filters, Ltd., Durham, UK
FMC Bioproducts, Rockland, ME 04841
IBF Biotechnics, Inc., Savage, MD 20763
ICN Biomedicals, Inc., Cleveland, OH 44128
Interfacial Dynamics Corp. (IDC), Portland, OR 97207
Kem-En-Tec Biotechnology Corp., Hellerup, Denmark
Kontes Biotechnology, Vineland, NJ 08360
Manville Filtration and Minerals, Denver, CO 80217
Memtek Corp., Billerica, MA 01821
Millipore Corp., Bedford, MA 01730
Nalge Company, Rochester, NY 14602
National Starch and Chemical Corp., Bridgewater, NJ 08807
Pall BioSupport Division, Glenn Cove, NY 11542
Pharmacia LKB Biotechnology Inc., Piscataway, NJ 08854
Phoenix Chemicals Ltd., New Zealand
Pierce Chemical, Rockford, IL 61105
Polysciences, Inc., Warrington, PA 18976
Protan Laboratories Inc., Richmond, WA 98073
Rainin Instrument Co., Inc., Woburn, MA 01801
Rohm Pharma GmbH, Darmstadt, FRG
Schleicher & Schuell, Inc., (S&S), Keene, NH 03431
Separation Industries, Metuchen, NJ 08840
Sepracor Inc., Marlborough, MA 01752
Seva Biochemicals, Westbury, NY 11590
Sterling Organics Ltd., Newcastle-upon-Tyne, UK
Sterogene Biochemicals, San Gabriel, CA 91775
Tessek Co., Mountain View, CA 94041
UOP, Inc., San Diego, CA 92121
Whatman Biosystems Inc., Clifton, NJ 07014

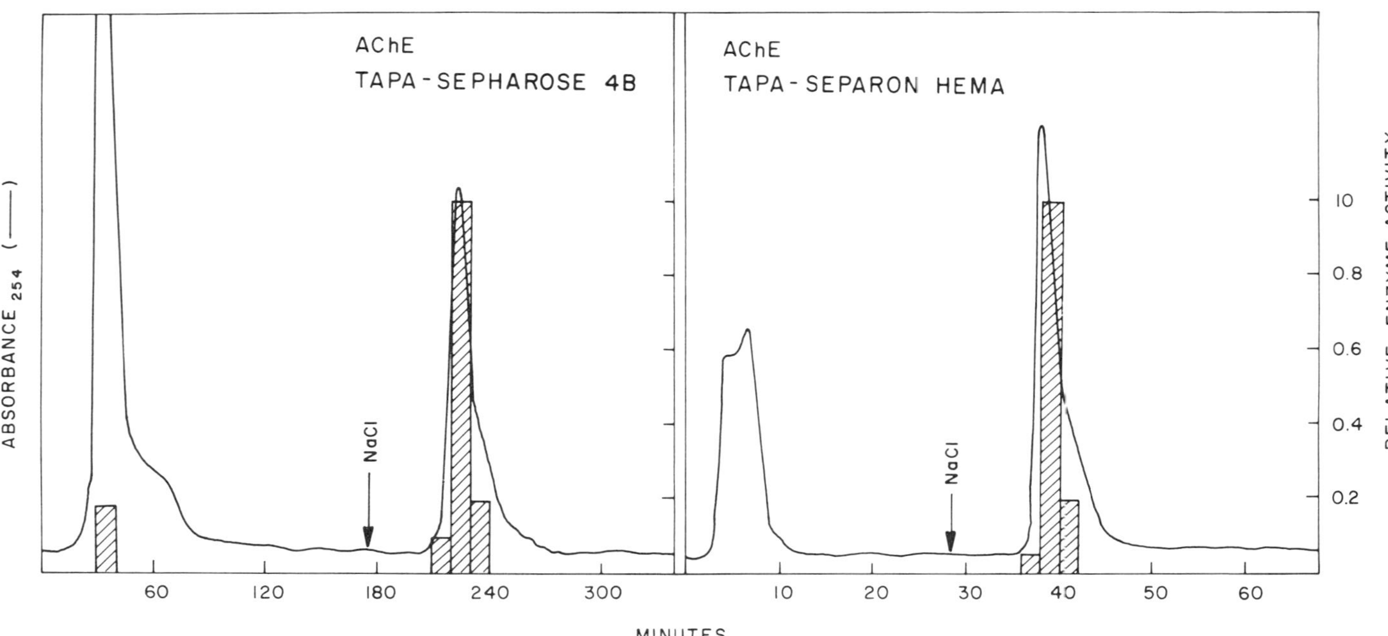

Figure 2 Purification of acetylcholinesterase (AChE) on a soft agarose (Sepharose 4B) or a pressure stable polymeric (Separon HEMA) affinity chromatography support. Enzyme was eluted from the columns by applying buffer containing 1 M NaCl. (From Ref. 4.)

HEMA support (up to 5000 psi), it can be packed into a HPLC column and eluted at flows up to 4 ml/min. The Sepharose support could not be run under pressure without compression, and the maximum (gravity) flow through the column was approximately 0.5 ml/min. As a result of the higher flow rate and rigidity of the HEMA support, purification of the enzyme was accomplished in 30–40 min in comparison to over 200 min using the Sepharose column. This example illustrates the advantages of new, stable polymeric supports for rapid purification of biomolecules. Such supports are also proving more stable for protein immobilization and the use of protein-immobilized polymeric supports in purification and enzyme reactors.

Silica-based supports are also stable to high flow rates and pressure. They are, however, susceptible to alkaline pH instability, and even prolonged use in neutral pH aqueous solutions can lead to leaching of immobilized ligands from the supports. One of the first attempts to circumvent these problems was the coating of controlled pore glass beads with glycerol or similar hydrophilic phases to provide a stable support surface chemistry resistant to alkaline pH (e.g., Glycophase glass beads made by Corning and marketed by Pierce Chemical). Newer silica-based supports utilize coatings of (proprietary) polymers both to protect the silica base material and to provide a reactive surface for protein immobilization.

Stabilization also refers to the shelf life of activated supports and is especially important for preactivated supports (see below). Except for activation chemistries such as cyanogen bromide, N-hydroxysuccinimide and carbonyldiimidazole, most of the activation chemistries now used for commerical support products have shelf lives in excess of 1 year stored at 4–25°C.

4. Protein Capacity

Studies have been carried out with most commercially available supports to determine binding capacities for common proteins such as albumin, IgG, trypsin, and other enzymes. In general, most supports will bind from 2 to 50 mg protein per gram of support. While some supports claim to bind as high as 170 mg protein/g (5), such high binding capacities may result in steric interference problems, loss of protein activity, and changes in the net surface charge of the support (6,7).

Even at low protein binding capacities, the amount of protein bound may not correlate with the amount of activity recovered after immobilization. For example, Figure 3 illustrates the immobilization of three enzymes on cyanogen bromide–activated Sepharose at varying pH. While in the cases of peroxidase and glucose oxidase the amount of protein bound correlates with recovered activity after immobilization, this is not the case

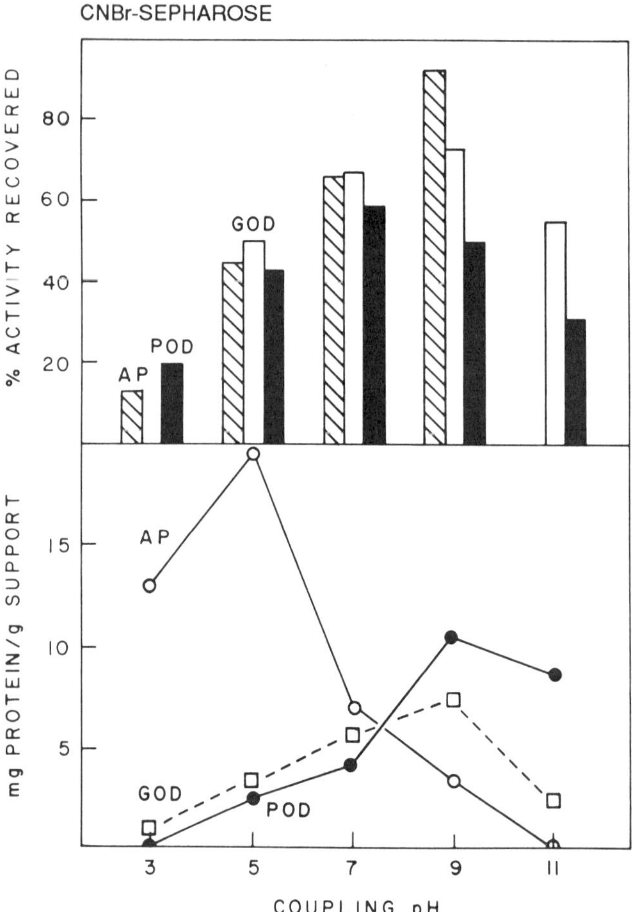

Figure 3 Amounts of protein and activity recovered after immobilizing alkaline phosphatase (AP), peroxidase (POD), and glucose oxidase (GOD) on CNBr-activated Sepharose 4B. The percentage activity recovered is relative to the activity of the free enzyme prior to immobilization. (From ref. 8.)

with alkaline phosphatase (8). The loss in AP activity may be due to the coupling pH (see below).

The protein capacity of commercially available supports largely depends on the surface density of available linking groups. All commercial supports will differ in the concentration of these groups but, in general, the majority of the supports contain from 50 to 200 μmol of activated group or available binding group per gram of support. Some supports may have amounts of binding groups as low as 1 to 5 μmol/g, while others may have very high concentrations (e.g., as much as 1500 μmol/g for epoxy-activated Separon HEMA supports, Table 1).

The amount of protein bound to a support will also directly depend on the coupling reaction conditions used, e.g., time, temperature, and pH. While the optimal coupling conditions have been determined for most commercial supports for common proteins, these same conditions may not hold true for other proteins, and optimization studies may be necessary. The example presented in Figure 3 illustrates that the optimal binding pH suggested by the manufacturer does not hold in the case of alkaline phosphatase.

B. Support Chemistries

The surface modification of immobilization and separation supports has been reviewed both in this text and elsewhere (1,2,9–11). A number of these are used for the preparation of commercial supports and will be considered here. The coupling reactions and conditions given below have been drawn from information provided by the support manufacturers and from the literature.

Table 1 lists two general categories of commercially available supports: *preactivated supports,* which allow direct, spontaneous coupling of the protein to the support; and *functional group supports,* which contain groups (such as amino, carboxylic, sulfhydryl, and hydroxyl) which are available for chemical coupling to the protein and other ligands.

1. Preactivated Supports

These supports are the easiest to use for protein (or other ligand) immobilization. They are purchased, either wet or dry, and after initial washing steps are ready for direct coupling with the ligand in a buffer. In most cases, coupling is completed in less than 24 hours and can be carried out at room temperature. The major preactivated support chemistries are as follows.

Cyanogen Bromide (CNBr) Activation. One of the first commercially available preactivated supports was CNBr-activated Sepharose. The CNBr reacts with the hydroxyl groups on the surface of the support to form the cyclic imidocarbonate (Reaction 1). This active group can react with ligand primary amine group(s) to form *N*-substituted isoureas, *N*-substituted imidocarbonates, or *N*-substituted carbamates.

$$\left.\begin{array}{l}-OH\\-OH\end{array}\right. + CNBr \longrightarrow \left.\begin{array}{l}-O\\-O\end{array}\right\rangle C{=}NH + H_2N{-}R \longrightarrow \begin{cases}\begin{array}{l}-OH\\-O{-}\underset{\underset{NH}{\parallel}}{C}{-}NH{-}R\end{array}\\ \left.\begin{array}{l}-O\\-O\end{array}\right\rangle C{=}N{-}R\\ \begin{array}{l}-OH\\-O{-}\underset{\underset{O}{\parallel}}{C}{-}NH{-}R\end{array}\end{cases} \qquad (1)$$

Coupling to CNBr-activated supports needs to be carried out rapidly since the cyclic imidocarbonate active group is unstable in aqueous solutions. Typical coupling is at pH 8.5–9.5 in, e.g., carbonate buffer, although lower pH (≥7) may be used depending on the pKa of the ligand to be immobilized (see Table 3). With CNBr-activated and all amine-coupling supports, amine-containing buffers such as Tris/HCl should not be used since the buffer amino groups will compete with the ligand for the reactive

Table 3 Summary of Ligands Bound and Reaction Conditions for the Major Types of Commercially Available, Preactivated Supports

Active group	Ligands bound	Typical reaction conditions[a]
CNBr	$-NH_2$	pH 7–10, 2–18 h, 4–25°C
Aldehyde (Glutaraldehyde)	$-NH_2$	pH 3–10, 1–12 h, 4–25°C
Epoxy	$-NH_2$	pH 5–12, 4–72 h, 4–60°C
	$-OH$	
	$-SH$	
Hydrazide	$-NH_2$	pH 6–9, 2–12 h, 4–25°C
	$-CHO$	
N-Hydroxysuccinimide	$-NH_2$	pH 5–10, 1–4 h, 4–25°C
Carbonyl diimidazole	$-NH_2$	pH 8–10, 18–24h, 4–25°C
Divinylsulfone	$-NH_2$	pH 6–11, 2–24 h, 4–25°C
	$-OH$	
	$-SH$	
Tresyl	$-NH_2$	pH 7–9, 2–16 h, 4–25°C
Fluoromethylperidinium	$-NH_2$	pH 7–9, 2–8 h, 4–37°C
	$-SH$	

[a] In general, after ligand reaction with the support, the support is reacted with a low molecular weight amine (such as ethanolamine or glycine) or Tris/HCl buffer to quench any remaining active groups.

groups of the support. Coupling can take place at 4–25°C, although CNBr-activated agarose gels (such as Sepharose) are temperature sensitive and are best reacted at 4°C. After the reaction, any residual active groups on the support are blocked with ethanolamine, glycine, Tris/HCl buffer, or other appropriate low molecular weight amines.

There are a number of problems with the use of CNBr supports past the toxicity of the reagent. The support-ligand linkage formed can be partially hydrolyzed at pH <5 and >10 and at elevated temperatures leading to ligand leakage. Spontaneous leakage may also occur during normal use and aging of the support. Secondarily, isocyanates released during the support-ligand linkage reaction can interact with proteins leading to potential changes in their structure and function and/or the interduction of new antigenic sites. N-Substituted urea derivatives formed during the coupling reaction can act as bases and introduce an ion-exchange characteristic into the support. For these reasons, CNBr coupling is becoming less used for protein immobilization and CNBr-activated supports are offered by only a few commercial suppliers.

Glutaraldehyde Activation. One of the most-used methods for support preactivation in current commercial products is treatment of an amide- or primary amine-containing support with glutaraldehyde (e.g., 10 ml of 2.5–20% glutaraldehyde per gram of support for 2 h at room temperature). The glutaraldehyde adds directly to the support resulting in the addition of (at least) a 5-atom spacer arm and a reactive aldehyde end group (Reaction 2). The reaction of the aldehyde group with a primaary amine at pH 3–10 for 4–12 h at 4–25°C results in the formation of the Schiff base, which is then stabilized to the secondary amine by reduction with an agent such as sodium borohydride, sodium cyanoborohydride, or trimethylamine borate. The reducing agent may be added together with the ligand or as a separate step after the ligand-support reaction. The support is then reacted with an amine-containing buffer (such as 0.1–0.5 M Tris/HCl, pH 7) or 1 M ethanolamine to quench residual aldehyde groups.

$$\vdash CHO + H_2N-R \rightleftarrows \vdash CH{=}N-R \xrightarrow{H} \vdash CH_2-NH-R \qquad (2)$$

Proteins and ligands immobilized on glutaraldehyde-activated supports form stable bonds. The resulting supports have proved resistant to ligand leakage during long-term use.

Epoxy Activation. The reaction of hydroxyl-containing supports with epichlorohydrin or bisoxiranes, such as 1,4-butanediol diglycidyl ether, results in supports with active epoxy groups able to covalently link ligands with available amino, hydroxyl, or sulfhydryl groups (Reaction 3). The resulting alkylated amines, ethers, or thioethers are stable, and significant ligand leakage does not occur with time or use.

$$\vdash OCH_2-\underset{\displaystyle OH}{CH}-(CH_2)_n-\overset{\displaystyle O}{CH-CH_2} + H_2N-R \longrightarrow \vdash OCH_2\underset{\displaystyle OH}{CH}(CH_2)_n\underset{\displaystyle OH}{CH}CH_2-NH-R \quad (3)$$

Commercially available epoxy-activated supports usually contain higher numbers of activated groups than other preactivated supports, e.g., 200–1500 μmol epoxy groups/g support compared to 50–200 μmol active group/g for other supports. This appears due to both the high efficiency of the activation reaction and the need for such high active group concentrations for efficient ligand coupling.

Coupling to epoxy-activated supports is carried out at pH 5–12 for 4–72 h at 4–60°C. In addition, coupling may be carried out in buffers containing up to 50% by volume organic solvent (such as dioxane and dimethylformamide). Such coupling in organic solvents is necessary for many ligands and may be necessary for the immobilization of proteolipids and hydrophobic peptides.

Hydrazine Activation. Hydrazine activation of supports was originally used as a step toward azide group formation for reaction with primary amines (Reaction 4). Such azide-activated supports can be used to immobilize proteins, although the resulting supports appear less stable to ligand leakage than others such as aldehyde- and epoxy-activated supports.

$$\vdash CONH_2 + H_2N-NH_2 \longrightarrow \vdash CON_3 + H_2N-R \longrightarrow \vdash CONH-R \quad (4)$$

Most commercial supports using hydrazine activation now provide the support for reaction with aldehyde groups formed from the periodate oxidation of vicinyl hydroxyl groups of sugar groups in the ligand (e.g., antibodies, glycoproteins, and nucleic acids) to be immobilized (Reaction 5). This more recent use of hydrazine-activated supports is especially applicable to the directional immobilization of antibodies, e.g., in cases where the immobilized antibodies are to be used for purification or for the detection of an antigen for diagnostic applications (see Chapter 8).

$$\begin{bmatrix}\vdash CONHNH_2 \\ \vdash CONHNH_2\end{bmatrix} + (CH_2OH;\ O;\ HC{=}O;\ CH{=}O)-O-R \longrightarrow \begin{bmatrix}\vdash CONHN{=}CH \\ \vdash CONHN{=}CH\end{bmatrix}(CH_2OH;\ O;\ O-R) \quad (5)$$

For directional immobilization of antibodies to hydrazine supports, vicinyl hydroxyl groups on the sugar residues of the Fc region of the antibody are oxidized in the presence of sodium periodate for 1 h at room tenperature. This reaction does not appear to affect the antigen binding activity of the antibody. The oxidized antibody is then reacted with the hydrazine-activated support at pH 6–9 for 2–8 h at 4–25°C. This type of binding results in orientation of the immobilized antibody with its antigen binding region away from the support and thus able to better interact with antigen. As the result of this type of immobilization, suppliers of hydrazine-activated supports claim better than 90% retention of antibody binding activity after immobilization. The supports resulting also appear very stable to prolonged use.

While the use of hydrazine-activated supports for periodate-treated glycoproteins provides a means for directional immobilization, this is also possible using other activated supports. For example, Figure 4 illustrates the immobilization of human IgG on four preactivated supports. While the CNBr-Sepharose and CDI Reacti-Gel supports resulted in low retention of antibody activity after immobilization, the epoxy-activated Separon HEMA and glutaraldehyde-activated Ultrogel both resulted in high antibody activity retention after immobilization. Low activity retention was also found with two *N*-hydroxysuccinimide–activated supports and two other epoxy-activated supports with lower densities of active group (8).

N-Hydroxysuccinimide (NHS) and Carbonyldiimidazole (CDI) Activation. NHS and CDI activation represent two older methods to produce supports for coupling ligands with primary amine group(s). NHS is used to activate carboxyl-containing supports, while CDI is used for hydroxyl-containing supports. Reaction of the supports at appropriate pH (Table 3) results in spontaneous coupling between the supports and ligands (Reactions 6 and 7).

$$\vdash COOH + HO-N(succinimide) \longrightarrow \vdash COO-N(succinimide) + H_2N-R \longrightarrow \vdash CONH-R \qquad (6)$$

$$\vdash OH + (imidazolyl)N-C(=O)-N(imidazolyl) \longrightarrow \vdash OOC-N(imidazolyl) + H_2N-R \longrightarrow \vdash OOC-NH-R \qquad (7)$$

Some problems have been observed in the use of NHS and CDI supports. NHS can react with hydroxyl groups on the ligand to result in an unstable support–ligand ester linkage. The *N*-alkylcarbamate formed be-

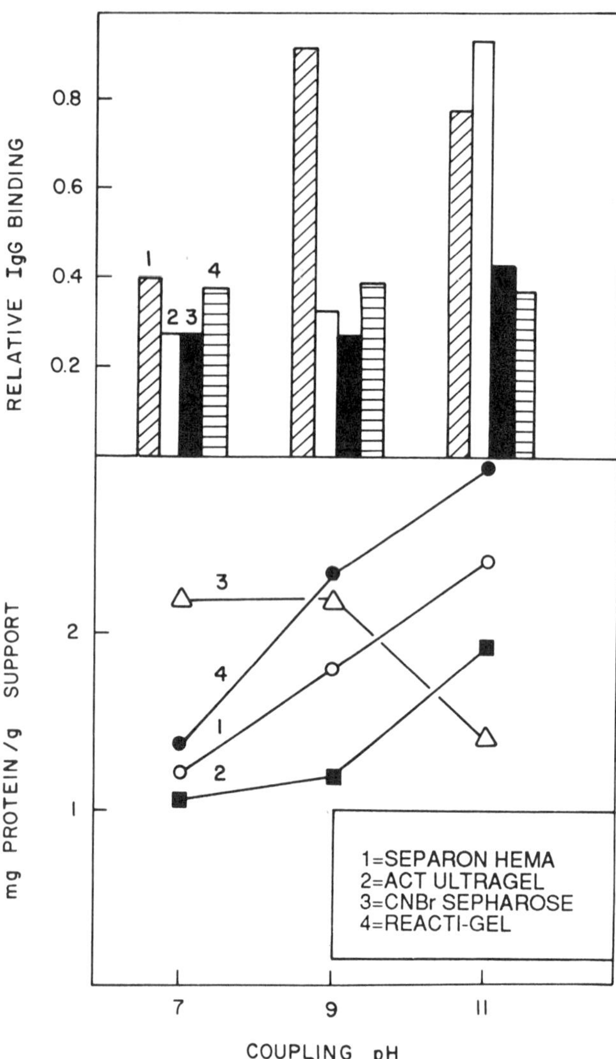

Figure 4 Immobilization of human immunoglobulin G (IgG) on four activated supports at varying pH. The relative IgG binding is the binding activity for antibody to IgG by the support in comparison to antibody binding of free IgG prior to immobilization. (From Ref. 8.)

tween the ligand and the CDI support is not stable at pH > 10 and may leak ligand with use similar to CNBr-activated supports. In addition, proteins can bind/interact with the imidazole leaving group during CDI coupling, resulting in possible changes in protein structure and function.

Divinyl Sulfone (DVS) Activation. Similar to bisoxiranes, DVS is a bifunctional cross-linking reagent. Treatment of a hydroxyl-containing support with DVS results in a vinylsulfonyl ethyl ether, which will react with ligands containing amino, hydroxyl, and sulfhydryl groups (Reaction 8). The vinyl group is more reactive than the oxirane group on epoxy-activated supports, and thus DVS-activated supports will couple a wider variety of ligands with higher efficiency.

$$\text{|}-OH + CH_2{=}CH-SO_2-CH{=}CH_2 \longrightarrow \text{|}-OCH_2CH_2SO_2CH{=}CH_2 + H_2N-R \longrightarrow \text{|}-OCH_2CH_2SO_2CH_2CH_2NH-R \quad (8)$$

DVS coupling is best carried out at pH 8–11, although coupling will occur at lower pH. Unfortunately, the ligand-support bond resulting is also unstable with time at alkaline pH. Thus, the use of a DVS support requires optimization of the reaction conditions. The support resulting from DVS coupling is, however, very stable at acidic and neutral pH.

Tresyl Activation. Organic sulfonyl chlorides can activate support hydroxyl groups. Tosyl chloride was one of the first materials to be used for support activation. It has now been replaced in commercial supports by tresyl chloride (2,2,2-trifluoroethanesulfonyl chloride). The resulting tresyl-activated support can couple ligands containing amino and sulfhydryl groups (Reaction 9). The resulting supports are relatively stable but may show ligand leakage at pH < 4 and > 10.

$$\text{|}-CH_2OH + CF_3CH_2SO_2Cl \longrightarrow \text{|}-CH_2OSO_2CH_2CF_3 + H_2N-R \longrightarrow \text{|}-CH_2NH-R \quad (9)$$

Other Activated Supports. New supports activated with fluoromethylperidinium (FMP) are available for spontaneous coupling of ligands containing amino and sulfhydryl groups. While the exact reaction mechanism for FMP coupling is still undefined, questions have been raised about potential interaction of proteins with methyl pyridone, a postulated leaving group during FMP coupling.

One product listed in Table 1 is representative of a biologically activated rather than a chemically activated support. This is an aviden-coated glass bead product (from Kontes Biotechnology) which directionally immobilizes periodate oxidized antibodies pretreated with hydrazine-biotin. The hydrazine and aldehyde groups of the oxidized sugar group form a covalent bond (see Reaction 5), and the biotin binds to the avidin on the support. Questions remain, however, concerning the stability of the noncovalent biotin-avidin bond in this support with use and time.

2. Nonactivated Supports

Nonactivated, commercially available supports include primarily those containing a functional group which can be covalently coupled to ligands and microcarriers intended primarily for cell immobilization but also applicable to protein immobilization.

The primary coupling groups offered on commercial supports are carboxyl, amino, and sulfhydryl groups. The supports are most often prepared by introducing bifunctional spacer arms into the base support such as aminocaproic acid, hexanediamine, or ethanolamine. The resulting supports can then be used to couple ligands using the appropriate chemistry. For example, the most widely used coupling chemistry is carbodiimide coupling of the support carboxyl or amino groups to the amino or carboxyl group(s) of the ligand (Reaction 10). The reaction is carried out at pH 4.6–6 for 2–18 h at 4–25°C and may use either water-soluble (e.g., EDC, *N*-ethyl-*N'*-(3-dimethylaminopropyl)carbodiimide HCl) or organic-soluble (e.g., DCC, dicyclohexylcarbodiimide) carbodiimide. The resulting peptide bond is stable with both use and time.

$$\left|\!-NH(CH_2)_5COOH + R'N{=}C{=}NR'' \longrightarrow \left|\!-\underset{\displaystyle NH-R''}{COOC}{=}N-R'\right.\right. \qquad (10)$$

$$+ \; H_2N-R \longrightarrow \left|\!-CONH-R\right.$$

One filter product listed in Table 1 also contains spacer groups which may be activated for ligand binding. The Transa-Bind filters of Schleicher & Schuell contain one of two spacer arms ending in primary aromatic amines (aminophenylthioether- and aminobenzyloxymethyl-celluloses) which may be converted to their diazo derivatives by diazotization. The resulting activated support will spontaneously couple ligands containing hydroxyl groups.

A number of microcarriers are listed in Table 1 which are applicable to ether cell or protein immobilization. Those listed are representative of available products and do not include all such products. Other microcarriers and their use are discussed in other chapters of this text.

Some microcarrier characteristics are of interest for comparison purposes to the supports discussed above. The carriers have a base of glass, plastic, dextran, collagen, agarose, alginate, carragenan, chitosan, cellulose, or other materials common to spearation and immobilization supports. The base particles of the carriers are usually 50–500 μm in size with capacities of 3–200 ml cells/g support or 0.1–1 ml cells/ml settled bed volume. In most cases, for application to protein immobilization, the surface of the microcarriers would need to be activated by reaction with an appropriate activation reagent such as glutaraldehyde, epichlorohydrin, DVS, etc. While these microcarriers have not been used extensively for protein immobilization, improvements in their stability and their ease of use makes them attractive candidates for protein immobilization.

IV. CONCLUSION

Commercially available supports provide a means to rapidly access and evaluate immobilization technologies and applications. During the past 10 years, significant advances in the stability, capacity, and ease of use of these supports have taken them from mostly unreliable, high-cost specialty products to routinely available, dependable, and cost-effective immobilization media. Ongoing advances in the development of new polymeric supports and better activation chemistries promise to provide even better supports for protein immobilization in the next 5 years. The use of these supports for immobilization of proteins will be largely dependent on proving their applicability for protein immobilization and the awareness on the part of immobilized protein users of the existence and applications of these supports.

REFERENCES

1. P. Cuatrecasas and C. B. Anfinsen, *Meth. Enzymol. 22:*345 (1971).
2. C. R. Lowe, *An Introduction to Affinity Chromatography,* North-Holland Publishing Co., Amsterdam, 1979.
3. I. M. Chaiken, M. Wilchek, and I. Parikh (Eds.), *Affinity Chromatography and Biological Recognition,* Academic Press, New York, 1983.
4. R. F. Taylor and I. G. Marenchic, *J. Chromatogr. 317:*193 (1984).
5. O. Hannibal-Friedrich, M. Chun, and M. Sernetz, *Biotechnol. Bioeng. 22:*157 (1980).
6. O. Valentova, J. Turkova, R. Lapka, J. Zima, and J. Coupek, *Biochim. Biophys. Acta 403:*192 (1975).
7. K. Babor and L. Rexova-Benkova, *Coll. Czechoslov. Commun. 45:*617 (1980).

8. R. F. Taylor, *Anal. Chim. Acta 172:*241 (1985).
9. J. Porath, *Meth. Enzymol. 34:*13 (1974).
10. P. V. Sundaram and F. Eckstein (Eds.), *Theory and Practice in Affinity Techniques,* Academic Press, New York, 1978.
11. H. H. Weetall, *Sep. Purifica. Met. 2:*199 (1973).

5

Determination of Coupling Yields and Handling of Labile Proteins in Immobilization Technology

Bo Mattiasson and Rajni Kaul

University of Lund, Lund, Sweden

I. INTRODUCTION

Reports on chemically immobilized proteins first appeared around 20 years ago. Since then, innumerable systems have been studied, and many of them are now used routinely. Initially, most of the interest was focused on the immobilization of enzymes, but with the introduction of various binding immunoassays, immobilized antibodies found a great commercial interest. In many cases the preparations have been characterized to some degree, but often it has been difficult to compare data between different reports, since each author has performed the characterization in his own way. A working party on immobilized biocatalysts appointed by the European Federation of Biotechnolgy has published some guidelines defining the minimum parameters required for a satisfactory characterization of the immobilized biocatalysts, so that adequate information is obtained for general comparisons and for designing processes for large-scale purposes (1) (Table 1).

This chapter summarizes the methods used to characterize the proteins attached to insoluble matrices or to soluble polymers. These include mainly the measurements of total and active bound protein. Most of the methods have been discussed in detail earlier (2, 3), and thus will not be elaborated here. Since a lot of emphasis has been put on enzymes, we shall also devote most of the chapter to enzymatic systems. It should, however, be kept in mind that the techniques presented in many cases will also be applicable to

Table 1 Some Guidelines for the Characterization of an Immobilized Biocatalyst

1. General description of the biocatalyst, reaction scheme, matrix, and immobilization technique.
2. Preparation of the immobilized biocatalyst—immobilization conditions, dry weight yield and activity left in supernatant.
3. Physical/chemical characterization—biocatalyst size and shape, behavior in column and stirred vessel.
4. Kinetics of the immobilized vs. free biocatalyst, including storage and operational stability.

Source: Ref. 1.

noncatalytic proteins. As immobilization is an established method both in enzyme- and immunotechnology, the strategies applied to immobilize labile proteins have also been discussed.

II. DETERMINATION OF BOUND PROTEIN

Estimation of the amount of bound protein is important for ascertaining the coupling efficiency of a matrix, and also the extent of protein modification during immobilization since it is well known that all the protein that gets bound is not active.

A. Difference Between Protein Added and Unbound Protein

The immobilization procedure is usually followed by a rigorous washing routine to eliminate the last traces of the non–firmly bound protein. Quantitation of the bound protein by measurement of the difference between the amount of protein put into the immobilization mixture and that recovered in the washing solution has been performed in several cases by ultraviolet (UV) absorbance studies, and also by colorimetric determination, e.g., using Lowry's method. But the determination of low protein concentrations in relatively large volumes of wash solutions generally used to remove the adsorbed material gives, more or less, a rough estimate of the protein that is bound. UV absorbance measurements cannot, however, be applied when other UV-absorbing materials are present in the immobilization mixture. Lowry's method has been shown not to be applicable for systems using polyacrylamide supports, because of the interference by the residual monomer (4). The use of nitrogen-containing buffers may also lead to errors in such estimations.

A more accurate method has, therefore, been suggested to be a direct determination of the carrier bound protein.

B. Amino Acid Analysis

One of the most precise and a widely adopted method to date for determination of carrier bound protein has been the quantitative amino acid analysis of the protein-carrier conjugate after an exhaustive acid hydrolysis. A small amount (a few milligrams) of the conjugate is washed with acetone and dried prior to hydrolysis with 6 M HCl for 24 h at 110°C in vacuo. The hydrolysate is evaporated to dryness, the residue dissolved in a suitable buffer, and applied to the amino acid analyzer. The quantity of the protein is then calculated from the molar amounts of certain selected amino acids (or even one) and their known residue numbers (5).

One has to remember that the amino acids involved in the covalent linkage between enzyme and carrier may give lower yields and thus should not be used as a basis for the calculation of the bound protein. Analyses of alanine and/or leucine has been suggested as being useful for quantitation (6). Even the high concentration of ammonia released from the samples of acrylamide carriers were not seen to disturb the determinations of these two amino acids. Errors arising due to sample dilution etc. can be sufficiently corrected if a known amount of norleucine (internal standard) is added prior to hydrolysis, which is analyzed together with the other amino acids, and then compared with the expected value.

The free amino acid content can also be measured by applying specific colorimetric methods (7).

Gas chromatography (GC) has been used for quantifying amino acids in immobilized trypsin (8). This procedure requires the derivatization of the starting material in order to convert the amino acids into more volatile, less polar compounds that are suitable for GC. Hence, the amino acids obtained after the acid hydrolysis of the insoluble complex were converted to *N*-trifluoroacetyl-*n*-butyl esters for their analysis. For comparison, known amounts of an inner standard (e.g., 4-amino benzoic acid) and/or one of the amino acids (e.g., aspartic acid) were also added to the hydrolysate.

C. Other Methods

Provided the immobilized protein is the only UV-absorbing entity, it may be possible to perform UV measurements on the insoluble complexes. To avoid the interference by the support, the absorbance is measured by suspending the insoluble material in 50% sucrose or 85% glycerol. This helps in eliminating the changes in refractive index in the system. It is advisable to have an absorption spectrum over a broad wavelength range (instead of only at 280 nm) or at any other wavelength where the protein in question absorbs and compare it with the absorbance contributed by a control matrix taken through the same coupling and washing procedures, but without the addition of the protein.

Analysis of the nitrogen content of the immobilized preparation by Kjeldahl's method for determining the amount of bound protein can be useful when high amounts of protein have been immobilized. A parallel determination of nitrogen on a control matrix is suggested (5). The method, however, is not applicable for matrices such as polyacrylamide, DEAE cellulose, collodion, etc., whose nitrogen content greatly exceeds that of the protein introduced.

It may sometimes be possible to measure the bound protein by its deliberate release from the matrix. This approach is generally taken for proteins that are reversibly bound to the supports either by ionic, hydrophobic, affinity, or cleavable covalent bonds. In case of Sepharose bound G actin, the protein was dissociated from the support by digestion with pronase and then measured by developing with ninhydrin (9). Such a procedure may have a drawback in that all the bound protein may not be accessible to enzyme attack, and varying amounts of the protein may be liberated depending on the type of protein bound and its susceptibility to pronase action.

Proteins bound to supports which are easily solubilized can be assayed by different methods after solubilization of the complex. For example, agarose bound affinity ligands have been quantified after dissolving the support in 50% acetic acid (10).

A convenient way to determine the amount of bound protein is to add a small amount of radioisotope-labeled protein to the preparation prior to immobilization. Provided no dramatic changes in protein properties have taken place in the radiolabeling step, one can calculate the coupling yield by measuring the percentage of the isotope retained in the immobilized preparation.

Some of the methods discussed above may not be applicable in situations when the carrier itself is a protein, or two or more proteins are co-immobilized. In the latter case, the different proteins may not have the same coupling yield, i.e., one protein may bind preferentially to the support as compared to the other. In such cases, therefore, measurements based on groups characteristic to the immobilized protein may be exploited. An example is seen in the estimation of the glucose oxidase content in an immobilized trypsin–glucose oxidase system (11). Flavin adenine dinucleotide, the prosthetic group of the enzyme, was dissociated by treatment with trichloroacetic acid, and the amount of the nucleotide released was measured fluorimetrically. As trypsin was in 1000-fold excess, the amino acid analysis of the hydrolyzed conjugate gave a close estimation of the trypsin content.

With collagen as the support material for immobilization of enzymes, advantage has been taken of the fact that most enzymes contain tryptophan and cysteine, while collagen does not. Thus, specific methods for tryptophan and cysteine analysis were employed successfully for measuring the bound protein (12, 13).

Covalent attachment of soluble polymers such as polyethylene glycol (PEG) to proteins is also regarded as a part of some immobilization strategies. PEG modification has generally been carried out via the primary amino groups on the proteins; and the amount of the modified protein has been determined by measuring the free amino groups by reacting with trinitrobenzenesulfonic acid (14) or with fluorescamine (15). The latter procedure has been shown to be more sensitive and reproducible.

III. DETERMINATION OF THE ACTIVITY OF BOUND PROTEIN

The amount of enzymatic activity retained in an immobilized preparation is often measured by determining the maximal reaction rate per certain amount of the catalyst under defined experimental conditions, and comparison with the activity of the native enzyme. The amount of activity remaining in solution or desorbed by washing should also be considered. Various procedures used to assay enzymes have been detailed by Mattiasson and Mosbach (2). In several cases, the enzyme is allowed to interact with the substrate, aliquots of the reaction mixture are removed at different time intervals, and the product and/or the substrate is measured by spectrophotometry, GC, HPLC, etc., after filtering off the gel particles.

On the other hand, some measurements require the enzyme to be present during the assay. Spectrophotometric methods commonly used for assaying enzymes have been adapted in several ways for such measurements, so as to avoid the problems of heterogeneity posed by the immobilized preparations. Provided the matrix is not too dense (e.g., Sepharose), the enzymatic reaction can be followed in the spectrophotometer by keeping the enzyme particles continuously stirred in the cuvette (16). A number of commercially available spectrophotometers are equipped with a stirring device, or such units can be obtained as accessories. In situations when it is not possible to place the enzyme-matrix in the light path of the photometer for activity determinations, the reaction is allowed to take place in a column or a batch reactor, and a part of the clear product containing solution is intermittently or continuously pumped through a flow cuvette for spectrophotometric analyses (Fig. 1)(2).

In the case of an enzyme reaction in a column, recirculation of the eluate under conditions of low substrate conversions (1–2%) gives rise to a continuous process, and the absorbance of the recirculating fluid may be mea-

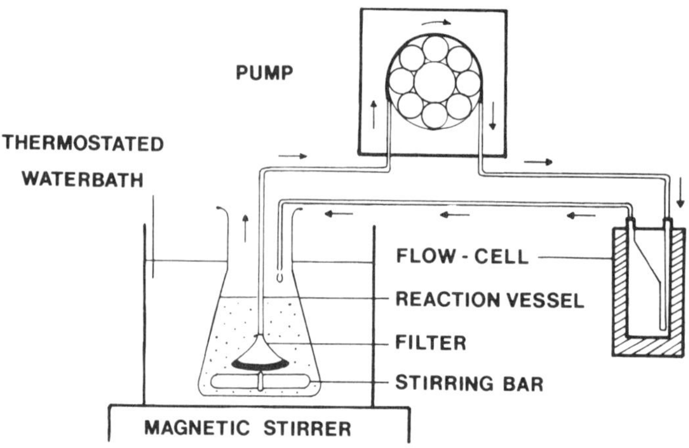

Figure 1 Continuous spectrophotometric analysis of the supernatant from a stirred batch reactor. (Reproduced with permission from Ref. 2.)

sured continuously by integrating the system with a spectrophotometer (17). Such a continuous system is identical to a batch reactor if the column bed is shortened.

Fluorescent procedures for determining enzyme activity are usually several orders of magnitude more sensitive than colorimetric methods and have replaced the latter wherever possible. Therefore, the assays for immobilized enyzmes can also be adapted for fluorometric analyses using a different flow cell. For example, a modified fluorometric technique developed in Guilbault's laboratory (18) for the determination of glucose in plasma by immobilized glucose oxidase and peroxidase can easily be used for assaying other insoluble enzymes. In this case, the fluoromicrophotometer is provided with a stirring motor beneath the cell compartment so that the contents of the cuvette can be continuously stirred. A special stirrer has been constructed that restricts the immobilized enzyme in a nylon tube, preventing its dissipation throughout the solution and, at the same time, allowing the reaction mixture to come in contact with the enzyme easily (Fig. 2).

Spectrophotometric methods cannot be used in cases where immobilized enzyme preparations are highly opaque. However, the opacity of insoluble preparations does not pose a problem for measurement techniques such as electrode-monitoring, conductivity measurements, or titrimetry. Monitoring the consumption or liberation of an ion or a gaseous compound by means of an electrode constitutes one of the most convenient methods for measuring the activity of a bound enzyme. The analytically useful range of the ion-selective electrodes is generally between 10^{-1}–10^{-5} M. However,

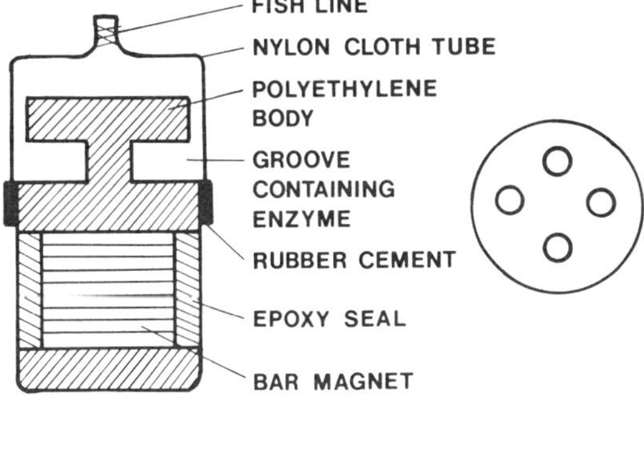

Figure 2 Cross-sectional and top views of the immobilized enzyme stirrer device for fluorometric analyses. (Reproduced with permission from Ref. 18.)

there are several electrodes that are useful for much lower concentrations. Since the response of these electrodes is logarithmic, the precision of the instruments is constant over their dynamic range. For example, an NH_4^+ sensitive electrode has been used for assaying activity of urease immobilized on glass (19).

Bouin et al. (20) have used an oxygen electrode for the determination of enzyme activities in an immobilized glucose oxidase–catalase system. For the assay of glucose oxidase, the decrease in oxygen in solution was measured in a reaction medium containing 13.9 mM D-glucose in a citrate-phosphate buffer, pH 5.5. Catalase was assayed by following the production of oxygen in the buffer containing 0.5 mM H_2O_2. The initial concentration of oxygen in the reaction medium was reduced by bubbling nitrogen through the buffer and substrate solutions.

Messing has employed differential conductivity as an assay method for immobilized glucose oxidase-catalase (21) and urease (22). The differential mode of measurement offered an advantage of detecting very small changes in conductivity with respect to background contributions.

Titrimetric methods have been used for measurements of enzyme catalyzed reactions producing or consuming protons. The esterase activity of immobilized trypsin has been determined titrimetrically (5). The method has been found to be simple and reproducible, although the sensitivity is lower than obtained with spectrophotometry. However, it is a realistic alternative in cases with excess substrate and high enzyme activities. Titrimetry cannot be used in conjunction with charged matrices and strongly buffering substances.

A. Determination of Kinetic Constants

The effects of the microenvironment and the immobilization process itself cause the kinetic parameters of the immobilized enzyme to frequently differ from those of the soluble enzyme. Diffusional restriction is the most dominant effect influencing the kinetics of the immobilized enzyme. When assaying for the level of enzyme activity in an immobilized preparation, it is important to operate with the smallest diffusional restrictions possible. Diffusional constraints may be eliminated to a great extent by decreasing the particle size, increasing the stirring or flow rate, and/or increasing the substrate concentration in the bulk solution, thus making it possible to measure, more or less, a true value of V_{max} and K_m of the immobilized enzyme (1).

Kinetic measurements on immobilized enzymes are also dependent on enzyme loading and the microenvironment of the active enzyme. In cases of excess loading of enzyme onto a matrix, it must be realized that the characteristics measured relate only to a fraction of the active enzyme molecules. The reaction rate in such cases is strictly dependent on the diffusion of substrate. Systems with an excess of unused catalytic capacity will appear to be insensitive to changes in the external medium.

The microenvironment around the immobilized enzyme is significantly altered if a charged matrix is used for immobilization, or if the enzyme itself catalyzes proton production or consumption giving rise to local proton concentrations different from that in the bulk solution (2). This leads to pH and ionic effects on the activity of the preparation, which may be reduced or eliminated by increasing the ionic strength of the surrounding medium. The effect of proton production or consumption is also controlled by diffusion of product between the site of catalysis and the surrounding medium. This may be diminished by reducing the size of the enzyme particles.

B. Active Site Titration

Active site titration is a more accurate and, hence, a preferred method for the determination of active protein, but it is applicable for only a limited number of enzymes. The basis for this method lies in the specific stoichiometric reactions of enzymes with active site directed reagents. A molar amount of the product equivalent to the molar concentration of the active enzyme is liberated in a quick burst reaction.

1. Immobilized Trypsin

For active site titration of immobilized trypsin, *p*-nitrophenyl *p*′-guanidinobenzoate (NPGB) was used as the titrant (23, 24), and the *p*-nitrophenol formed was determined spectrophotometrically. The initial burst of the

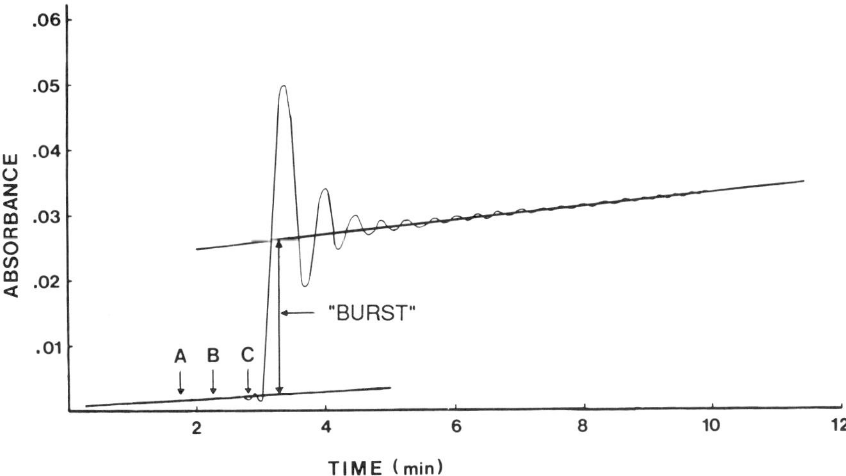

Figure 3 A recorder trace from active site titration of glass bound trypsin with *n*-nitrophenyl-*p*′-guanidinobenzoate (NPGB) in a recirculation system. A) peristaltic pump off; B) enzyme reactor plugged into the system; C) pump restarted. (Reproduced with permission from Ref. 23.)

product was estimated by extrapolating the absorbance increase to zero reaction time (Fig. 3).

In a method using simultaneous titrations, the determination of active site concentration of Sepharose-trypsin was carried out by measuring its loss of activity towards its substrate, *N*-benzoyl-L-arginine ethyl ester (BAEE), as it was being titrated with NPGB (24). The residual BAEE activity was plotted against the quantity of NPGB added, and at zero activity the value gave the molarity of active sites.

Active site titration of immobilized trypsin with soyabean trypsin inhibitor (STI) was carried out to study the interaction of a gel bound enzyme with large molecules (24). The loss of activity of Sepharose-trypsin towards different substrates appeared to be directly related to the molar amount of STI added. However, the immobilized enzyme exhibited 12–15% residual activity toward small molecular weight substrates like BAEE, which was attributed to the active sites not reached by STI. With chymotrypsinogen as the substrate, the immobilized enzyme was found to be completely inhibited.

Electron spin resonance (ESR) spectroscopy has also been used to determine the concentration of active trypsin covalently bound to porous glass (25). The enzyme was spin-labeled with 1-oxyl-2,2,6,6-tetrametyl-4-piperidinyl methylphosphonofluoridate. The concentration of active sites which were inhibited was determined by hydrolyzing the spin label from a fixed weight of beads with a known volume of 0.5 M NaOH and comparing the final free spin label concentration with that of a standard.

2. Immobilized Chymotrypsin

Fluorometric titration has been used to determine the concentration of active sites of immobilized chymotrypsin (26). This technique has been reported to be more sensitive than the spectrophotometric method. The scattering of the exciting light by the gel particles does not appear to interfere with the signal observed. The amount of active chymotrypsin present in the Sephadex-bound enzyme was determined by reacting it with 4-methylumbelliferyl-P(N_1N_1N-trimethylammonium) cinnamate (MUTMAC) in a fluorescence cuvette. The fluorescence due to liberated 4-methyl umbelliferone was followed with time, and extrapolated to zero time. The concentration was calculated from a calibration curve of 4-methyl umbelliferone in the same buffer.

3. Immobilized Lactate Dehydrogenase

Equilibrium binding of ^{14}C-labeled NAD^+ to glass bound lactate dehydrogenase was shown to provide a measure of the number of active sites of immobilized lactate dehydrogenase (27). The immobilized enzyme was equilibrated for 1 h at room temperature with ^{14}C-NAD^+ solution of various concentrations. After equilibration, the NAD^+ concentration in the supernatants was determined by scintillation counting. An identical amount of untreated glass beads equilibrated with NAD^+ solution under the same conditions served as a control. The amount of NAD^+ bound to the enzyme beads was calculated by subtracting the amount in solution from the total NAD^+ added. The data was then analyzed using a Scatchard plot. The computer fitted curve provided the number of active sites per enzyme tetramer.

4. Other Proteins

Binding site concentrations of immobilized antibodies, lectins, and other binding proteins can easily be determined by labeling techniques. A preparation of immobilized binding-protein, such as an antibody preparation, is first characterized for its protein content. Then the binding capacity is measured by titrating with molecules which are enzyme-, fluorescent-, or radioisotopically labeled. By successive additions of labeled ligand a saturation level is reached. This gives the molar concentration of binding sites and can be used together with the protein determination studies to calculate the relative binding site efficiency of the immobilized protein in relation to that of the protein in free solution.

IV. STABILITY OF IMMOBILIZED PROTEINS

The concept of stabilization was for a long time the main driving force for immobilizing enzymes, since the economic feasibility of many enzyme-catalyzed biotransformations depends critically on the active lifetime of the catalyst. Initially, the majority of stabilization studies involved enzymes available at reasonable prices, or those catalyzing simple hydrolytic reactions. This limited, of course, the number of potential candidates to be studied.

The observed operational stabilization of an enzyme is generally the sum of the effects of stabilization at the molecular level and the loading of an excess amount of the enzyme. The latter causes the process to be diffusion limited rather than to be kinetically controlled as mentioned earlier. This appears to lead to a buffering effect: As the enzyme molecules present in the outer layer lose activity, other resting molecules come into operation, thus resulting in an apparent stabilization. Since the process is diffusion controlled, a relatively constant activity is observed.

When an enzyme is stabilized towards thermal denaturation, or other factors like proteases, inhibitors, etc., upon immobilization, it is the result of stabilization at the molecular level. The most commonly used method for studying molecular stabilization has been to expose the enzyme preparation for a fixed period of time to a stressing situation—elevated temperature, organic solvent, proteases, etc., followed by transfer of the enzyme to a standardized test solution where its residual activity is measured. This procedure, however, does not differentiate between true stabilization and the effects of an operational stabilization. Furthermore, if the inactivation of the enzyme is reversible, reactivation may take place when reverting back to the standardized conditions. For instance, removal of denaturants from the sample of "irreversibly" denatured Sepharose bound trypsin and α-chymotrypsin resulted in regeneration of 70–100 % of the catalytic activity (28).

Covalent binding of a protein to an inert support leads to its stabilization owing to multipoint attachment, resulting in increased rigidity of the structure. On the other hand, it can also lead to destabilization due to immobilization induced conformational changes. In an immobilized preparation, a population of protein molecules having different stabilities is thus quite possible. For example, immobilization of *Streptomyces griseus* proteases to succinamidopropyl porous glass via their α and/or ϵ amino groups resulted in an increased stability of the attached enzymes as measured by urea denaturation (29). The degree of stabilization was seen to be related to the number of potential binding sites on the proteases.

Differential scanning calorimetry (DSC) studies on the transition behavior of the immobilized enzyme molecules subjected to heat denaturation have also shown a correlation between the degree of stabilization in terms of a changed transition temperature and the number of covalent bonds between the enzyme molecules and a Sepharose matrix (30). Similar conclusions were drawn from fluorescence studies on immobilized chymotrypsin (31), where it was shown that multiple cross-linkages cause changes in conformation and conformational flexibility. The same study provided evidence that there was a significant difference in conformation between the immobilized enzyme and its heat-treated counterpart.

Electron paramagnetic resonance (EPR) spectroscopy studies on immobilized monoclonal antibodies have indicated some alterations in the antigen-binding site (32). These conformational changes may reflect the stabilization achieved, but so far no clear correlations have been demonstrated.

EPR spectroscopy has also been applied together with measurements of catalytic activity and the quantity of active immobilized protein for studying the deactivation in 50% *n*-propanol of α-chymotrypsin immobilized on CNBr-Sepharose 4B (33). These analyses have shown that the relative quantities of the two active forms of the immobilized enzyme (designated as chymotrypsin A and B) change as a result of exposure to alcohol, with the relative quantity of B form increasing with time. The enzyme immobilized last was the first to deactivate, probably because the molecules binding to the surface initially occupied certain preferred sites which perturb the native enzyme structure relatively little, and the subsequent enzyme attachment occurs at different, less favorable sites, which cause more substantial disruption of enzyme molecular structure. These more perturbed molecular forms are apparently more vulnerable to further unfolding and inactivation in the presence of *n*-propanol.

Stabilization against proteolytic attack was one of the interesting observations made while studying bound enzymes, and has also been achieved by coupling soluble polymers, such as dextrans, to the protein molecules (34, 35).

Some enzymes have been shown to display improved stability in various organic solvents. One way to obtain active and stable enzyme preparations has been to adsorb the enzyme onto a solid support, and then to dehydrate the preparation before introducing the organic solvent (36). In a subsequent step small amounts of water are added. The stabilization obtained has been ascribed to the fact that there is too little water available to allow protein denaturation. It appears in any case that the amount of water present plays a crucial role in the stabilization process (37). (See Chapter 6 for an extensive treatment of immobilized enzymes in organic solvents.)

V. LABILE ENZYMES

Labile enzyme molecules have been used only to a limited degree when operational stability was the goal. A reason for this is that the immobilization process takes time, and during this period inactivation occurs. Such inactivation could potentially be avoided by adopting different approaches.

A. Reversible Immobilization

One way to get around the problem of enzyme instability is to reversibly immobilize the enzymes on the support via biospecific interactions (38). This technology can, in principle, exploit extensive experience from affinity chromatography provided no enzyme inhibitors are used. Affinity binding may be performed using a range of interactions from the rather unspecific to very specific ones (Table 2). For example, carboxypeptidase A immobilized via monoclonal antibodies was seen to retain full catalytic activity, in contrast to the covalently bound enzyme preparations, which had considerably lower activity than that to be expected from the amount of enzyme bound (40).

Biospecific interactions are preferably mediated via groups on the protein located far from the active site. Otherwise, inhibitory effects may be foreseen. This potential effect of inhibition is exploited in several of the homogeneous immunoassays that are on the market today. The basic idea behind reversible biospecific immobilization is that enzyme is added in an amount suitable for the application of interest. When the activity drops or after the application is successfully terminated, the support is rinsed and new enzyme added. When an enzyme or any protein molecule does not contain a suitable group for affinity binding, chemical modification may be performed to introduce the appropriate groups (41–43). Furthermore, gene technologists have modified proteins by synthesizing various reporter groups on them, in order to purify them in easier way (44, 45). These groups may as well be used to reversibly immobilize the enzyme (46).

Table 2 Affinity Pairs Used for Reversible Immobilization

avidin–biotin[a]
antibody–hapten[a]
antibody–antigen[a]
protein A–Fc region of IgG[a]
lectin–carbohydrate[a]
enzyme–inhibitor (carboanhydrase–sulfonamide)[b]

[a] From Ref. 38.
[b] From Ref. 39.

Table 3 Affinity Interactions Used for Binding of Enzymes

Ligand used	Enzyme bound	Ref.
concanavalin A	glucose oxidase	48
	peroxidase	48
	invertase	48
	glycosylated catalase	48
	tyrosinase	48
	carboxypeptidase Y	47
anti-albumin	catalase–albumin	41
	penicillinase–albumin	41
	invertase–albumin	41
	phenol oxidase–albumin	41
	glucose oxidase–albumin	41
monoclonal antibody	carboxypeptidase A	40,49

Chemical coupling of affinity bound protein to the support could provide a way to immobilize a protein in a preferred molecular arrangement. Turkova et al. (47) formed an affinity complex of carboxypeptidase Y and concanavalin A bound to Spheron particles, and in a subsequent step covalently linked the lectin and the enzyme.

Most reports on affinity binding have so far dealt with analytical applications (38) (Table 3). If the affinity ligands are available at a reasonable cost, affinity binding may prove to be a suitable method for labile enzymes on a large scale as well.

Immobilized antibodies may sometimes be labile, or extremely low yields may be obtained upon covalent immobilization. The use of biospecific interactions between Protein A and the Fc-region of the antibodies has made it possible to use such labile preparations for immobilization (50, 51). However, protein A does not bind to all subclasses of antibodies. This has limited its use to some extent. It has recently been reported that Streptococci produce a protein G, which has a broader binding specificity and thus may be useful for a wider range of antibody subclasses (52).

B. Protection of the Active Site During Immobilization

Loss of enzyme activity during immobilization is, at times, ascribed to the oxidation of an essential SH group, loss of cofactor, or the involvement of essential amino acid residues in binding to the support. To increase the yields in such cases, it may be advisable to include an inhibitor, substrate or an effector during immobilization (53–55). As a result, the essential groups are blocked during the coupling procedure and thus will not be modified.

In some cases, however, the objective may not be achieved, probably due to the effector/substrate/inhibitor inducing a conformational change and thereby exposing sensitive groups on the protein to the immobilization conditions. For example, in coupling a yeast β-galactosidase, a decreased yield was observed when substrate was present in the coupling solution (56). This observation was interpreted to be a result of a conformational change induced by substrate.

Another effect resulting from immobilization in the presence of an effector is that the protein may be immobilized in a certain conformation. Upon multipoint attachment it might be possible to freeze the conformation so that the protein will function as if the effector was present even when it is not. This situation is mainly of interest for oligomeric proteins. Studies using this approach have been reported with glutamate dehydrogenase (53), phosphofructokinase (54), as well as hemoglobin (57).

C. Immobilization of Subunits

Still another approach for immobilization of labile proteins is to immobilize stable monomers of oligomeric proteins. Missing, more labile subunits are biospecifically bound at a later stage without any chemical modification. This approach also makes it possible to investigate whether the immobilized subunit is enzymatically active on its own, or if the quaternary structure is essential for the expression of the activity. Furthermore, it provides an opportunity to produce hybrid oligomers between subunits from different species (27,58,59).

D. Immobilization of Zymogens

A special approach that may be useful for the immobilization of proteolytic enzymes is to couple their zymogens and then to activate the insoluble preparation (60). This may reduce the self-degradation phenomenon common among such enzymes. For example, immobilization of chymotrypsinogen A through covalent attachment to an insoluble matrix has provided a means for the study of regeneration of biologically functional tertiary structure from completely reduced, denatured zymogen chains under conditions that would minimize interactions leading to nonspecific aggregation and formation of intermolecular disulfide bridges (61).

VI. IMMUNOLOGICAL REACTIVITY AND IMMUNOGENICITY

Medical applications of proteins may raise problems of immunological character. When a nonself protein or a modified self protein is administered into the body, an immunological reaction is inevitable. Several approaches to get over this problem have been presented which include

Immobilization of the protein and its use in an extracorporeal shunt outside the body.

Encapsulation or entrapment of the protein so that there is no direct exposure of the proteins to the body fluids. A selective membrane will only allow small molecules to penetrate. This has been shown to be feasible in enzyme therapeutic applications (*62*).

Modification of the surface properties of the protein in order to reduce the immunogenicity.

The last approach involves attachment of nonimmunogenic polymers to the protein, such that the end product remains soluble. Modification of proteins with PEG molecules decreases their immunoreactivity towards antibodies against respective proteins (63–66). The modified proteins have been shown to be extremely stable in blood, and possess an extended clearance time as compared with that of the native proteins. An appropriate balance between degree of modification and the remaining biological activity is an important parameter when dealing with protein modification. When studying the immunological response to such preparations, one can foresee that it may not necessarily have the same sensitivity as the native protein. It is thus extremely important to look for antibodies raised to the modified protein in the animal after some time of repeated treatment. In the cases reported in literature, no antibody formation was observed (64, 65). The reason for this loss of immunogenicity may, of course, be that the modified protein is inert or that the modifying groups are not attached via bonds that are stable enough, are lost, and are cleared from the body before stimulating antibody formation. In the latter case, antigenic determinants start to appear after some time, and this brings up the possibility of antibody production to the original, unmodified protein.

Studies are also being carried out on the use of nonimmunogenic macromolecules such as fatty acids and α,β-poly (2-hydroxyethyl)-DL-aspartamide (PHEA) to chemically modify allergens (66). The modified allergens do not show any reactivity with IgE antibodies produced in response to inhaled, native allergen molecules. This approach would also suppress the production of IgE antibodies, bringing about the tolerance to the allergen, e.g., ovalbumin chemically modified with a fatty acid or with PHEA, specifically inhibited production of IgE to ovalbumin (66). The mechanism of this phenomenon is still not clear.

The whole area of conjugate formation between polymeric structures and proteins is very interesting and, in recent years, has found many new areas of application in biotechnology such as enzymatic catalyses in organic solvents (60, 66, 67) and competitive binding assays in aqueous two-phase systems (68).

ACKNOWLEDGMENTS

The financial support of The National Swedish Board for Technical Development is gratefully acknowledged. The authors also wish to thank Anja Broeders for help with the figures.

REFERENCES

1. The Working Party on Immobilized Biocatalysts within the European Federation of Biotechnology, Guidelines for the characterization of immobilized biocatalysts, *Enzyme Microb. Technol., 5:*304–307 (1983).
2. B. Mattiasson and K. Mosbach, *Methods Enzymol. 44:*335 (1976).
3. D. Gabel, and R. Axén, *Methods Enzymol. 44:*383 (1976).
4. P.S. Bunting and K. J. Laidler, *Biochem. 11:*4477 (1972).
5. Y. Levin, M. Pecht, L. Goldstein, and E. Katchalski, *Biochem. 3:*1905 (1964).
6. M.-R. Kula, in *Characterization of Immobilized Biocatalysts* (K. Buchholz, ed.), Dechema Monograph, vol. 84, Verlag Chemie, Weinheim, 1979, p. 182.
7. W. F. Line, A. Kwong, and H. H. Weetall, *Biochim. Biophys. Acta 242:*194 (1971).
8. E. Ehrentahl, J. Schlünsen, and G. Manecke, in *Characterization of Immobilized Biocatalysts* (K. Buchholz, ed.), DECHEMA Monograph, vol. 84, Verlag Chemie, Weinheim, 1979, p. 188.
9. P. D. Chantler and W. B. Gratzer, *FEBS Lett. 34:*10 (1973).
10. D. Failla and D. V. Santi, *Anal. Biochem. 52:*363 (1973).
11. S. Gestrelius, B. Mattiasson, and K. Mosbach, *Eur. J. Biochem. 36:*89 (1973).
12. A. Eskamani, T. Chase, Jr., J. Freudenberger, and S. G. Gilbert, *Anal. Biochem. 57:*421 (1974).
13. W. R. Veith and K. Venkatasubramanian, *Methods Enzymol. 44:*243 (1976).
14. A. F. S. A. Habeeb, *Anal. Biochem. 14:*328 (1966).
15. S. J. Stocks, A. J. M. Jones, C. W. Ramey, and D. E. Brooks, *Anal. Biochem. 154:*232 (1986).
16. J. S. Mort, D. K. K. Chong, and W. W.-C. Chan, *Anal. Biochem. 52:*162 (1973).
17. J. R. Ford, A. H. Lambert, W. Cohen, and R. P. Chambers, *Biotechnol. Bioeng. Symp. 3:*267 (1972).
18. S. W. Kiang, J. W. Kuan, S. S. Kuan, and G. G. Guilbault, *Clin. Chem. 22:*1378 (1976).
19. H. H. Weetall and L. S. Hersh, *Biochim. Biophys. Acta 185:*464 (1969).
20. J. C. Bouin, M. T. Atallah, and H. O. Hultin, *Methods Enzymol. 44:*478 (1976).
21. R. A. Messing, *Biotechnol. Bioeng. 16:*897 (1974).
22. R. A. Messing, *Biotechnol. Bioeng. 16:*1419 (1974).

23. J. R. Ford, R. P. Chambers, and W. Cohen, *Biochim. Biophys. Acta 309:*175 (1973).
24. R. J. Knights and A. Light, *Arch. Biochem. Biophys. 160:*377 (1974).
25. L. J. Berliner, S. T. Miller, R. Uy, and G. P. Royer, *Biochim. Biophys. Acta 315:*195 (1973).
26. D. Gabel, *FEBS Lett. 49:*280 (1974).
27. I. C. Cho and H. Swaisgood, *Biochim. Biophys. Acta 334:*243 (1974).
28. V. V. Mozhaev and K. Martinek, *Enzyme Microb. Technol. 4:*229 (1982).
29. F. C. Church, G. L. Catignani, and H. E. Swaisgood, *Enzyme Microb. Technol. 4:*313 (1982).
30. A. C. Koch-Schmidt, and K. Mosbach, *Biochem. 16:*2101 (1977).
31. D. Gabel, I. Z. Steinberg, and E. Katchalski, *Biochem 19:*4661 (1971).
32. E. J. Fernandez, F. B. Fernandez, R. B. Jagoda, and D. S. Clark, in *Separation, Recovery, and Purification in Biotechnology* (J. A. Asenjo and J. Hong, eds.), ACS, Washington, DC, vol. 314, 1986, p. 208.
33. D. S. Clark and J. E. Bailey, *Biotechnol. Bioeng. 26:*1090 (1984).
34. J. J. Marshall, *Trends Biochem. Sci. 3:*79 (1978).
35. J. J. Marshall, *ACS Symposium Series 123:*125 (1980).
36. A. Zaks and A. M. Klibanov, *Science 224:*1249 (1984).
37. P. Adlercreutz and B. Mattiasson, *Biocatalysis* 2:99 (1987).
38. B. Mattiasson, *J. Appl. Biochem. 3:*183 (1981).
39. R. Horowitz and G. M. Whitesides, *J. Amer. Chem. Soc. 100:*4632 (1977).
40. B. Solomon, R. Koppel, G. Pines, and E. Katchalski-Katzir, *Biotechnol. Bioeng. 28:*1213 (1986).
41. B. Mattiasson, *FEBS Lett. 77:*107 (1977).
42. M. Wilchek, *J. Solid Phase Biochem. 5:*193 (1980).
43. M-H. Remy and D. Thomas, *Enzyme Microb. Technol. 4:*381 (1982).
44. S. J. Brewer and H. M. Sassenfeld, *Trends Biotechnol. 3:*119 (1985).
45. M. Uhlen, B. Nilsson, B. Guss, M. Lindberg, S. Gatenbeck, and L. Philipson, *Gene 23:*369 (1983).
46. B. Mattiasson, *Methods Enzymol.* 137:647 (1988).
47. J. Turkova, M. Fusek, J. J. Maksimov, and Y. B. Alakhov, *J. Chromatogr. 376:*315 (1986).
48. B. Mattiasson, and C. A. K. Borrebaeck, *FEBS Lett. 85:*119 (1978).
49. B. Solomon, R. Koppel, and E. Katchalski-Katzir, *Biotechnol. 2:*709 (1984).
50. A. Forsgren, and J. Sjöquist, *J. Immunol. 97:*822 (1966).
51. H. Hjelm, *Acta Universitatis Uppsaliensis 213* (thesis) (1977).
52. L. Björck and G. Kronvall, *J. Immunol. Methods 133:*969 (1984).
53. J. H. Julliard, C. Godinot, and D. C. Gautheron, *FEBS Lett. 14:*185 (1971).
54. B. Mattiasson, S. Gestrelius,, and K. Mosbach, in *Enzyme Engineering,* Vol. 2 (E. K. Pye and L. B. Wingard, Jr., eds.), Plenum Press, New York, 19XX, p. 181.
55. L. Bulow and K. Mosbach, *Biochem. Biophys. Res. Commun. 107:*458 (1982).
56. A. Dahlqvist, B. Mattiasson, and K. Mosbach, *Biotechnol. Bioeng. 15:*395 (1973).

57. E. Antonini and M. R. R. Fanelli, *Methods Enzymol. 44:*538 (1976).
58. W. W.-C. Chan, *Biochem. Biophys. Res. Commun. 41:*1198 (1970).
59. W. W.-C. Chan, H. Schutt, and K. Brand, *Eur. J. Biochem. 40:*533 (1973).
60. A. Matsushima, M. Okada, and Y. Inada, *FEBS Lett. 178:*275 (1984).
61. H. R. Horton and H. E. Swaisgood, *Methods Enzymol. 44:*516 (1976).
62. J. Campbell and T. M. S. Chang, in *Biomedical Applications of Immobilized Enzymes and Proteins* (T. M. S. Chang, ed.), Vol. 2, 1977, p. 281.
63. P.S. Pyatak, A. Abuchowski, and F. F. Davis, *Res. Commun. Chem. Pathol. Pharmacol. 29:*113 (1980).
64. A. Abuchowski, T. van Es, N.C. Palczuk, and F. F. Davis, *J. Biol. Chem. 252:*3578 (1977).
65. A. Abuchowski, J. R. McCoy, N. C. Palczuk, T. van Es, and F. F. Davis, *J. Biol. Chem. 252:*3582 (1977).
66. Y. Inada, T. Yoshimoto, A. Matsushima, and Y. Saito, *Trends Biotechnol. 4:*68 (1986).
67. Y. Inada, K. Takahashi, T. Yoshimoto, A. Ajima, A. Matsushima, and Y. Saito, *Trends Biotechnol. 4:*190 (1986).
68. B. Mattiasson, *Methods Enzymol. 92:*498 (1983).

II

APPLICATION OF IMMOBILIZED PROTEINS

6

Immobilized Enzymes in Organic Solvents

Atsuo Tanaka and Takuo Kawamoto

Kyoto University, Kyoto, Japan

I. GENERAL CONCEPT

In the case of bioconversions of lipophilic compounds, it is desirable to carry out enzymatic reactions in mixtures of water and suitable organic cosolvents or in appropriate organic solvent systems, provided that the catalytic activities of enzymes are maintained in such reaction systems. The use of organic solvents can improve the poor solubility in water of substrates and/or other reaction components of hydrophobic nature.

Hydrolytic enzymes, which require water as one of the coreactants in hydrolytic reactions, can catalyze synthetic reactions and group exchange reactions when the concentration of water in the systems is low. If the water fraction in a reaction mixture is reduced by introducing organic solvents, it is possible to synthesize useful hydrophobic compounds more efficiently by shifting unfavorable reaction equilibria in desired directions. However, attempts to make free, native enzymes function in organic solvent systems have not been successful in general. Enzymes as well as other proteins maintain their structural conformations through intramolecular interactions among side chains of the component amino acids. Hydrophobic side chains also contribute to such interactions. Enzyme molecules in aqueous solutions have both hydrophilic domains in contact with water and hydrophobic domains folded inside the molecules. When the polarity of the media surrounding the enzyme molecules is reduced by introducing organic solvents, the hydrophobic domains are liable to disperse, resulting in the unfolding of the molecules. Furthermore, hydrophobic interactions between the enzyme and substrate molecules are also disrupted. These characteristics indicate that suitably hydrated states of enzyme molecules should be maintained to keep the enzymes active and stable even when organic solvents are utilized in enzymatic reaction systems. Therefore, the choice of a solvent having an adequate polarity becomes important. In general, nonpolar solvents are suitable for use with enzymes because of their decreased ability to remove water from enzyme molecules.

Another problem in carrying out enzymatic reactions in organic solvent systems is poor solubility or poor dispersibility of enzymes in the solvents. This brings about apparent low or no activity of enzymes.

Although chemical modification of enzymes (1) and construction of reverse micelles (2) are important means to protect enzymes from distortion by organic solvents and to solubilize or disperse them in organic solvents, immobilization seems to be the most promising approach for enzyme stabilization and enzyme dispersion in the presence of organic solvents (3). The conformational structure of enzyme molecules immobilized on or in rigid supports becomes more resistant against distortion, and thus, unfolding of the higher dimensional structures of enzymes which can lead to loss of catalytic activity becomes much more difficult than in the cases of unimmobilized counterparts. Immobilization is also able to "fix" the water surrounding enzyme molecules, which is essential for stabilization and activity of the enzymes. Furthermore, immobilization permits the enzymes to be used repeatedly or continuously. Topics in this chapter, therefore, will be focused on the applications of immobilized enzymes (and microbial cells). Such topics have also been discussed in review papers published hitherto (4–8).

II. ENZYMES AND CATALYTIC ACTIVITIES

Different kinds of enzymes and microbial cells that contain the enzymes have been immobilized and applied to various types of reactions in the presence of organic solvents. These include lipases and esterases, proteases, dehydrogenases or oxidases, hydroxylases, etc., which have been used for hydrolysis, esterification, ester exchange, peptide synthesis, peptide exchange, dehydrogenation, oxidation, hydroxylation, etc.

A. Lipases and Esterases

Lipases and esterases are useful not only for hydrolysis of esters but also for esterification, interesterification (transesterification), and, in some cases, peptide synthesis. Among 50 Celite-adsorbed preparations of lipases, lipoprotein lipases, esterases, and cholesterol esterases from different sources, 22 preparations were found to be very active in an organic solvent in the synthesis of a terpenoid ester, and some of the others also had moderate enzymatic activity under the same conditions, while the rest were inactive (9). These results indicate that a wide variety of enzymes do retain their catalytic activities in organic solvents after appropriate immobilization. Several examples of the applications of immobilized enzymes and microbial cells are shown in Table 1. Yeast cells having esterase activity were employed for the stereoselective hydrolysis of menthyl ester (10). Stereoselective esterification and interesterification were applied for the optical resolution of alcohols (11,13,14). Regiospecific interesterification of triglyceride with saturated fatty acids is useful for the reformation of fats having a low melting point (15–17). In addition to immobilized biocatalysts, free enzymes have also been applied to a variety of stereoselective reactions in the presence of organic solvents, although direct comparisons of activity between the immobilized and free enzymes were not made (19).

B. Proteases

α-Chymotrypsin bound to polyacrylamide gel (20) and adsorbed on porous glass beads (21) was utilized in organic solvent systems for the synthesis of an amino acid ester. Various kinds of proteases, belonging to the groups of serine-, sulfhydryl-, carboxyl-, and metallo-proteinases, have been applied to the synthesis of peptide linkages based on their substrate specificities (22,23,25). These include trypsin, chymotrypsin, subtilisin, *Achromobacter* protease I, carboxypeptidase Y, papain, and thermolysin, all of which are active even in the presence of organic solvents. Trypsin and *Achromobacter* protease I are especially useful for the conversion of porcine insulin to human insulin (30–32), and thermolysin for the synthesis of a precursor of the practical sweetener aspartame (34,35). Table 2 summarizes several examples of proteases which have been applied to peptide synthesis and other reactions in the presence of organic solvents in free or immobilized states.

C. Dehydrogenases and/or Oxidases

Reactions with dehydrogenases are more complicated than those with hydrolytic enzymes because pyridine nucleotides are involved in the reactions as coenzymes or cosubstrates. Therefore, microbial cells, living or dead, are often utilized as catalysts to oxidize or reduce the substrates and to regenerate reduced or oxidized pyridine nucleotides. Several examples are summarized in Table 3.

Table 1 Several Examples of Immobilized Lipases and Esterases Used in the Presence of Organic Solvents

Application	Solvent	Substrate	Product	Enzyme (source) or microorganism	Ref.
Stereoselective hydrolysis	*n*-Heptane[a]	*dl*-Menthyl succinate	*l*-Menthol	*Rhodotorula minuta* cells	10
Stereoselective esterification	Cyclohexane[a] etc.	*dl*-Menthol + acid	*l*-Menthyl ester	Lipase (*Candida cylindracea*)	11
Stereoselective interesterification	Methyl acetate	2-Substituted 1,3-propanediol + methyl acetate	2-Substituted 1,3-propanediol monoacetates	Carboxyl esterase (porcine pancreas)	12
	2-Butanol[b] etc.	Tributyrin + 2-butanol etc.	Butyrate esters	Lipase (*C. cylindracea*)	14
	1-Butanol	Methyl 2-(*p*-chlorophenoxy)-propionate + 1-butanol	Butyl 2-(*p*-chlorophenoxy)-propionate	Lipase (*C. cylindracea*)	13
	Methyl propionate[a]	3-Methoxy-1-butanol etc. + methyl propionate	Propionate esters	Carboxyl esterase (hog liver)	14
Regiospecific interesterification	Petroleum ether	Palm oil + saturated fatty acids	Reformed palm oil	Lipases (*Aspergillus niger, Rhizopus niveus*)	15
	n-Hexane[a]	Olive oil + saturated fatty acids	Reformed olive oil	Lipases (*Geotrichum candidum, R. delemar*)	15–17
Peptide synthesis	Ethyl acetate	Carbobenzyloxyphenylglycine + alanine propyl ester etc.	Various peptides	Lipase (*C. cylindracea*)	18

[a]Water-saturated.
[b]Aqueous buffer-organic solvent two-phase system.

Table 2 Several Examples of Proteases Used in the Presence of Organic Solvents

Enzyme	Solvent	Application	Ref.
Chymotrypsin	Ethanol-glycerol (1:1) (90%),[a] Chloroform	Esterification of *N*-acetyl-L-tyrosine	20*,21*
	Dimethyl sulfoxide	Hydrolysis of *p*-nitrophenyl acetate	27
	n-Octane	Transesterification of *N*-acetyl-L-amino acid esters	24
	Ethyl acetate[b]	Synthesis of *N*-acetyl-L-tryptophanyl-L-leucine amide	25
Trypsin	1-Propanol (92%)[a]	Hydrolysis of lysyl dipeptides	28*
	Dimethyl sulfoxide	Hydrolysis of *p*-nitrophenyl acetate	27
	Dimethylformamide (50%)[a]	Synthesis of human insulin from desoctapeptide-insulin	29
	Dimethylformamide-ethanol (1:1) (60%)[a]	Synthesis of human insulin from desalanine-insulin	30
	Dimethylformamide (60%)[a]	Conversion of porcine insulin to human insulin	31
Achromobacter protease I	Dimethylformamide-ethanol (1:1) (60%)[a]	Synthesis of human insulin from desalanine-insulin	23,32
Subtilisin	*n*-Octane	Transesterification of *N*-acetyl-L-amino acid esters	24
Thermolysin	Ethyl acetate[b]	Synthesis of *N*-(benzyloxycarbonyl)-L-aspartyl-L-phenylalanine methyl ester	22,33*, 34*,35
	Ethyl acetate[b]	Synthesis of destyrosine-leucine enkephalin	36

[a]Cosolvent in aqueous buffer.
[b]Aqueous buffer-organic solvent two-phase system.
*Immobilized proteases.

Table 3 Several Examples of Oxido-Reduction Reactions Carried out by Immobilized Biocatalysts in the Presence of Organic Solvents

Enyzme or microbial cells	Solvent	Application	Ref.
Arthrobacter simplex cells	Methanol (10%)[a]	Δ^1-Dehydrogenation of hydrocortisone	37–39
Nocardia erythropolis cells	Toluene, carbon tetrachloride	Dehydrogenation of 3β-hydroxysteroid	40
Nocardia rhodochrous cells	Benzene-*n*-heptane (1:1),[b] chloroform-*n*-heptane (1:1),[b] carbon tetrachloride[c]	Dehydrogenation of 3β-hydroxysteroids	41–43
	Benzene-*n*-hexane (4:1)[b]	Dehydrogenation of 17β-hydroxysteroids	44
	Benzene-*n*-heptane (1:1 or 4:1)[b]	Δ^1-Dehydrogenation of steroids	44,45
Pseudomonas testosteroni cells	Methanol (10%)[a]	Δ^1-Dehydrogenation of cortexolone	46
β-Hydroxysteroid dehydrogenase	Ethyl acetate[c]	Dehydrogenation of 17β-hydroxysteroid	47
3α-Hydroxysteroid dehydrogenase	Ethyl acetate[c]	Dehydrogenation of androsterone	47
20β-Hydroxysteroid dehydrogenase	Ethyl acetate[c]	Reduction of cortisone	47
Alcohol dehydrogenase	Isopropyl ether	Stereospecific dehydrogenation of alcohols	48
	Isopropyl ether	Stereospecific reduction of carbonyl compounds	48
Yeast cells	Many	Stereospecific reduction of carbonyl compounds	Many

[a]Cosolvent in aqueous buffer.
[b]Water-saturated.
[c]Aqueous buffer-organic solvent two-phase system.

Dehydrogenation or oxidation of 3β-hydroxy-Δ^5-steroids (e.g., cholesterol) to the corresponding 3-keto-Δ^4-steroids (e.g., cholestenone) was investigated extensively in organic solvent systems or in water–organic solvent two-phase systems with free cells of *Nocardia* sp. as a model of bioconversion of hydrophobic compounds (49). Immobilized cells of *Nocardia rhodochrous* are also active in dehydrogenation of 3β-hydroxysteroids and 17β-hydroxysteroids, and, in the presence of a suitable electron acceptor, Δ^1-dehydrogenation of several steroids, in organic solvent systems (41–45). Δ^1-Dehydrogenation of hydrocortisone to form prednisolone is usually carried out with immobilized living or acetone-dried cells of *Arthrobacter simplex* in the presence of low concentrations of organic cosolvents (37–39).

Horse liver alcohol dehydrogenase adsorbed on glass beads was applied to the reduction of (±)-2-phenylpropionaldehyde to (−)-2-phenylpropanol and of (±)-2-chlorocyclohexanone to (+)-*trans*-2-chlorocyclohexanol in isopropyl ether. In this case, ethanol was used as cosubstrate to regenerate NADH. The same enzyme preparation was useful to oxidize (±)-*trans*-3-methylcyclohexanol and (±)-*cis*-2-methylcyclopentanol to yield (−)-3-methylcyclohexanone and (+)-2-methylcyclopentanone, respectively, in the presence of isobutylaldehyde as cosubstrate (48). Alcohol dehydrogenases from different sources and microbial cells, especially yeast cells, are important catalysts for stereospecific reduction of ketones and aldehydes to alcohols and vice versa, although most of the reactions have been performed with free biocatalysts or in the absence of organic solvents. Dried baker's yeast is also a useful biocatalyst for the reduction of carbonyl compounds of different structures.

D. Hydroxylases

Hydroxylase systems are usually composed of two or more protein components and thus their reconstruction in immobilized systems with purified components is not practical. Therefore, microbial and plant cells, especially living cells, are applied to these systems.

Conversion of 1,7-octadiene to 7,8-epoxy-1-octene and/or 1,2-,7,8-diepoxyoctane was successfully achieved by incorporating a high concentration of cyclohexane with free cells of *Pseudomonas oleovorans* (50) or *Ps. putida* (51). Epoxidation of propene and l-butene by immobilized cells of *Mycobacterium* sp. was facilitated in a two-phase system containing an organic solvent of a low polarity and a high molecular weight, such as *n*-hexadecane (52).

Hydroxylation of steroids was carried out with immobilized living fungal or bacterial cells in the presence of low concentrations of appropriate or-

Table 4 Several Examples of Immobilized Microbial Cells Used for Steroid Hydroxylation in the Presence of Organic Solvents

Microorganism	Cosolvent[a]	Application	Ref.
Corynebacterium sp.	Dimethyl sulfoxide (15%)	9α-Hydroxylation of 4-androstene-3,17-dione	53
Rhizopus nigricans	Ethanol (1%)	11α-Hydroxylation of progesterone	54
Rhizopus stolonifer	Methanol (2.5%)	11α-Hydroxylation of progesterone	55
Aspergillus ochraceus	Ethanol (0.9%)	11α-Hydroxylation of progesterone	56
Curvularia lunata	Methanol (2%)	11β-Hydroxylation of cortexolone	57
Curvularia lunata	Dimethyl sulfoxide (2.5%)	11β-Hydroxylation of cortexolone	59
Streptomyces roseochromogenes	*N*,*N*-Dimethylformamide (0.25 %)	16α-Hydroxylation of dehydroepiandrosterone	61
Sepedonium ampullosporum	*N*,*N*-Dimethylformamide (0.65 %)	16α-Hydroxylation of estrone	80

[a]Percent in aqueous media or buffer.

ganic cosolvents (Table 4). When the activities of the immobilized cells become low after repeated reactions, the hydroxylation systems can be restored by incubating the immobilized cells in nutrient media containing suitable inducers. Hydroxylation of 4-androstene-3,17-dione to 9α-hydroxy-4-androstene-3,17-dione (53), of progesterone to 11α-hydroxyprogesterone (54–56), of cortexolone (Reichstein's Compound S) to hydrocortisone (57–60), and of dehydroepiandrosterone to 16α-hydroxydehydroepiandrosterone (61) has been reported.

E. Miscellaneous

A flavoprotein, D-oxynitrilase, adsorbed on ion-exchange cellulose catalyzed the synthesis of D-α-hydroxynitriles from aldehydes and hydrocyanic acid (62). Thus, D-(+)-mandelonitrile was synthesized from benzaldehyde and hydrogen cyanide in cold 50% methanol.

Gel-entrapped cells of *Enterobacter aerogenes* are a potent catalyst in the synthesis of adenine arabinoside from uracil arabinoside and adenine. Introduction of dimethyl sulfoxide at 40% improved the solubilities of adenine and adenine arabinoside, permitting a high productivity of the product (63). The enzyme system in the immobilized cells was stable over 35 days of the reaction when the reaction was carried out at 60 °C in the presence of 40% cosolvent.

Photochemical and chemical debromination of *meso*-1,2-dibromostilbene was performed in ethyl acetate with combination of Sepharose-bound alcohol dehydrogenase, L-lactic dehydrogenase or L-alanine dehydrogenase and the respective substrates as an electron donor system (64). Acid phosphatase bound on glass beads was examined for the hydrolytic activity of *p*-nitrophenyl phosphate in water–organic cosolvent systems (65).

Polyethylene glycol-modified horseradish peroxidase (66) and catalase (67) were active in benzene. Modified catalase was found to have a much higher activity in benzene than in an aqueous system. There are many reports on the application of free enzymes in the presence of organic solvents: for example, synthesis of urea by urease in the presence of 50% organic cosolvents (68), oxidation of phenols by polyphenyl oxidase in chloroform (69), depolymerization of lignin by horseradish peroxidase in organic solvents containing 5% aqueous buffer (70), and conversion of geranyl pyrophosphate to monoterpene olefins by (+)-α-pinene cyclase in hexane (71). A variety of enzymes and cells proved to have specified activities in organic solvent systems which had not been observed in the aqueous reaction systems.

III. METHODOLOGIES

A. Reaction Systems

Organic solvents are often introduced into enzymatic reaction systems to shift reaction equilibria (72,73) and to solubilize reactants or, in some cases, to increase the permeability of cells. When the organic solvents are miscible in water, homogeneous reaction systems can be constructed regardless of the concentration of organic cosolvents dependent on the solubility of the reactants, the stability and solubility of each enzyme and/or the dispersibility of the cells. In the case of organic solvents with low water solubility, three alternate systems can be employed. These are: water-organic solvent two-phase (biphasic) systems, organic solvent systems containing water at less than saturated concentrations, and aqueous systems containing organic solvents at less than saturated concentrations. As most enzymes and intact microbial cells are, in general, not soluble or dispersible in organic solvents, two-phase systems are most often used for free biocatalysts (5). An interesting alternate method to make enzymes soluble and active in organic solvent systems is to modify enzymes with proper modifiers. For example, polyethylene glycol-modified enzymes showed excellent activities in organic solvent systems (1).

Dispersion of biocatalysts in organic solvents is easily achieved by immobilization. Immobilization often stabilizes biocatalysts by interaction of the supporting materials with biocatalysts and also by entrapping water around the immobilized biocatalysts. Immobilized biocatalysts can be applied to any reaction systems mentioned above. Aqueous systems containing low concentrations of water-insoluble or water-immiscible organic solvents were effective to increase the permeability of gel-entrapped plant cells permitting the excretion of hydrophobic intracellular metabolites (74). Reactions with living cells are often carried out in aqueous systems containing low concentrations of water-miscible organic solvents.

A reaction system with immobilized biocatalysts should be selected depending on the properties of the biocatalysts, the solubilities of the reactants, the type of reactor, etc.

B. Immobilization Methods

At present, various immobilization techniques are available (75) (see Chapters 2, 3, and 4). Principles of these methods are: (1) carrier-binding through covalent linkage, electrostatic force, physical adsorption, biologically specific affinity, etc.; (2) crosslinking by using bi- or multifunctional reagents; and (3) entrapment in gel matrices, microcapsules, etc. Since every method can be applied to the immobilization of biocatalysts to be used in the presence of organic solvents, the method should be selected according to the specific approach, considering the kind of biocatalyst, type of bioreaction, and type of bioreactor.

IV. FACTORS AFFECTING IMMOBILIZED ENZYME REACTIONS

To achieve efficient bioreactions in organic solvents, it is essential to construct adequate systems, considering the hydrophobicity of the reactants, the hydrophilicity-hydrophobicity balance of supports, especially of gel materials for entrapment, and the polarity of organic solvents.

When enzymes and microbial cells are immobilized by entrapment in gel matrices, the hydrophilicity-hydrophobicity balance of the gels seriously affects the diffusion of substrates and, subsequently, the effective activities of the entrapped biocatalysts. The effect of gel hydrophobicity has been investigated systematically by using photo-cross-linkable resin prepolymers and urethane prepolymers of different properties as gel materials (76). One of the most representative examples of these effects is the conversion of 3β-hydroxy-Δ^5-steroids to the corresponding 3-keto-Δ^4-steroids by *Nocardia rhodochrous* cells in nonpolar organic solvent systems. Highly hydrophobic substrates, such as cholesterol, were oxidized in water-saturated benzene-*n*-heptane (1:1 by volume) by the cells entrapped in hydrophobic gels but not by those entrapped in hydrophilic gels (41). Figure 1 shows the relationship between the relative activity of the entrapped cells and the partition coefficient of cholesterol (the ratio of cholesterol concentration between the gel and the external solvent, which is measured in the absence of the biocatalysts), both of which changed depending on the hydrophobicity of gels. The abscissa shows the mixing ratio of the hydrophobic urethane prepolymer PU-3 and the hydrophilic prepolymer PU-6 used for the entrapment of the cells. Only the cells entrapped with high ratios of PU-3 exhibited catalytic activity paralleling a high partition coefficient of the substrate, although the entrapped cells retained their catalytic activity in any cases. These results indicate that high affinity between the gel and the substrate is essential for the efficient catalytic activity of the immobilized biocatalysts. Water contents of gels also affect the affinity or diffusion of hydrophobic substrates and, subsequently, the activities of the entrapped biocatalysts (77). However, a critical amount of water is essential to make enzymes active, probably due to maintenance of the native conformation of enzyme molecules.

When the hydrophobicity of substrates decreases, the hydrophobicity of gels is not necessarily a limiting factor for the enzymatic reactions. For example, dehydroepiandrosterone, with a carbonyl group at its C17 position and less hydrophobic than cholesterol, can be converted to 4-androstene-3,17-dione by the *Nocardia* cells entrapped in hydrophilic gels (41). How-

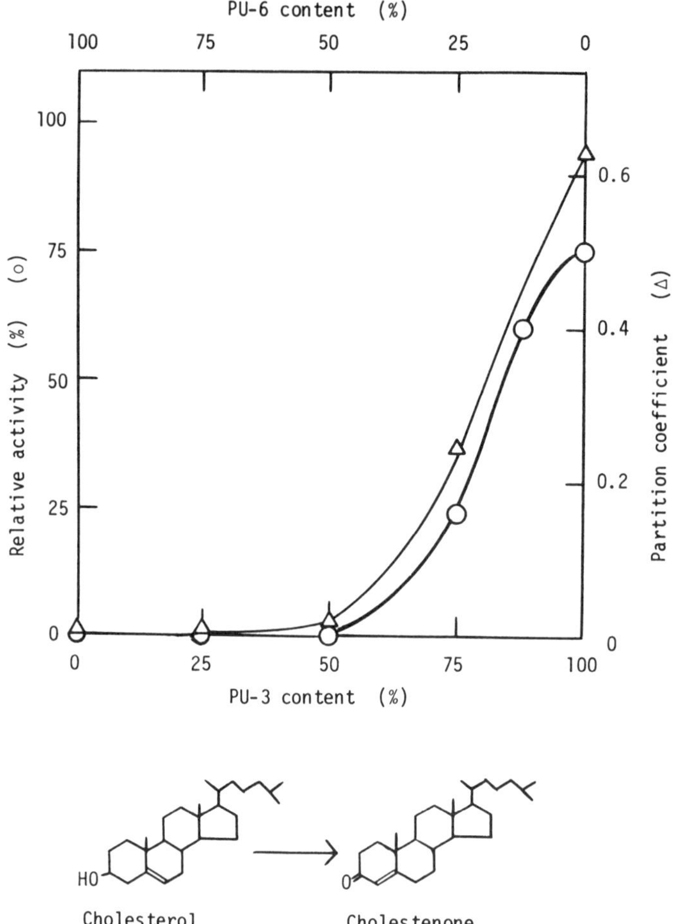

Figure 1 Effect of hydrophobicity of polyurethane gels on relative activity of cholesterol transformation by *Nocardia rhodochrous* and partition coefficient of cholesterol in water-saturated benzene-*n*-heptane (1:1 by volume) (41). The activity of the free cells was expressed as 100%. (○), Relative activity; (△), partition coefficient.

ever, the hydrophobic gels were still more effective, along with high partition coefficients of the substrate, than the hydrophilic gels in this system when the reaction was carried out in water-saturated benzene-*n*-heptane.

Such differences in the partition coefficients of substrates dependent on the properties of gels were also successfully applied to the control of diverse conversion of testosterone in the presence of a hydrophilic electron acceptor

Table 5 Effects of Gel Hydrophobicity on Bioconversions Catalyzed by Entrapped Biocatalysts

			Hydrophilic gel		Hydrophobic gel		
Reaction	Biocatalyst	Solvent[a]	ENT[b]	Pu-6[c]	ENTP[d]	PU-3[e]	Ref.
			Relative activity (%)[f]				
Oxidation of cholesterol	*Nocardia rhodochrous* cells	Benzene-*n*-heptane (1:1)	0	0	100	74	41
Oxidation of cholesterol	*Nocardia rhodochrous* cells	Chloroform-*n*-heptane (1:1)	78	76	100	83	42
Oxidation of dehydroepiandrosterone	*Nocardia rhodochrous* cells	Benzene-*n*-heptane (1:1)	72	67	100	98	41
Oxidation of pregnenolone	*Nocardia rhodochrous* cells	Chloroform-*n*-heptane (1:1)	100	95	92	59	42
Oxidation of testosterone	*Nocardia rhodochrous* cells	Benzene-*n*-heptane (4:1)	14	12	100	73	44
Hydrolysis of menthyl succinate	*Rhodotorula minuta* cells	*n*-Heptane	—	80	—	100	10
Interesterification of triglyceride	Lipase (*Rhizopus delemar*)	*n*-Hexane	27	—	100	—	17
Esterification of citronellol	Lipase (*Candida cylindracea*)	Cyclohexane	1	6	77	100	9

[a]Water-saturated.
[b]Hydrophilic photo-cross-linked gel.
[c]Hydrophilic urethane gel.
[d]Hydrophobic photo-cross-linked gel.
[e]Hydrophobic urethane gel.
[f]The highest activity obtained in each reaction was expressed as 100%.

to yield Δ^1-dehydrotestosterone or 4-androstene-3,17-dione (44). A similar effect of the gel hydrophobicity was also observed in the optical resolution of *dl*-menthol (10), ester exchange of triglyceride (17), esterification of citronellol (9), synthesis of oleate esters (78), etc. (Table 5).

In addition to gel hydrophobicity and substrate hydrophobicity, the polarity of the reaction solvents also had a marked effect on the conversion of steroids (42). As mentioned above, *N. rhodochrous* cells entrapped in hydrophilic gels could not oxidize cholesterol in a nonpolar solvent, e.g., benzene-*n*-heptane (1:1 by volume). However, if benzene is replaced by chloroform to increase the polarity of the solvents, the hydrophilic gel-

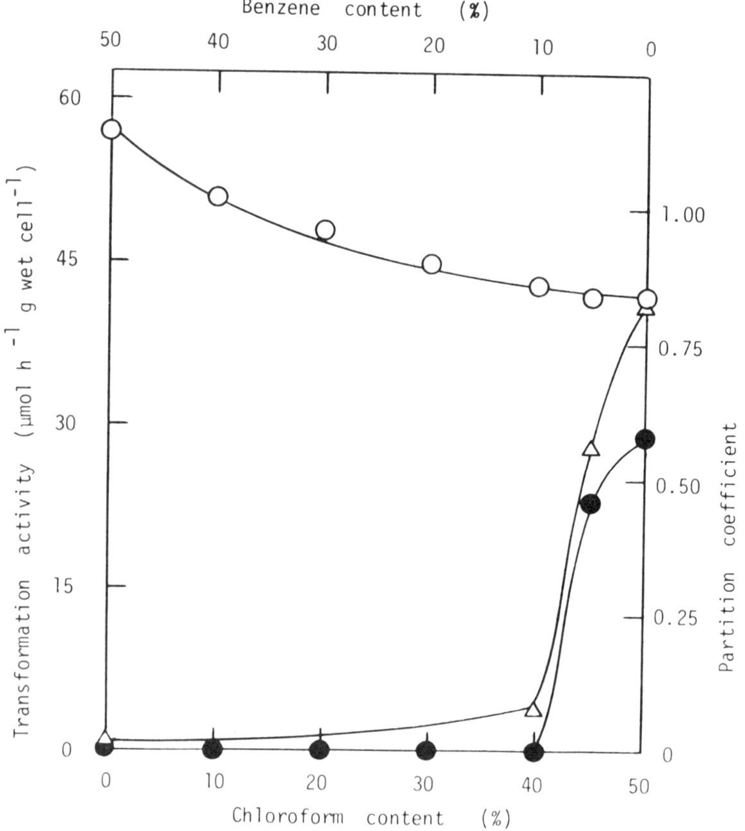

Figure 2 Effect of solvent polarity on cholesterol transformation by free and hydrophilic gel-entrapped *Nocardia rhodochrous* cells (42). Solvents employed were prepared by mixing benzene and chloroform in different ratios keeping the content of *n*-heptane constant (50% by volume). (○), Activity of free cells; (●), activity of gel-entrapped cells; (△), partition coefficient of cholesterol.

Table 6 Effects of Organic Solvents on Bioreactions

Organic solvent	Oxidation of cholesterol[a]	Esterification of menthol[b]	Esterification of citronellol[c]
		Relative activity (%)[d]	
Acetone	0	0	0
Benzene	84	2	2
Carbon tetrachloride	100	15	14
Chloroform	62	0	0
Cyclohexane	—	100	61
Dioxane	—	0	0
Ethanol	0	—	—
Ethyl acetate	40	—	—
n-Hexane	—	32	63
Isooctane	—	76	100
Methanol	0	0	0
Methylene chloride	28	—	—
Methyl isobutyl ketone	—	0	0
Tetrahydrofuran	—	—	0
Toluene	100	—	—

[a]Free cells of *N. rhodochrous* (42). Mixtures of organic solvents and *n*-heptane (1:1 by volume).
[b]Polyurethane-entrapped *C. cylindracea* lipase (11).
[c]Celite-adsorbed *C. cylindracea* lipase (9).
[d]The highest activity obtained in each reaction was expressed as 100%.

entrapped cells become active in agreement with the increased partition coefficients of cholesterol (Fig. 2) (42). In a water-saturated mixture of chloroform and *n*-heptane (1:1 by volume), little effect of gel hydrophobicity was observed on the conversion of cholesterol and dehydroepiandrosterone, differing from the results obtained in the nonpolar solvent system. However, enhancement of solvent polarity lowered the activity and stability of the free cells and, subsequently, those of the entrapped cells. Similar effects from solvent polarity were observed on other steroid conversions (41), esterification of menthol (11) and citronellol (9), hydrolysis of menthyl ester (10), etc. (Table 6).

V. APPLICATION EXAMPLES

Although a variety of reactions have been achieved successfully in the presence of organic solvents with immobilized biocatalysts as mentioned above, several examples mainly carried out in the authors' laboratory will be introduced in some detail.

A. Stereoselective Hydrolysis and Esterification

Stereoselective or stereospecific hydrolysis and esterification with lipases or esterases are useful methods to prepare optically active acids or alcohols from the racemic compounds.

l-Menthol, a compound having a peppermint flavor and useful in the food and pharmaceutical industries, can be obtained by stereoselective hydrolysis of an appropriate ester or by stereoselective esterification of chemically synthesized *dl*-menthol (79). The stereoselective hydrolysis of *dl*-menthyl succinate (39 mM) by gel-entrapped cells of *Rhodotorula minuta* var. *texensis* (Fig. 3) was carried out successfully at 30°C in water-saturated *n*-heptane, and the reactants were analyzed by gas chromatography (10). The ammonium salt of *dl*-menthyl succinate is water-soluble, and its stereoselective hydrolysis can be achieved in an aqueous system by using the free cells which possess esterase activity. However, due to poor solubility in aqueous buffers, the *l*-menthol formed accumulated on the surface of the yeast cells, thus decreasing the catalytic activity. To prevent the accumulation of *l*-menthol on the cell surface, various kinds of water-miscible organic solvents were tested as cosolvents. However, the hydrolytic activity of the cells was reduced in the presence of organic cosolvents. The combination of gel-entrapped cells and water-saturated *n*-heptane was finally employed to obtain a homogenous reaction system after many attempts to carry out the reaction effectively. Although the effect of gel hydrophobicity was not so remarkable as in the case of the steroid conversions described above, the activity of the gel-entrapped cells increased along with increased gel hydrophobicity. Figure 4 shows the comparison of operational stability between the free cells and PU-3-entrapped cells over repeated reactions. The half-life of the free cells was 2 days in a *n*-heptane-aqueous buffer two-phase system, while that of the entrapped cells was estimated to be 63 days

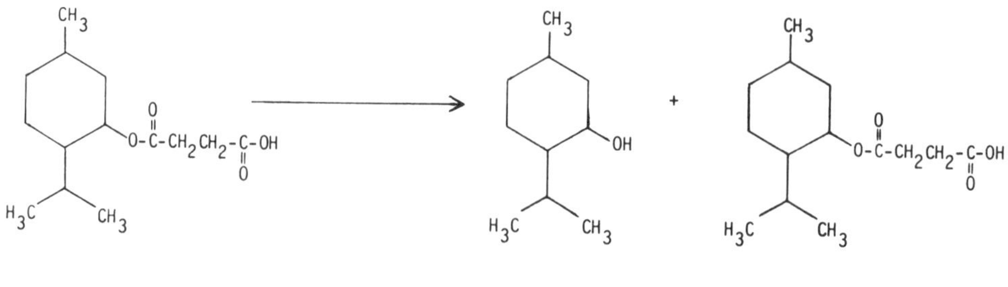

Figure 3 Stereoselective hydrolysis of *dl*-menthyl succinate catalyzed by *Rhodotorula minuta* var. *texensis* cells (10).

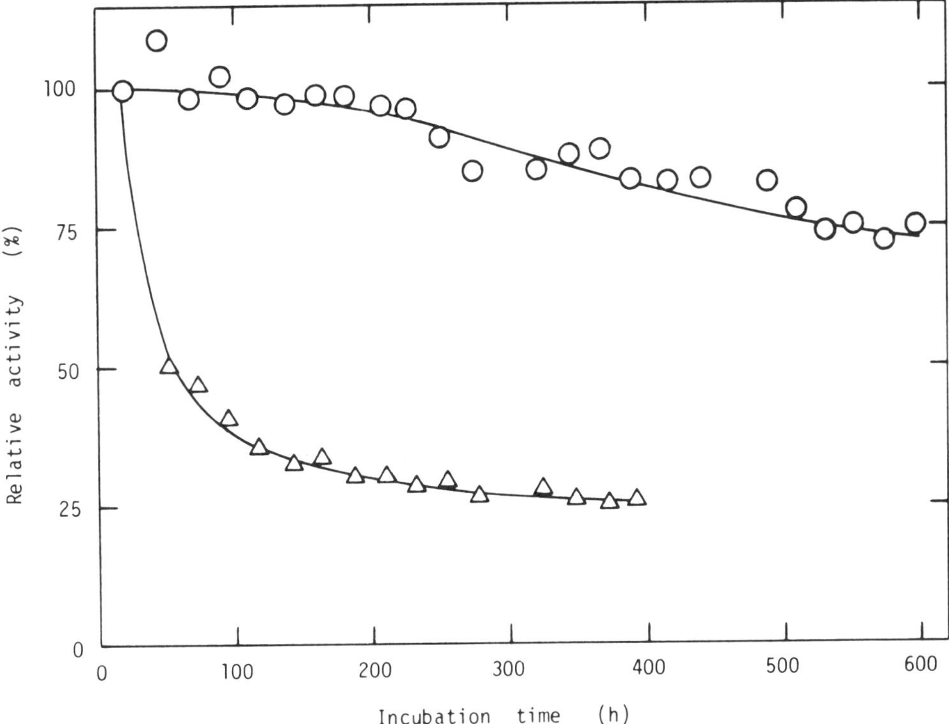

Figure 4 Repeated use of free and polyurethane gel-entrapped *Rhodotorula minuta* cells in hydrolysis of *dl*-menthyl succinate (10). Each reaction was carried out for 24 h in water-saturated *n*-heptane (entrapped cells) or in aqueous buffer-*n*-heptane two-phase system (free cells). The activity at each first reaction was expressed as 100%. (○), Gel-entrapped cells; (△), free cells.

in water-saturated *n*-heptane. Thus, immobilization greatly improved the operational stability of the hydrolytic enzyme in the yeast cells. The optical purity of the product was also constantly maintained at 100% even after a long-term operation.

Stereoselective esterification of *dl*-menthol (130 mM) with an appropriate acyl donor (100 mM) was performed with polyurethane-entrapped lipase from *Candida cylindracea* in an appropriate water-saturated organic solvent system at 30°C (11). As a solvent, cyclohexane and isooctane were found to be suitable, while polar solvents inactivated the enzyme (Table 6). 5-Phenylvaleric acid was chosen as the acyl donor based on the stereoselectivity and the yield of *l*-menthyl ester. Longer-chain acids served as good substrates for esterification but with a low stereoselectivity, while shorter-chain acids were not good substrates. Not only *C. cylindracea* lipase but also porcine pancreas lipase catalyzed the stereoselective reaction under the conditions mentioned above, while fungal lipases were ineffective.

B. Modification of Glyceride

Reformation of olive oil to cacao butter-like fat has been tried by exchanging oleic acid moieties in positions 1 and 3 of the triglyceride with saturated fatty acids (such as stearic acid) by 1 and 3 position-specific lipase from *Rhizopus delemar* (Fig. 5) (16). In this reaction, it is very important to

$$\begin{array}{l} H_2C\text{-}O\text{-}CO\text{-}R_1 \\ \quad | \\ HC\text{-}O\text{-}CO\text{-}R_2 \\ \quad | \\ H_2C\text{-}O\text{-}CO\text{-}R_3 \end{array} + 2\ X\text{-}COOH \longrightarrow \begin{array}{l} H_2C\text{-}O\text{-}CO\text{-}X \\ \quad | \\ HC\text{-}O\text{-}CO\text{-}R_2 \\ \quad | \\ H_2C\text{-}O\text{-}CO\text{-}X \end{array} + R_1\text{-}COOH + R_3\text{-}COOH$$

Figure 5 Regiospecific ester exchange of triglyceride catalyzed by *Rhizopus delemar* lipase (17).

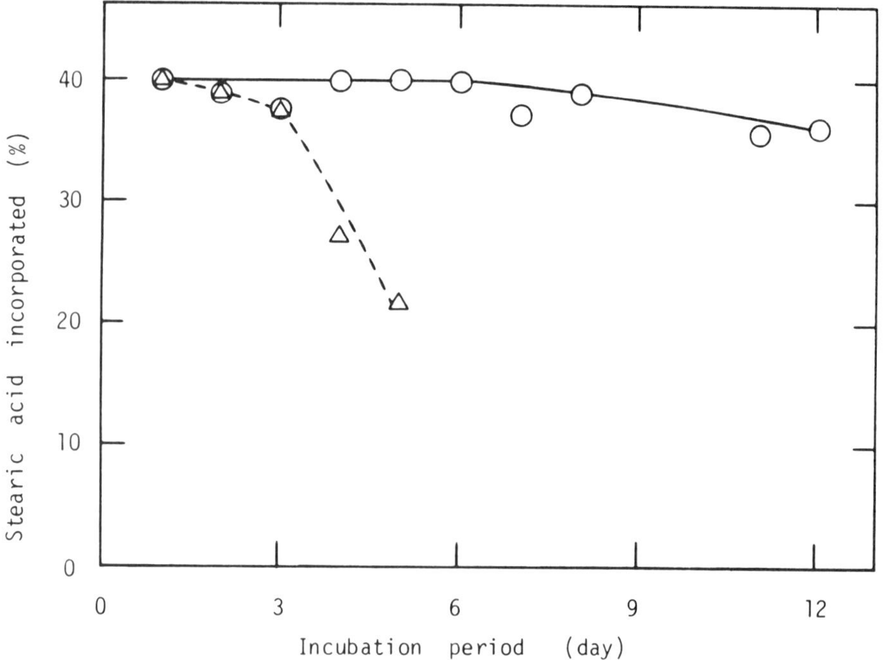

Figure 6 Repeated use of lipase preparations for interesterification of triglyceride (17). Each reaction was carried out for 24 h in water-saturated *n*-hexane. (○), Hydrophobic gel-entrapped Celite-adsorbed lipase; (△), Celite-adsorbed lipase.

control the water content in the reaction system, because the ester exchange reaction cannot be initiated without water, while high water content results in hydrolysis of the ester. Therefore, water-saturated *n*-hexane was selected as the reaction solvent, taking into consideration the activity and stability of the enzyme as well as the solubility of the reactants. To provide water in the vicinity of the enzyme, lipase (5 mg) was first adsorbed on a suitable porous support (0.25 g), such as Celite, with a controlled amount of water (200 μl). When the Celite-adsorbed lipase was entrapped with different prepolymers (0.5 g), the enzyme entrapped with a hydrophobic photo-cross-linkable resin prepolymer showed the highest activity of interesterification: about 75% of that of the lipase preparation simply adsorbed onto Celite. On the other hand, the hydrophilic gel-entrapped enzyme exhibited only low activity. These results again indicate the usefulness of hydrophobic gels in the bioconversions of lipophilic compounds in nonpolar solvents. Entrapment markedly enhanced the apparent operational stability of lipase. For example, in one experiment the enzyme lost only 10% of the original activity after 12 batches of reaction (operational period, 12 days) (Fig. 6) (17).

C. Regio- and Stereospecific Dehydrogenation and Hydroxylation of Steroids

N. rhodochrous cells catalyzed 3β-hydroxysteroid dehydrogenation, 17β-hydroxysteroid dehydrogenation, and, in the presence of an artificial electron acceptor, Δ^1-dehydrogenation of steroids in organic solvent systems (Table 7) (41,42,44). To carry out these reactions efficiently, correct selection of the organic solvents (depending on the hydrophobicity of substrates) and the gels (depending on the solvent polarity) is very important, as described above. The catalytic activities of the entrapped cells are stable and remain high in organic solvents of a low polarity and, therefore, the application of hydrophobic gels is in general very effective.

Δ^1-Dehydrogenation of hydrocortisone to yield prednisolone by acetone-dried cells of *Arthrobacter simplex* was carried out in the presence of 10% methanol. Addition of an electron acceptor, phenazine methosulfate, to the reaction system was very useful for the stabilization of the enzyme system and stimulation of the reaction. Immobilization was also effective to stabilize the catalytic system (the half-life of the free cells is about 10 days; that of the entrapped cells, more than 24 days) (38,39).

Hydroxylation of steroids was catalyzed by immobilized living or growing microbial cells in the presence of organic cosolvents (Fig. 7, Table 4). Entrapped spores (3.6×10^5/g gel) of *Curvularia lunata* were germinated in situ by incubating for 50–60 h at 30°C in 50 ml of a nutrient medium

Table 7 Dehydrogenation/Oxidation of Steroids Catalyzed by Gel-Entrapped Cells of *Nocardia rhodochrous* in Organic Solvent Systems

Reaction	Solvent[a]	Substrate	Product
3β-Hydroxysteroid dehydrogenation	Benzene-*n*-heptane (1:1)	Cholesterol	Cholestenone
	Benzene-*n*-heptane (1:1)	β-Sitosterol	β-Sitostenone
	Benzene-*n*-heptane (1:1)	Stigmasterol	Stigmastenone
	Benzene-*n*-heptane (1:1)	Dehydroepiandrosterone	4-Androstene-3,17-dione
	Chloroform-*n*-heptane (1:1)	Pregnenolone	Progesterone
17β-Hydroxysteroid dehydrogenation	Benzene-*n*-heptane (4:1)	Testosterone	4-Androstene-3,17-dione
	Benzene-*n*-heptane (4:1)	Δ^1-Dehydrotestosterone	1,4-Androstadiene-3,17-dione
	Benzene-*n*-heptane (1:1)	β-Estradiol	Estrone
Δ^1-Dehydrogenation	Benzene-*n*-heptane (4:1)	Testosterone	Δ^1-Dehydrotestosterone
	Benzene-*n*-heptane (1:1)	17β-*O*-Acetyl-testosterone	17β-*O*-Acetyl-Δ^1-dehydrotestosterone
	Benzene-*n*-heptane (1:1)	4-Androstene-3,17-dione	1,4-Androstadiene-3,17-dione

[a]Water-saturated.

Source: Refs. 41,42,44.

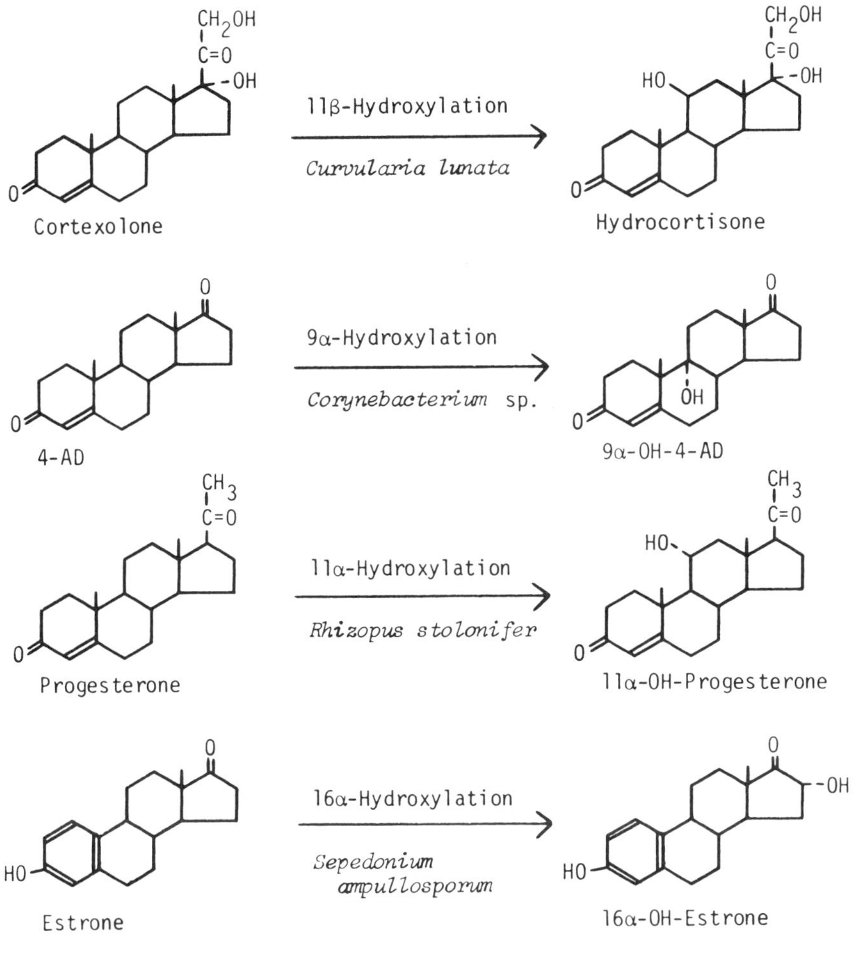

Cortexolone → 11β-Hydroxylation, *Curvularia lunata* → Hydrocortisone

4-AD → 9α-Hydroxylation, *Corynebacterium* sp. → 9α-OH-4-AD

Progesterone → 11α-Hydroxylation, *Rhizopus stolonifer* → 11α-OH-Progesterone

Estrone → 16α-Hydroxylation, *Sepedonium ampullosporum* → 16α-OH-Estrone

Figure 7 Hydroxylation of steroids catalyzed by synthetic gel-entrapped microbial cells in the presence of organic cosolvents. See Table 4 for solvents used.

containing 2 mg/ml of cortexolone (Reichstein's Compound S) and the resulting gel-entrapped mycelia were used for the 11β-hydroxylation of cortexolone (20 mg) in 20 ml of an aqueous buffer containing 14 mg of Tween 80 (58,59). Introduction of dimethyl sulfoxide at 2.5% to the reaction system was not effective for dissolving the substrate but was effective in solubilization of the product, thus stimulating the reaction by the entrapped mycelia. When the catalytic activity of the entrapped mycelia decreased during reaction, the hydroxylation system could be reactivated by incubating the gel-entrapped cells in a nutrient medium containing cortexolone as an inducer as described above. By this method, the entrapped cells were able to be used repeatedly over 50 batches of reaction for 100 days under the conditions mentioned above (Fig. 8).

9α-Hydroxylation of 4-androstene-3,17-dione was catalyzed by the gel-entrapped cells of *Corynebacterium* sp. (53). In this case, however, the product, 9α-hydroxy-4-androstene-3,17-dione, was further metabolized by the cells. Thus, the cells entrapped in hydrophobic gels showed a low production rate: about one tenth of that with the hydrophilic gel-entrapped cells. Introduction of dimethyl sulfoxide at 15% stimulated the hydroxylation reaction 1.3 times or more. The product could be extracted from the vicinity of the cells regardless of the properties of the gels by using 15% dimethyl sulfoxide. In other experiments, 11α-hydroxylation of progesterone by *Rhizopus stolonifer* (55) and 16α-hydroxylation of estrone by *Sepedonium ampullosporum* (80) were carried out in the presence of 2.5% dimethyl sulfoxide and 0.65% *N,N*-dimethylformamide, respectively.

As shown by these examples of steroid hydroxylation, high concentrations of organic cosolvents cannot be used because the cells must remain viable to catalyze the complicated reactions involved.

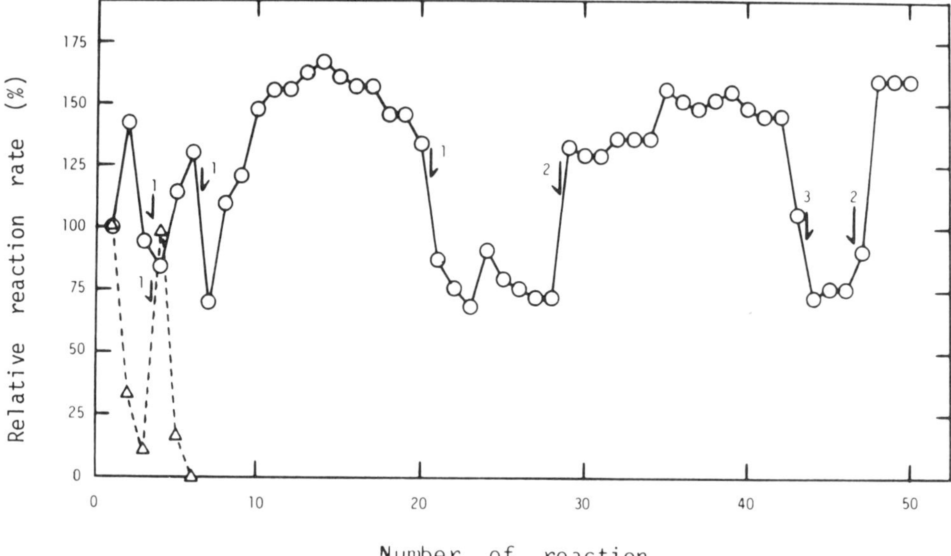

Figure 8 Repeated use of *Curvularia lunata* mycelia in 11β-hydroxylation of cortexolone (59). Each reaction was carried out for 48 h in the presence of 2.5% dimethyl sulfoxide or methanol. Arrows indicate the reactivation of mycelia in nutrient media under appropriate conditions (numbers indicate different incubation conditions). (○), Photo-cross-linked gel-entrapped mycelia; (△), free mycelia.

D. Peptide Synthesis

One example of the application of immobilized enzymes to peptide synthesis is the production of *N*-(benzyloxycarbonyl)-L-aspartyl-L-phenylalanine methyl ester (NBAPME), the precursor of the synthetic sweetener, aspartame. Thermolysin immobilized by different methods was applied to the peptide synthesis using *N*-(benzyloxycarbonyl)-L-aspartic acid and L-phenylalanine methyl ester as the reactants. The enzyme physically adsorbed onto Amberlite XAD-7 or XAD-8 was found to have a high activity and was used for the peptide synthesis in a packed column reactor with ethyl acetate as a solvent (33). Ethyl acetate was also found to be the best solvent among several organic solvents which can dissolve both substrates (34). The stability of the enzyme adsorbed on Amberlite XAD-7 was compared in a plug flow type reactor and a stirred tank reactor, and the latter was revealed to be suitable for the synthesis of NBAPME from 80 mM *N*-(benzyloxycarbonyl)-L-aspartic acid and 120 mM L-phenylalanine methyl ester in ethyl acetate. The reactor (total volume, 25 ml; working volume, 10 ml) was successfully operated for over 300 h with a yield of about 90% at a space velocity of 0.95 h^{-1}.

VI. CONCLUSION

At present, applications of biochemical processes to chemical, food, and pharmaceutical indusdtries are attracting worldwide attention. For this purpose the synthesis and conversion of a variety of lipophilic compounds or compounds with low water solubility must be attempted biochemically and such bioreactions appear best carried out in the presence of organic solvents. In the case of the use of biochemical processes for producing commodity petrochemicals, the substrates and/or products themselves are usually organic solvents. Thus, biocatalysts must maintain their activities toward such unconventional substrates and unfavorable conditions. In these cases the properties of supports used for immobilizing biocatalysts will seriously affect the efficiency of the reactions and the stability of the biocatalysts. New types of supporting materials, especially gel materials, which are easy to handle and whose properties are easily controlled, need to be developed to allow extensive applications of immobilized biocatalysts in industries.

At the same time, it is important to discover novel biocatalysts that retain high activities even in the presence of organic solvents. Not only screening from natural sources but also construction of new enzymes by protein engineering techniques are essential for this purpose. Such organic solvent-tolerant biocatalysts could become potent catalysts in organic solvent systems if they are immobilized properly and used in a suitable reactor.

Investigations on protein structure in organic solvents together with those on the role of water in enzymatic reactions are also important to design modified enzymes and to keep enzymes active and stable under unfavorable conditions. The supports from different scientific fields are necessary to carry out the bioreactions extensively in the presence of organic solvents.

REFERENCES

1. Y. Inada, K. Takahashi, T. Yoshimoto, A. Ajima, A. Matsushima, and Y. Saito, *TIBTECH 4:*190 (1986).
2. P. L. Luisi and C. Laane, *TIBTECH 4:*153 (1986).
3. A. M. Klibanov, *Anal. Biochem. 93:*1 (1979).
4. L. G. Butler, *Enzyme Microb. Technol. 1:*253 (1979).
5. M. D. Lilly, *J. Chem. Tech. Biotechnol. 32:*162 (1982).
6. G. Carrea, *TIBTECH 2:*102 (1984).
7. A. Tanaka and S. Fukui, in *Enzymes and Immobilized Cells in Biotechnology* (A. I. Laskin, ed.), Benjamin/Cummings, Menlo Park, CA, 1985, p. 149.
8. S. Fukui and A. Tanaka, *Endeavour 9:*10 (1985).
9. T. Kawamoto, K. Sonomoto, and A. Tanaka, *Biocatalysis 1:*137 (1987).
10. T. Omata, N. Iwamoto, T. Kimura, A. Tanaka, and S. Fukui, *Eur. J. Appl. Microbiol. Biotechnol. 11:*199 (1981).
11. S. Koshiro, K. Sonomoto, A. Tanaka, and S. Fukui, *J. Biotechnol. 2:*47 (1985).
12. G. M. R. Tombo, H.-P. Schär, X. Fernandez i Busquets, and O. Ghisalba, *Tetrahedron Lett. 27:*5707 (1986).
13. B. Cambou and A. M. Klibanov, *Biotechnol. Bioeng. 26:*1449 (1984).
14. B. Cambou and A. M. Klibanov, *J. Am. Chem. Soc. 106:*2687 (1984).
15. A. R. Macrae, *J. Am. Oil Chem. Soc. 60:*291 (1983).
16. T. Tanaka, E. Ono, M. Ishihara, S. Yamanaka, and K. Takinami, *Agric. Biol. Chem. 45:*2387 (1981).
17. K. Yokozeki, S. Yamanaka, K. Takinami, Y. Hirose, A. Tanaka, K. Sonomoto, and S. Fukui, *Eur. J. Appl. Microbiol. Biotechnol. 14:*1 (1982).
18. J. R. Matos, J. B. West, and C.-H. Wong, *Biotechnol. Lett. 9:*233 (1987).
19. G. Kirchner and A. M. Klibanov, *J. Am. Chem. Soc. 107:*7072 (1985).
20. R. G. Ingalls, R. G. Squires, and L. G. Butler, *Biotechnol. Bioeng. 17:*1627 (1975).
21. A. M. Klibanov, G. P. Samokhin, K. Martinek, and I. V. Berezin, *Biotechnol. Bioeng. 19:*1351 (1977).
22. K. Oyama and K. Kihara, *CHEMTECH 1984:*100 (1984).
23. K. Morihara, *TIBTECH 5:*164 (1987).
24. A. Zaks and A. M. Klibanov, *J. Am. Chem. Soc. 108:*2767 (1986).
25. M. Bergmann and H. Frankel-Conrat, *J. Biol. Chem. 119:*707 (1937).
26. A. N. Semenov, I. V. Berezin, and K. Martinek, *Biotechnol. Bioeng. 23:*355 (1981).

27. A. A. Klyosov, N. Van Viet, and I. V. Berezin, *Eur. J. Biochem. 59:*3 (1975).
28. H. H. Weetall and W. P. Vann, *Biotechnol. Bioeng. 18:*105 (1976).
29. K. Inouye, K. Watanabe, K. Morihara, Y. Tochino, T. Kanaya, J. Emura, and S. Sakakibara, *J. Am. Chem. Soc. 101:*751 (1979).
30. K. Morihara, T. Oka, and H. Tsuzuki, *Nature 280:*412 (1979).
31. A. Jonczyk and H.-G. Gattner, *Hoppe-Seyler's Z. Physiol. Chem. 362:*1591 (1981).
32. K. Morihara, T. Oka, H. Tsuzuki, Y. Tochino, and T. Kanaya, *Biochem. Biophys. Res. Commun. 92:*396 (1980).
33. K. Oyama, S. Nishimura, Y. Nonaka, K. Kihara, and T. Hashimoto, *J. Org. Chem. 46:*5241 (1981).
34. K. Nakanishi, T. Kamikubo, and R. Matsuno, *Bio/Technology 3:*459 (1985).
35. K. Nakanishi, Y. Kimura, and R. Matsuno, *Eur. J. Biochem. 161:*541 (1986).
36. K. Nakanishi, Y. Kimura, and R. Matsuno, *Bio/Technology 4:*452 (1986).
37. S. Ohlson, P. O. Larsson, and K. Mosbach, *Biotechnol. Bioeng. 20:*1267 (1978).
38. K. Sonomoto, A. Tanaka, T. Omata, T. Yamane, and S. Fukui, *Eur. J. Appl. Microbiol. Biotechnol. 6:*325 (1979).
39. K. Sonomoto, I.-N. Jin, A. Tanaka, and S. Fukui, *Agric. Biol. Chem. 44:*1119 (1980).
40. P. Atrat, E. Huller, and C. Horhold, *Z. Allg. Mikrobiol. 20:*79 (1980).
41. T. Omata, T. Iida, A. Tanaka, and S. Fukui, *Eur. J. Appl. Microbiol. Biotechnol. 8:*143 (1979).
42. T. Omata, A. Tanaka, and S. Fukui, *J. Ferment. Technol. 58:*339 (1980).
43. J. M. C. Duarte and M. D. Lilly, in *Enzyme Engineering,* Vol. 5 (H. H. Weetall and G. P. Royer, eds.), Plenum Press, New York, 1980, p. 363.
44. S. Fukui, S. A. Ahmed, T. Omata, and A. Tanaka, *Eur. J. Appl. Microbiol. Biotechnol. 10:*289 (1980).
45. T. Yamane, H. Nakatani, E. Sada, T. Omata, A. Tanaka, and S. Fukui, *Biotechnol. Bioeng. 21:*2133 (1979).
46. H. S. Yang and J. F. Studebaker, *Biotechnol. Bioeng. 20:*17 (1978).
47. G. Carrea, F. Colombi, G. Mazzola, and P. Cremonesi, *Biotechnol. Bioeng. 21:*39 (1979).
48. J. Grunwald, B. Wirz, M. P. Scollar, and A. M. Klibanov, *J. Am. Chem. Soc. 108:*6732 (1986).
49. B. C. Buckland, P. Dunnill, and M. D. Lilly, *Biotechnol. Bioeng. 17:*815 (1975).
50. R. D. Schwartz and C. J. McCoy, *Appl. Environ. Microbiol. 34:*47 (1977).
51. S. Harbron, B. W. Smith, and M. D. Lilly, *Enzyme Microb. Technol. 8:*85 (1986).
52. L. E. S. Brink and J. Tramper, *Biotechnol. Bioeng. 27:*1258 (1985).
53. K. Sonomoto, N. Usui, A. Tanaka, and S. Fukui, *Eur. J. Appl. Microbiol. Biotechnol. 17:*203 (1983).
54. I. S. Maddox, P. Dunnill, and M. D. Lilly, *Biotechnol. Bioeng. 23:*345 (1981).
55. K. Sonomoto, K. Nomura, A. Tanaka, and S. Fukui, *Eur. J. Appl. Microbiol. Biotechnol. 16:*57 (1982).

56. V. Bihari, P. P. Goswami, S. H. M. Rizvi, A. W. Khan, S. K. Basu, and V. C. Vora, *Biotechnol. Bioeng. 26:*1403 (1984).
57. S. Ohlson, S. Flygare, P.-O. Larsson, and K. Mosbach, *Eur. J. Appl. Microbiol. Biotechnol. 10:*1 (1980).
58. K. Sonomoto, M. M. Hoq, A. Tanaka, and S. Fukui, *J. Ferment. Technol. 59:*465 (1981).
59. K. Sonomoto, M. M. Hoq, A. Tanaka, and S. Fukui, *Appl. Environ. Microbiol. 45:*436 (1983).
60. T. K. Mazumder, K. Sonomoto, A. Tanaka, and S. Fukui, *Appl. Microbiol. Biotechnol. 21:*154 (1985).
61. Y. Y. Chun, M. Iida, and H. Iizuka, *J. Gen. Appl. Microbiol. 27:*505 (1981).
62. W. Becker and E. Pfeil, *J. Am. Chem. Soc. 88:*4299 (1966).
63. K. Yokozeki, S. Yamanaka, T. Utagawa, K. Takinami, Y. Hirose, A. Tanaka, K. Sonomoto, and S. Fukui, *Eur. J. Appl. Microbiol. Biotechnol. 14:*225 (1982).
64. R. Maidan and I. Willner, *J. Am. Chem. Soc. 108:*1080 (1986).
65. H. Wan and C. Horvath, *Biochim. Biophys. Acta 410:*135 (1975).
66. K. Takahashi, H. Nishimura, T. Yoshimoto, Y. Saito, and Y. Inada, *Biochem. Biophys. Res. Commun. 121:*261 (1984).
67. K. Takahashi, A. Ajima, T. Yoshimoto, and Y. Inada, *Biochem. Biophys. Res. Commun. 125:*761 (1984).
68. L. G. Butler and F. J. Reithel, *Arch. Biochem. Biophys. 178:*43 (1977).
69. R. Z. Kazandjian and A. M. Klibanov, *J. Am. Chem. Soc. 107:*5448 (1985).
70. J. S. Dordick, M. A. Marletta, and A. M. Klibanov, *Proc. Natl. Acad. Sci. USA 83:*6255 (1986).
71. C. J. Wheeler and R. Croteau, *Arch. Biochem. Biophys. 248:*429 (1986).
72. K. Martinek, A. N. Semenov, and I. V. Berezin, *Biochim. Biophys. Acta 658:*76 (1981).
73. K. Martinek and A. N. Semenov, *Biochim. Biophys. Acta 658:*90 (1981).
74. P. Brodelius and K. Mosbach, *Adv. Appl. Microbiol. 28:*1 (1982).
75. K. Mosbach (Ed.), *Methods in Enzymology,* Vol. 135, Academic Press, New York, 1987.
76. S. Fukui and A. Tanaka, *Adv. Biochem. Eng./Biotechnol. 29:*1 (1984).
77. K. Sonomoto, Y. Hosokawa, and A. Tanaka, *Ann. N.Y. Acad. Sci. 501:*343 (1987)
78. S. Fukui, A. Tanaka, and T. Iida, in *Biocatalysis in Organic Media* (C. Laane, J. Tramper, and M. D. Lilly, eds.), Elsevier, Amsterdam, 1987, p. 21.
79. S. Fukui and A. Tanaka, *Methods Enzymol. 136:*293 (1987).
80. J. M. Kim, K. Sonomoto, A. Tanaka, and S. Fukui, *Ann. Rep. ICME,* Japan, *6:*173 (1983).

7

Immobilized Enzyme Electrodes as Biosensors

George G. Guilbault

University of New Orleans
New Orleans, Louisiana

J-Michel Kauffmann and Gaston J. Patriarche

Pharmaceutical Institute, Free University of Brussels (U.L.B.)
Brussels, Belgium

I. INTRODUCTION

In the large and expanding area of protein (enzyme) immobilization technology, the field devoted to enzyme-immobilized sensors has found a number of important applications in clinical laboratories, fermentation processes, and pollution monitoring. The interdisciplinary structure of biosensor development contributes to its widespread appeal and constitutes certainly a powerful and stimulating aspect of research. Within the biosensor research field, the immobilized enzyme-based electrode technology is currently undergoing active experimentation to provide reliable, sensitive, accurate, easy to handle, and low-cost probes for the demanding environment.

Several excellent reviews dealing with the analytical performances of electrochemical biosensors have contributed to the illustration of their versatility and usefulness (1–17). Due to the enormous proliferation of literature on electrochemical biosensors, some of them have already found wide-scale commercialization (1,8,6,12,16). Today, biosensor-based electrode analyzers able to monitor about 10 different substances have been commercialized, and considerable efforts are being carried out to improve their performance with respect to reliability and stability and to try to extend their scope of applications (1–8).

The attractive and interesting aspect of biocatalyst-based electrodes relies upon their dual advantages of the specificity of the biocatalysts and the sensitivity of electrochemical detection. To achieve the former, isolated or purified enzymes have been successfully employed since the introduction of the concept of "enzyme-electrodes" in 1962 (87). In the late 1970s, the idea of utilizing the enzymatic properties of natural living cells, vegetable tissues, and animal tissues has emerged. Presently, a marked interest is directed towards the development of bacterial and tissue-based electrodes in order to achieve enhanced lifetimes and higher catalytic activities (4,10,13,18–20). This chapter will be devoted to the description of the most popular immobilization modes and sensors used for enzyme electrodes. Moreover, the performances of the resulting devices with respect to selectivity, sensitivity, and stability will be reported. Finally, several applications of enzyme-based electrodes will be reviewed. Examples of practical applications will cover more specifically the monitoring of organic compounds of biomedical interest. Obviously, enzyme electrodes are not limited to organic substrate determination; they have also successfully been utilized for the study of enzymatic activities per se (1,20).

II. GENERAL CONCEPT

An enzyme electrode is obtained by immobilizing a thin layer of enzyme on the surface of an electrochemical sensor. The resulting probe can operate without pretreatment of the sample since accuracy is achieved independently of the color and turbidity of the solution. The substrate(s) to be monitored diffuses into the enzyme layer where the catalytic reaction occurs generating a product (● or ○) or consuming a reactant (△), which can be detected electrochemically (Fig. 1).

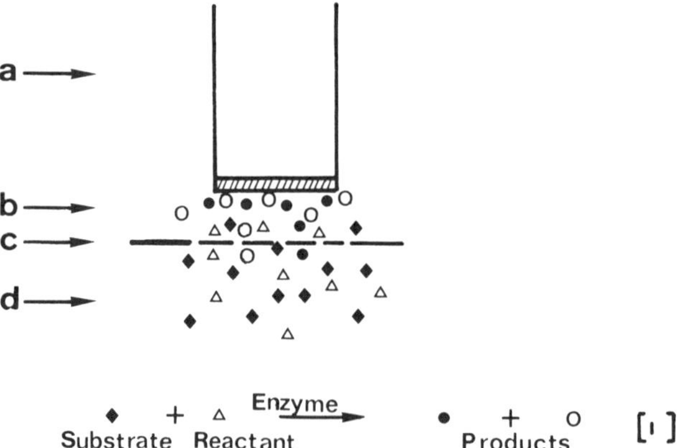

Figure 1 Schematic structure of a biocatalyst electrode. a) electrode; b) biocatalyst layer; c) permselective membrane; d) solution.

The electroactive species is monitored potentiometrically or amperometrically, and the electrochemical signal can be correlated back to the concentration of the substrate (or enzyme). In the potentiometric type sensor, a membrane (glass, solid state, liquid) selectively extracts a charged species into the membrane phase generating a potential difference between the internal filling solution and the sample solution (enzyme layer). This potential is, ideally, proportional to the logarithm of the analyte concentration (activity). Among the potentiometric ion-selective electrodes, the gas-sensing probes are of particular usefulness (22,23). Outstanding selectivity can be achieved by simply casting a gas selective membrane onto an ion-selective electrode (polypropylene, teflon) (Fig. 2). Interferences can only result from dissolved species in the sample, which can diffuse through the synthetic membrane, producing a pH change in the thin layer of internal filling solution.

Amperometric sensors monitor a current flow, at a selected fixed potential, between the working electrode and the reference electrode. The sensor exhibits a linear response versus the concentration of the substrate. Either the reactant or the product must be electroactive (oxidable or reducible) at the electrode surface. In some cases redox mediators must be introduced to allow fast electron-transfer reactions (5,24–26).

The response of an enzyme electrode can be measured by a steady-state (i.e., equilibrium) method measuring millivolts or microamperes, or by a kinetic measurement, in which the rate of change of the electrochemical signal is monitored as a function of time (89,127,143,150). The unique

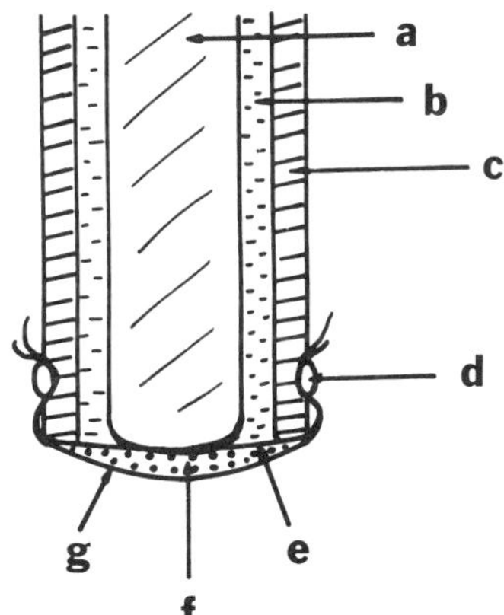

Figure 2 Cross-sectional view of a potentiometric gas-sensing enzyme electrode. a) glass pH electrode; b) internal filling solution; c) electrode outer jacket; d) "O" ring; e) gas selective membrane; f) enzyme layer; g) dialysis membrane.

properties of enzymes as selective biocatalysts have long been known, but their use in bioelectrocatalysis was introduced only in the late 1960s (27). The problems associated with the use of enzymes, i.e., the lack of pure enzyme preparations, the instability of enzymes, and impossibility of reuse, have limited their utilization. Physical or chemical binding of enzymes to solid supports has circumvented those major hindrances and facilitated their use as heterogeneous catalysts in reaction mixtures or in perfused beds (1,28,29). With some specific requirements, the modes of enzyme immobilization at the electrode surface follow the general guidelines of protein immobilization technology (see Chapters 2 and 3). The most widely used procedures will be presented in the following section.

III. IMMOBILIZATION METHODS

The techniques may be divided into covalent and noncovalent procedures with all the refinements these notions imply. Using any technique, the purpose is always to facilitate a physical localization of the enzyme in a thin layer across the electrode surface. The technique which consists of trapping a soluble enzyme in a thin film over the electrode surface by using some type of membrane layer(s) is easy, but does not produce a real immobilized enzyme electrode. However simple the procedure, the resulting electrodes may be utilized successfully. The electrodes obtained show, however, a low stability and necessitate the use of a large excess of enzyme. Electrodes with physically or chemically immobilized enzymes exhibit greater stability, fewer interferences, and reusability. In the usual practice, three methods are preferred: (1) to entrap the enzyme in an inert polymer matrix over the electrode, (2) chemical cross-linking of the enzyme by bifunctional reagents, to itself and to other macroscopic particles (protein), and holding a thin layer of the enzyme over the electrode with some type of membrane(s), and (3) the chemical coupling of the enzyme either directly onto the electrode surface or to water-insoluble membranes maintained over the electrode. The choice of the appropriate mode of immobilization is of significance since the chemically immobilized enzyme is exposed to molecules on all sides, whereas the entrapped enzyme is isolated from large molecules which cannot diffuse into the matrix. Hence, the type of immobilized enzyme will differ in its kinetics and in the kind of interferences observed. For the assay of large substrates, e.g., such as protein with proteolytic enzymes, an attached enzyme must be used and not an entrapped enzyme. Either form of enzyme could be used for the assay of small substrates such as glucose. Table 1 shows how diversified the immobilization techniques are that have been proposed for glucose electrodes. This is typical for every type of immobilized enzyme described in the

Table 1 Various Glucose Electrodes

Modes of immobilization	Modes of detection	Ref.
Physical entrapment between membranes	Amperometry	87–91,140
Physical entrapment into inert supports		
Gelatin	Amperometry	93,94
Polyacrylamide	Amperometry	89,95–99
	Potentiometry	100–102
Poly(vinylchloride)	Potentiometry	103
Conducting polymers	Amperometry	104
Polyacrylamide-graphite	Amperometry	105
Adsorption onto the electrode surface	Amperometry	106,107
Chemical cross-linking with proteins		
Glutaraldehyde-collagen	Amperometry	108
Glutaraldehyde-BSA	Potentiometry	102,103,109
	Amperometry	92,93,99,110–115,138,141
Glutaraldehyde-gelatin	Amperometry	92,116,117,154
Covalent linking to water-insoluble carriers		
Nylon mesh	Amperometry	118–120
Polyvinylalcohol-collagen	Amperometry	121
Collagen	Amperometry	122–124, 141, 146
Cellulose acetate	Amperometry	125
Teflon	Amperometry	126,127
Pig small intestine	Amperometry	128,142
Electrode surface	Potentiometry	102,109,129
	Amperometry	130–135,156,186

literature. To date no universal immobilization technique has emerged which is applicable to all cases.

A. Physical Entrapment

Inclusion of the enzyme into a gel, such as starch or polyacrylamide, is a physical and mild method of protein fixation, where the polymer is allowed to form cross-links in the presence of the enzyme.

Several enzymes have been entrapped in polyacrylamide gels, namely, glucose oxidase (GOD), catalase, lactic dehydrogenase (LDH), amino acid oxidase, and glutamic dehydrogenase (GLDH). The gels were found to show little loss of activity after 3 months of storage at 0–4°C (96). Entrapped glucose oxidase has been coupled with an oxygen electrode for the determination of glucose in blood and the decrease of oxygen was monitored:

$$\text{Glucose} + O_2 \xrightarrow[\text{oxidase}]{\text{glucose}} H_2O_2 + \text{gluconic acid} \quad (1)$$

The immobilization of several enzymes in starch, polyacrylamide, and silicone rubber was investigated for comparative purposes (30). The optimal method for immobilization appears to be physical entrapment in a polyacrylamide gel. The silicone rubber polymerization is too rough on the enzyme, resulting in loss of about 80% of activity. The starch-gel entrapped enzyme is too weakly held, and much of the enzymatic activity is lost due to leeching. Polyacrylamide provides better stability to the enzymes, particularly in the case of urease (80 days). However, some enzyme is lost during acrylamide polymerization (10–25% loss), and riboflavin and $K_2S_2O_8$ must be used as catalysts for the polymerization. These may later interfere with the assay depending on the mode of detection used. The technique has recently been successfully applied to the determination of urea in diluted plasma using a urease-immobilized glass pH electrode (31). The direct coating of the electrode surface of a glass pH electrode with a thin layer (50 μm) of entrapped enzyme has been reported (32). The enzyme entrapment was done by controlled chemical cross-linking of a prepolymerized linear chain of polyacrylamide-hydrazide by glyoxal. Acetylcholinesterase, urease, and penicillinase have been successfully immobilized, and the sensors exhibit a surprisingly long shelf life (more than 6 months for acetylcholinesterase) as well as short response times and wide linear responses.

Poly(vinyl alcohol) gels have also been utilized to entrap urease; the resulting membranes were mounted over a platinum anode. The pH-dependent hydrazine oxidation at the electrode was used to monitor urea in a rapid and sensitive procedure (33). Glucose oxidase has been immobilized in plasticized poly(vinyl chloride), either alone or with peroxidase, and the layers were coated onto an iodide ion-selective electrode. The performances of the immobilized PVC membranes have been compared under identical conditions to GOD chemically immobilized with glutaraldehyde-bovine serum albumin membranes. The PVC membranes exhibit higher sensitivity, but suffer from lower lifetimes, i.e., 7 days compared to a stability of longer than 2 weeks for the chemically immobilized GOD (103).

These gels are nonelectroconductive, thus making the electron transfer between the enzyme and electrode very difficult. To promote proximity between the enzyme active site and the conducting surface of the electrode, some promising techniques have been recently suggested (104,105,139).

Thin particles of graphite have been incorporated into the enzyme-polyacrylamide gel and an electron transfer promoter, namely ferrocene, was loaded into the gel (105). Another way is to entrap the enzyme into conductive polymers (104). Polypyrrole has been chosen as such a polymer. The film is electrochemically grown on a platinum electrode from aqueous solutions containing pyrrole and glucose oxidase. Despite some uncertainties concerning the mechanism of this type of immobilization, the procedure permits the realization of an enzyme electrode in a one-step process, the technique is simple, and it allows reproducible enzyme immobilization.

B. Chemical Immobilization

1. Cross-Linking with Bifunctional Reagents

Immobilization of enzymes with bifunctional reagents may be carried out by several different methods: direct cross-linking of enzymes or cross-linking of enzymes with inert proteins (albumin, gelatin, collagen). Glutaraldehyde as the bifunctional reagent and bovine serum albumin as the inert material are most often used (Eq. 2).

$$\begin{array}{ccccc} \mathrm{CHO} & & \mathrm{Enzyme} & & \mathrm{H} \\ | & & | & & | \\ (\mathrm{CH_2})_3 & + & \mathrm{NH_2} & \longrightarrow & \mathrm{C=N-Enzyme} \\ | & & & & | \\ \mathrm{CHO} & & \mathrm{NH_2} & & (\mathrm{CH_2})_3 \\ & & | & & | \\ & & \mathrm{Albumin} & & \mathrm{C=N-Albumin} \\ & & & & | \\ & & & & \mathrm{H} \end{array} \tag{2}$$

The technique is simple, fast, and allows control of the physical properties and particle size of the final product. Common reagents, BSA, glutaraldehyde, and buffer, are required for membrane formation. Several other bifunctional reagents have been employed, with varying degrees of success. The most common ones include: hexamethylene diisocyanate, trichloro-5-triazine, and diphenyl-4,4′-dithiocyanate-2,2′-disulfonic acid.

2. Covalent Binding to Water-Insoluble Matrices

This technique is the most widely used mode of enzyme insolubilization since it provides high stability, improved catalytic efficiency, and the possibility of using a wide range of carriers, including electroconductive ones. Carrier composition is of primary importance since it determines the type of enzyme attachment and the durability of the resulting probe. Corrosion or dissolution of the support can shorten the enzyme's operational half-life, either by enzyme loss or by deactivation of the enzyme caused by inhibition of the catalytic site with soluble corrosion products. Moreover, the carriers are chosen by their properties of solubility, functional groups, surface area, swelling, and hydrophobic or hydrophilic nature.

Essentially, three types of carriers have been used: inorganics, natural polymers, and synthetic polymers. The binding reactions must proceed under conditions which do not cause enzyme denaturation. Linkage is carried out using the functional groups on the enzyme which are not essential for its catalytic activity. The amino acid residues which are the most suitable for covalent binding are the α- and ε-amino groups, the β- and γ-carboxyl groups, and the phenol ring of tyrosine. There are many methods for covalent coupling of enzymes to water-insoluble carriers (1,28,34). The insolubilization of the enzyme onto the activated matrix via glutaraldehyde is a simple, inexpensive, and frequently used technique (Eqs. 3,4). Enzymes have been attached to a nylon matrix (38), a pig intestine (36,128), to the hydrophobic membrane of a gas selective sensor (37), or onto controlled pore glass (CPG) beads (35):

ACTIVATION OF THE SUPPORT:

$$\text{▨}-NH_2 + OH-(CH_2)_3-CHO \rightarrow \text{▨}-N=\overset{\displaystyle H}{\overset{|}{C}}-(CH_2)_3-CHO \quad (3)$$

LINKING TO ENZYME:

$$\xrightarrow{E-NH_2} \text{▨}-N=\overset{\displaystyle H}{\overset{|}{C}}-(CH_2)_3-\overset{\displaystyle H}{\overset{|}{C}}=N-E \quad (4)$$

Generally, the carriers must be activated to be suitable for covalent attachment of the enzyme molecule, i.e., it is necessary to introduce or liberate a functional group which undergoes a coupling reaction with the enzyme molecule under mild conditions.

Arginase and urease have been simultaneously immobilized onto controlled pore glass beads and deposited on the surface of an ammonia-sensing probe for the determination of arginine (34). In the electrode assembly, the enzyme beads are trapped between a nylon membrane and the ammonia-sensing membrane. Stability of the resulting probe stored at 4°C has been estimated to be at least 41 days. The immobilization procedure consists of an activation of the glass beads with an alkylamine derivative, and subsequent coupling of the enzymes with glutaraldehyde giving the following enzyme bead formula:

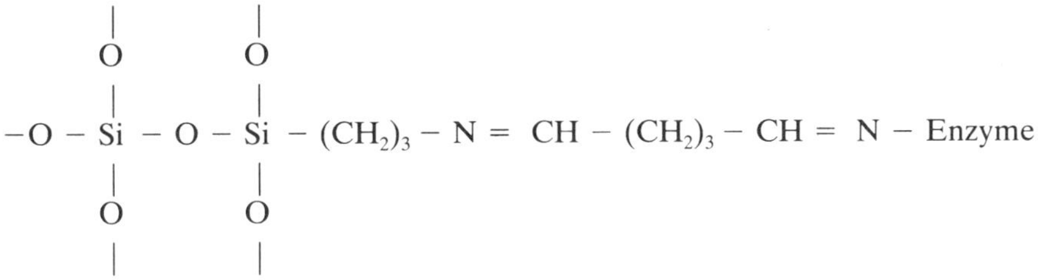

$$
\begin{array}{ccccccc}
 & | & & | & & & \\
 & O & & O & & & \\
 & | & & | & & & \\
-O - & Si & - O - & Si & - (CH_2)_3 - N = CH - (CH_2)_3 - CH = N - \text{Enzyme} \\
 & | & & | & & & \\
 & O & & O & & & \\
 & | & & | & & &
\end{array}
$$

The chemical binding of enzymes to a nylon net has been reported to be very simple, giving strong mechanically resistant membranes (118). The technique consists first of an activation of the nylon net by methylation; this is followed rapidly by treatment with lysine. The lysine acts as a spacer between the nylon and the enzyme. The chemical binding of the enzyme is finally effected using glutaraldehyde. The immobilized enzyme discs may be directly fixed to the sensor surface or stored in a phosphate buffer. The enzymes, glucose oxidase, ascorbate oxidase, cholesterol oxidase, galactose oxidase, urease, alcohol oxidase (118), and lactate oxidase (39) have been immobilized using this method and the respective enzyme electrode performance established.

Collagen membranes have also successfully been used to bind a variety of enzymes (40,122–124). The binding procedure is particularly mild because the enzymes never come in contact with the chemical reagents, thus avoiding all denaturation processes. Collagen carboxyl groups are activated in three successive steps involving methylation, hydrazide formation, and subsequent transformation into acylazides. After removal of all excess reagents by washing, the coupling of enzymes proceeds by simple immersion of the collagen film into the enzyme solution:

$$\text{Collagen - } CON_3 + H_2N \text{ - Enzyme} \longrightarrow \text{Collagen - CO-NH - Enzyme} \tag{5}$$

Recently, an extremely simple and fast procedure has been reported for glucose oxidase immobilization using commercially available polyamide preactivated membranes (41). The stability of the enzymic membranes was excellent, with more than 400 assays being performed within 50 days.

3. Direct Attachment onto the Electrode Surface

The linking of enzyme directly to a conductive support is a logic continuation of enzyme electrode improvement which has been developed progressively since the early eighties. This mode of enzyme fixation, thoroughly reviewed recently (42), is expected to permit improved electrocatalytic processes and faster responses. Metallic (44,100,102,109,129) and carbonaceous (43) attached enzyme electrodes have been shown to develop potentiometric responses to substances produced by the enzymatic reaction, but the signal is markedly dependent on the electrode surface redox states, and thus on electrode pretreatments (which are actually difficult to reproduce). The so-called amperometric enzyme electrodes are generally preferred. These devices generate electrocatalytic currents by the very thin layer of immobilized enzyme.

Physical Attachment. The presence of various functional groups on the surface of carbonaceous materials and their porosity make them effective in the adsorption of enzymes. However, the process is reversible and the probes exhibit poor stability and can lose their activity within a few days (42). Recently a novel enzyme immobilization procedure successfully produced, in a reproducible manner, a fairly stable and sensitive enzyme electrode (69). The technique consists of the entrapment or adsorption of the enzyme in an inert protein film electrochemically attached to the carbon electrode surface. Practically, after electrode activation, a protein sheeth (BSA) is electrochemically attached over the tip of a micro electrode, and then the enzyme (e.g., LDH) is fixed irreversibly by simply dipping the BSA-sheeted electrode in the appropriate enzymatic solution. The authors suggest not to use glutaraldehyde in this process. If glutaraldehyde is used, a loss of enzyme sensitivity (2.5×) occurs, possibly by some enzyme denaturation due to the cross-linking reagent. Pyruvate has been determined using the LDH electrode in the presence of the enzyme cofactor NADH using the normal pulse voltammetry mode, with a detection limit of 1 μM and an electrode stability of one week. Applications to cerebrospinal fluids have been realized.

Chemical Attachment. The covalent linking of a protein to a conductive matrix has already proven to be more suitable for preparation of stable enzyme electrodes (42,135). As a rule the initial step involves derivatization of the support to introduce appropriate functionalities. This step is considered to be critical to the final response of the enzyme layer and must therefore be controlled rigorously. The electrode surface may either be (1) oxidized by heat or by oxygen radio-frequency plasma, or chemically or electrochemically oxidized (cathodic and extreme anodic polarization cycles) to obtain highly reactive sites (carboxy, phenol, quinone-like structures), which can be linked covalently to enzymes directly or with a coupling agent, or (2) chemically modified by silanization or by introduction of an amine-terminated spacer arm. The coupling is generally realized by glutaraldehyde or carbodiimide, which can lead to intra- and interenzyme cross-linking in addition to the desired enzyme support cross-linking. The resulting layers maintain approximately 75% of the initial enzyme activity after a month's use. Carbonaceous substrates (graphite, glassy carbon) are generally preferred for such electrodes because of their inherently interesting chemical and electrochemical properties. However, excellent results have been reported on chemically modified platinum electrodes (46). Glucose oxidase has been successfully immobilized by cross-linking the enzymes with bovine serum albumin (BSA) using glutaraldehyde onto an electrochemically oxidized platinum surface previously silanized with 3-aminopropyltriethoxysilane:

$$\text{Pt} \,|\!\!-\text{O}-\underset{|}{\overset{|}{\text{Si}}}-(\text{CH}_2)_3-\text{NH}_2+\text{glutaraldehyde}+\text{BSA} + \text{Enzyme—NH}_2 \qquad (6)$$

Detection of the H_2O_2 liberated, by fixing the potential of the electrode at +0.7 V vs. S.C.E., gave a linear response from 0.1 μM to 2 mM. The electrode response is fast, and the stability longer than one month. Applications to human control serum were in agreement with the manufacturer's data. Working amperometrically, the sensors detect in a fast and sensitive manner the electroactive products liberated within the thin (monolayer) enzyme membrane. The stability of chemically attached enzyme electrodes ranges from 20 to 30 days.

IV. ELECTRODE PROBES

The choice of an appropriate electrochemical sensor is governed by several requirements including: (1) nature of substrate to be determined (ions or redox species), (2) shape of the final sensor (e.g., micro-electrodes), (3) selectivity, sensitivity, and speed of the measurements required, and (4) desired reliability and stability of the probe. The most frequently used sensors operate under potentiometric or amperometric modes. Amperometric enzyme electrodes, which detect the consumption of a specific product of the enzymic reaction, display an expanded linear response range and a larger apparent Michaelis-Menten constant (Km) than potentiometric enzyme electrodes. However, the latter, due to their unique configuration, offer generally greater selectivity than their amperometric counterparts.

A. Potentiometric Sensors

Ion-selective potentiometric sensors have gained general acceptance among scientists (47). Their use with enzymes to produce biosensors is a natural extension. Ion-selective electrodes sensing various ions, such as F^-, I^-, S^{--}, NH_4^+, CN^-, and H^+, have been employed successfully together with enzymes such as glucose oxidase, peroxidase, urease, catalase, et al. (48–51) Glucose has been determined with a fluoride electrode (51) and with an iodide ion-selective electrode coated with GOD, either alone or with horseradish peroxidase (HPR) (83,103). The iodide electrode potentiometrically monitors a decrease in background iodide levels, and permits the determination of glucose according to the following equations:

$$\text{Glucose} + O_2 \xrightarrow{\text{GOD}} \text{gluconic acid} + H_2O_2 \tag{7}$$

$$H_2O_2 + 2I^- + 2H^+ \xrightarrow[\text{HRP}]{\text{Mo(VI)}} I_2 + 2H_2O \tag{8}$$

When using an iodide sensor to measure glucose, several types of interferences may be encountered, e.g., (1) interferences at the iodide electrode (SCN^-, $S^=$, CN^-, Ag^+) and (2) interferences from oxidizable compounds present in blood (uric acid, tyrosine, ascorbic acid, Fe(II)), which compete in the oxidation of iodide to iodine in Eq. 8 (83). These compounds must thus be removed by sample pretreatment prior to analysis.

The pH sensor remains one of the most sensitive and selective commercially available ion-selective electrodes. However, when applied in a biosensor, these probes suffer from lack of sensitivity due to the buffer system of the samples normally analyzed (45,52). Nevertheless, successful applications have been realized (53) and more recently a urea-pH glass electrode operating in a differential mode with an uncoated pH electrode has been applied to the urea determination in diluted plasma samples (31). Innovation in the enzyme electrode field has been introduced by the use of metal–metal oxide pH sensors (antimony, palladium, iridium, etc.) in association with the catalyst (54,56,171). Such devices show great promise in the fabrication of cheap, disposable, multigate sensors with considerable potential for miniaturization. Although current investigations have already provided good results with enzyme-based pH transducers, these suffer from a difficulty in obtaining reproducible electrode surface states and may be subjected to interfering substances, which can reduce these states. (54)

Considerable selectivity has been achieved by casting the ion-selective electrodes with a synthetic membrane (teflon, polypropylene, etc.), thus allowing selective gas diffusion towards the sensor (22,23,50). Potentiometric ammonia and carbon dioxide selective gas sensing devices are the ones most frequently used for these electrodes because of their high selectivity. Figure 2 illustrates a typical biocatalytic potentiometric gas-sensing probe. The electrode consists of a combination of pH glasss and reference electrodes and a gas-permeable membrane. Positioned between the glass electrode and the membrane is a thin layer of sodium bicarbonate solution (carbon dioxide sensor) or ammonium chloride (NH_3 sensor). When the whole electrode is exposed to a solution of substrate to be analyzed, the enzymatic reaction produces a pH change. The gas diffuses both back to the sample and towards the pH sensor, passing through the enzyme and the gas-permeable membrane, then dissolves in the internal filling solution, thereby changing its pH and giving rise to the potentiometric response of the electrode system. Linear Nernstian responses are generally observed in the range of substrate concentration from 10 μM to 0.1 M with a response time ranging from 1 to 10 min and a baseline recovery time of the same order of magnitude. Despite some minor interferences which may affect their responses (22,57,58), potentiometric gas sensing probes offer real advantages when operating in complex biological fluids (9,10). Studies are presently underway to improve their performances (59,60) in order to use them in combination with a great number and range of biocatalysts (10,14,18).

B. Amperometric Sensors

Amperometric sensors monitor a current flow at a selected fixed potential between the working electrode and the reference electrode. Optimization of this type of sensor (in regard to stability), low background currents, and fast electron-transfer kinetics constitutes a vast field of research in electrochemistry. Platinum electrodes, due to their catalytic response towards hydrogen peroxide oxidation and oxygen reduction have so far mainly been used in combination with enzymes for enzyme electrode development. Due to an increasing interest in solid electrode-based sensors, the development of new carbon substrates, modified or not (glassy carbon, carbon paste), has added new dimensions to the research on amperometric enzyme electrodes (61–63). As a result, the high sensitivities (0.1 μM) reported earlier for glucose determination using glucose oxidase immobilized platinum electrodes (122) have recently been observed for various substrates using carbon-based electrodes, combined with a judicious choice of enzymatic substrate amplification systems (6,64,116,117) and chemical redox mediators. This new concept, consisting of a bienzyme system and a redox mediator, shows promise not only because of its increased sensitivity, but also due to a reduction of interfering or electrode fouling substance effects resulting from working at lower positive potentials (24,25,65–67).

Redox mediators have been introduced into enzyme electrodes to circumvent the problems associated with the sluggish redox behavior of proteins at the electrode surface. Many enzymes involved in oxidation and reduction reactions contain electroactive centers (haemin, flavin) surrounded by a protein matrix, preventing efficient electron transfer to electrodes. Consider for example, the NAD^+-dependent enzyme dehydrogenases which catalyze the oxidation of a substrate RH:

$$\begin{array}{ccc} \text{Enzyme-NAD}^+ & & \text{Enzyme-NADH} \\ \nwarrow & & \nearrow \\ \text{RH} & \longleftrightarrow & \text{R} \end{array} \qquad (9)$$

Ideally, the reaction should be followed by direct oxidation of the reduced NADH-dependent enzyme at the electrode surface. This possibility of direct electrochemical oxidation of NADH by solid electrodes has been demonstrated by a number of biosensor applications (66,68). However, the coenzyme is generally so well embedded in the enzymic structure that the change transfer between the redox enzyme and the electrode requires high overvoltages to overcome the reaction barrier or steric hindrance exhibited by the protein. Moreover, the NADH oxidation is accompanied by nonspecific side reactions, including some NAD^+ decomposition. In addition, the reaction products often cause severe electrode surface fouling, which can cause slow, heterogeneous electron transfer. If we consider, in a second example, the flavoprotein oxidases, similar surface problems are encountered. In the classical glucose sensor, the reaction may be divided into the following steps:

Immobilized layer: $\text{glucose} + \text{FAD} \rightarrow \text{gluconolactone} + FADH_2$ (10)

$FADH_2 + O_2 \rightarrow \text{FAD} + H_2O_2$ (11)

Electrode: $H_2O_2 \rightarrow O_2 + 2H^+ + 2e$ (12)

The oxidation of glucose occurs enzymatically, and the flavin within the glucose oxidase is reduced. In turn, the presence of the natural electron acceptor oxygen permits the regeneration of the enzyme in its oxidized state, while oxygen is reduced to hydrogen peroxide. In the reaction, the two FAD (flavin adenine dinucleotide) molecules of glucose oxidase are embedded in a protein matrix of a molecular weight of about 150,000. Although the protein-free flavin prosthetic group shows diffusion controlled electron transfer at the electrode, in the presence of the protein the reaction becomes highly inhibited. For this reason, most of the glucose enzymes have been constructed to detect the product (H_2O_2) or the natural reactant (O_2) of the reaction. Although perfectly suitable for glucose monitoring in solution, these devices do suffer potential difficulties when operated in vivo or in nondiluted biological samples where the response is limited by physiological amounts of oxygen available. Moreover, hydrogen peroxide detection occurs at fairly high anodic potentials with associated problems of oxidation of interfering molecules. To solve the problems associated with slow electron transfer kinetics and molecular oxygen limitation, several ingenious strategies have been developed such as: (1) addition of an electron-transfer mediator to the immobilized enzyme layer (66,67,71, 94,105) directly onto the electrode surface (covalently or noncovalently) (65–67,134,146,147) or into the electrode material (105,106); (2) chemically or electrochemically preactivating the electrode surface (69); (3) immobilization of the enzyme into an electroconducting polymer that is deposited on the electrode surface (104); or (4) use of conducting organic salts as the electrode material (24,138–140,164). Today, a tremendous amount of research is devoted to the construction of redox mediated enzyme electrodes. The mediator provides redox coupling between the electrode and the redox center in the enzyme (see Eq. 13). In order to be efficient, a mediator should meet several requirements: it should not be pH-sensitive or autoxidizable, and it should exhibit fast interaction with the enzyme active site and electrochemical behavior. A compilation of mediator compounds for the electrochemical study of biological redox systems (70) as well as several reviews dealing with the acceleration of enzyme electrode reactions have been reported recently (27,63,66,67,139).

$$\text{Electrode} \;\; e^- \;\; \begin{matrix} \nearrow \text{Mediator OX} \nwarrow & \nearrow \text{RH} \\ \searrow \text{Mediator RED} \swarrow & \searrow \text{R} \end{matrix} \qquad (13)$$

V. ENZYME ELECTRODE CHARACTERISTICS

A. Biosensor Lifetime

The stability of an enzyme electrode can be characterized by both its storage and operational stability. Because of the hybrid structure of enzyme electrodes, factors affecting both electrode stability and enzyme stability have to be considered. Potentiometric ion-selective electrodes possess great stability, but in the case of gas-sensing devices care has to be taken to regularly replace the internal filling solution due to slow diffusion of sample solution through the gas selective membranes. Amperometric sensors are generally platinum, gold, or carbon based, and the surface stability can be regulated by appropriate chemical or electrochemical pretreatments. Their stability is rather limited due to rapid surface fouling caused by reaction products, especially when operating at high overvoltages. In order to diminish the extent of electrode surface fouling, redox mediators have been introduced, as already discussed above. Furthermore, it has been suggested to monitor the reaction kinetically; instead of waiting until an equilibrium current is reached, the rate of change in current ($\Delta i/\Delta t$) can be measured and equated with the concentration of substrate. Thus, the time during which the reaction occurs is short, minimizing the amount of end product formation. In regard to biocatalyst lifetime, the mode of immobilization and the source and extent of purification of the enzyme are of primary concern. Of the two methods used to immobilize an enzyme (covalent and noncovalent) the technique of chemical (covalent) immobilization offers the greatest stability. As a general rule, a "soluble" enzyme electrode is useful for about one week or 25–50 assays, and physically entrapped polyacrylamide electrodes perform satisfactorily for about 50–100 assays depending primarily on the degree of care exercised in the preparation of the polymer. Chemically attached enzymes can be kept indefinitely, if not used very much. The electrode has a lifetime of over one year for glucose oxidase, and greater than 4–6 months for 2-amino acid oxidase or uricase (30). One can expect about 200–1000 assays per electrode, depending on the immobilization technique. With amperometric sensors the high overvoltages used, as well as the product of the reaction, can inactivate the enzyme. It is thus advantageous to measure the electrode response kinetically, in order to minimize the time during which the enzyme membrane is in contact with high concentrations of substrate (172). Although the electrode can be stored at room temperature, it is recommended that all enzyme-based electrodes be kept in the refrigerator (around 4–5°C) and covered with a dialysis membrane to prevent any contamination by bacteria, which tend to degrade the enzyme and destroy its activity. Dialysis membranes prevent loss of enzyme from the electrode and keep bacteria from diffusing to the enzyme. Another factor affecting the stability of some enzyme electrodes is the leeching of a loosely bound cofactor from the active site, as demonstrated in the case of D-amino acid oxidase in a polyacrylamide membrane (223). The immobilized enzyme D-amino acid oxidase must be stored in the presence of its coenzyme, flavin adenine dinucleotide, in order to maintain its activity for periods longer than 3 weeks.

B. Response Time

Optimization of the response time of a bioelectrode encompasses a short electrode response time, a parameter often controlled by the immobilized enzyme layer rather than by the sensor. In enzyme electrodes highly active enzymes are kept in a thin layer near the sensor to obtain fast response times. In the mechanism of the response, (1) the substrate must diffuse through the solution to the membrane surface, (2) the substrate must diffuse through the membrane and react with the biocatalyst at the active site, and (3) the products formed must then diffuse to the electrode surface where they are measured. Mathematical models describing this effect have been thoroughly presented in the literature.

Many factors affect the speed of response as listed in Table 2. The enzyme concentration, either in a pure form in the immobilized layer or as

Table 2 Factors Affecting the Response Time of an Enzyme Electrode[a]

1. Stirring rate of solution
2. Concentration of substrate $10^{-1} > 10^{-3} > 10^{-5}$
3. Concentration of enzyme
4. pH optimum
5. Temperature (most effect on rate)
6. Dialysis membrane

[a]A fast response is defined as a low response time.

crude enzyme, will also affect the speed of response of the electrode. However, compromises need to be made between the increase in enzyme activity and the concomitant increase of membrane thickness, the latter affecting the rate of electrode response. As the amount of enzyme is increased, a shorter response time is observed, until an optimum level is reached. Further increase in the amount of enzyme tends to diminish the response time. This effect is due to a thickening of the membrane layer by the use of more weight of biocatalyst, resulting in an increase in the time required for the substrate to diffuse through the membrane. Hence, the degree of purification of the enzyme must be considered. As a rule, it is recommended to operate with the highest possible enzyme activity to obtain thin membranes and to ensure rapid kinetics.

C. Selectivity

As a rule of thumb, any biosensor will be only as good as its selectivity; possible interferences in the electron sensor and in the biocatalyst reaction must be taken into account.

1. Interferences in the Electrode Probe

Ideally, the sensor used to detect the products of, or reactants in, the biocatalyzed reaction should not react with other substances present in the sample being assayed. Practically, this requirement is not always met using either potentiometric or amperometric methods. For example, immobilized urease electrode probes operating with a cation glass sensor measuring the NH_4^+ produced have been reported to be inadequate for blood and urine assays due to the fact that they respond also to Na^+ and K^+ (72,73). However, improvements in selectivity have been observed by using a glass electrode sensor (74) or even better by using a solid antibiotic nonactin electrode as the sensor (53). This electrode has a selectivity for NH_4^+/K^+ of 6.5 and for NH_4^+/Na^+ of 7.5×10^{-2}, thus partially eliminating the response to these ions by the sensor. Considerable selectivity improvement has been achieved by using an ammonia electrode. The enzyme layer is placed directly onto the gas electrode thus causing a large buildup of product at the electrode surface producing linear calibration plots at various pH values (37,75).

Amperometric devices using solid electrodes operate at selected potentials but are still subject to electroactive interfering substances. The most widely used sensor, due to its inherent sensitivity toward oxygen reduction and hydrogen peroxide catalytic oxidation, is the platinum electrode. Biased at positive potentials, some interferences are observed in the presence of easily oxidizable compounds like ascorbic acid, uric acid, acetaminophen, glutathione, cysteine, and bilirubin (97,134,136). Carbon-based electrodes offer several advantages with respect to low background currents and large available potential ranges, but they still require some surface modifications in order to achieve a greater selectivity.

2. Interferences with the Biocatalyst Reaction

Interferences with biocatalyst reactions fall into two classes: substrates that can catalyze the reaction in addition to the compound to be measured, and substances that either activate or inhibit the enzyme.

The selectivity of the enzyme probe is, of course, highly related to the degree of purity of the available enzyme. With some enzymes, such as urease, the only substrate that reacts at a reasonable rate is urease; hence, the urease-coated electrode is specific (73,74). Uricase, likewise, acts almost specifically for uric acid (76), and aspartase for aspartic acid (1,137). Others, such as penicillinase, react with a number of substrates; for example, ampicillin, naficillin, penicillin G, penicillin V, cyclibillin, and dicloxacillin can all be determined with a penicillinase electrode (77,78).

Similarly, D-amino acid oxidase and L-amino acid oxidase are less selective in their responses (79,80). The former in an electrode gives good response to D-phenylalanine, D-alanine, D-valine, D-methionine, D-leucine, D-norleucine, and D-isoleucine; the latter electrode senses L-leucine, L-tyrosine, L-phenylalanine, L-tryptophan, and L-methionine. Alcohol oxidase responds to methanol, ethanol, and allyl alcohol (81,82). Therefore, when using electrodes of these enzymes, a separation must be performed if two or more substrates are present, or, alternatively, a combination of enzymes can be advantageously used (82). In the case of L-amino acid assays, the use of individual decarboxylative enzymes, each of which acts specifically with a different amino acid, is an attractive alternative (52). Enzyme electrodes of this type are known for L-tyrosine, L-phenylalanine, L-tryptophan, and others. Glucose oxidase acts with a large number of glucides (83): Glucose and 2-deoxyglucose are the main substrates, but cellobiose and maltose also react, probably because of the presence of other hydrolytic enzymes in the glucose oxidase preparation.

The activity of the enzyme can also be strongly affected by the presence of certain compounds acting as inhibitors. Fluoride ions reversibly inhibit the enzyme urease (84) and oxalate ions inhibit lactate oxidase (85). The major enzyme inhibitors are heavy-metal ions, such as Ag^{+}, Hg^{2+}, Cu^{2+}, and sulfhydryl-reacting organic compounds (p-chloromercuribenzoate and phenylmercury (II) acetate), which block the free S-H groups of many active centers, especially oxidases (89,101). However, an immobilized enzyme is not influenced as strongly by such inhibitors, because of the protective action of the immobilization matrix.

D. Influence of pH

Influence of pH on the response of the immobilized enzyme electrode must be considered both for its effect on the immobilized system and for its effect on the detector performance. Every enzyme has a maximum pH at which it is most active and a certain range of pH in which it demonstrates reactivity. When the enzyme is immobilized, the optimal pH may shift depending on the nature of the carrier (1). According to the Nernst equation, the potential at the electrode at a given reaction may be governed by the pH of the solution layer. A number of electrode redox processes are known to be pH-dependent, therefore, a rigorous control of pH is required for optimal response. The apparent pH in the vicinity of the immobilized biocatalytic layer may differ from the solution pH (86), thus, a high buffer capacity is generally suggested to minimize this effect.

For the fastest and most sensitive responses, one should work at the optimum pH. This is not always possible, because the electrode may not respond optimally at the pH of the enzyme reaction. Thus, a compromise between these two factors is generally necessary. One should, however, be careful not to be trapped into forcing the pH of the biocatalyst system to conform with the pH requirements of the sensor (1).

E. Quantitative Determinations and Limits of Detection

All enzyme electrodes measure substrate in the concentration range of 10^{-2} to 10^{-4} M, or as low as 10^{-5} M (1). Generally, reaction curves approximately Nernstian in the linear range, with a slope close to 59.1 mV/decade, are obtained. All curves level off at high substrate concentrations, as predicted by the Michaelis-Menten equation, which states that the reaction becomes independent of substrate at high substrate concentration. A leveling off of the curve at low substrate concentration is observed also, due to the limit of detection of the electrode sensor used. Higher sensitivities may be achieved when using amperometric devices (10^{-6} M, 10^{-7} M), which consume the biocatalyst reaction product, thereby expanding the linear response range, and displaying a larger apparent Michaelis-Menten constant (Km) than the potentiometric probes.

VI. ANALYTICAL APPLICATIONS

A. Glucose and Glucide Electrodes

Among the vast number of enzyme-based electrodes, the ones devoted to glucose monitoring have received most attention. The first description of an enzyme electrode was dedicated to the determination of glucose, and since then a wide variety of glucose enzyme electrodes have been developed and applied with varying success to glucose determination.

The principle upon which the majority of enzyme glucose electrodes operate is based on the following reactions:

$$\beta\text{-D-glucose} + O_2 \xrightarrow{\text{GOD}} \text{D-glucono-}\delta\text{-lactone} + H_2O_2 \quad (14)$$

$$\text{D-glucono-}\delta\text{-lactone} + H_2O \longrightarrow \text{D-gluconic acid} \quad (15)$$

The oxidation of glucose is specifically catalyzed by the enzyme glucose oxidase (GOD) in the presence of oxygen as electron acceptor. To follow the reaction in Eq. 14, the probe must detect the formation of hydrogen peroxide or gluconic acid, or the depletion of oxygen dissolved in solution. Amperometric sensors monitoring the product (H_2O_2) or oxygen depletion have been preferentially applied, each type exhibiting performances and drawbacks inherent to their detection mode. As illustrated schematically in Figure 3, the construction of a glucose enzyme Clark type electrode is a rapid and simple technique, requiring just the entrapment of the enzyme

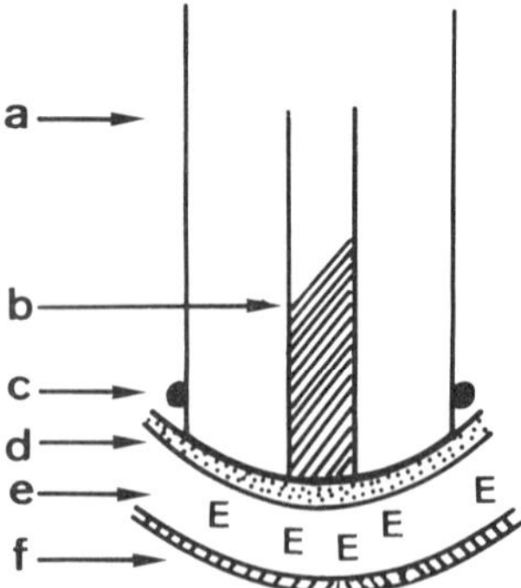

Figure 3 Cross-sectional view of an amperometric enzyme sensor (Clark-type). a) electrode body; b) working electrode; c) reference electrode; d) inner permselective membrane; e) enzyme layer; f) outer membrane.

between an inner permselective membrane and an outer membrane (collagen, polycarbonate, cellulose acetate, or polyurethane) cast at the tip of the amperometric probe. When the probe is immersed in a solution of glucose, the latter diffuses through the outer membrane into the enzyme layer and the reactions shown in Eqs. 14 and 15 proceed. The inner permselective membrane, which must be extremely thin (1–2 μm) to allow rapid response, is selected depending on the nature of the product to be detected. An inner hydrophobic membrane (polyethylene) allows the selective permeation of oxygen to the platinum cathode. The consumption of oxygen in the enzyme layer is monitored amperometrically providing a measurement of glucose concentration. A cellulose acetate or polyurethane membrane allows selective diffusion of H_2O_2 towards the platinum anode, where it undergoes oxidation; the current produced is directly proportional to the H_2O_2 generated by the enzymatic reaction and, in turn, to the glucose concentration in the sample.

The permselectivity towards products or reagents of the enzyme reaction is the most important function of the membrane of an enzyme electrode. For example, in glucose electrodes, the membrane covering the probe needs to be permeable to hydrogen peroxide or oxygen, but must also exclude other electrically active substances such as uric acid, ascorbic acid, acetaminophen, phenols, etc. The immobilization of the enzyme constitutes the key factor in determining the stability and signal enhancement of the enzyme-based electrode. As already reported in the preceding sections, glucose oxidase may be immobilized over an electrode in many different ways (some of these are listed in Table 1). Although no universal technique has emerged, to date covalent enzyme immobilization procedures offer stabilities longer than several months and provide the greatest success in construction of practical probes. Several GOD-based electrodes have already been constructed and applied to monitor glucose in vivo or in complex media (Table 1 and references therein). Most of these are platinum-based electrodes poised at a selected potential to monitor either oxygen depletion (cathodic) or hydrogen peroxide production (anodic). The cathodic measurement of molecular oxygen concentration is of particular interest because the electrode response is generally not affected by interfering substances in contrast to the anodic monitoring of H_2O_2, where oxidizable substances may interfere (144). Although these devices operate successfully in solution, they often suffer difficulties when applied to glucose determination in whole blood, in vivo, or in fermentation broths where oxygen concentration is usually low and is subject to large variations. Electrode response fluctuation and nonlinear range of current responses at high glucose concentration are observed. To solve the oxygen deficit problem, redox mediators have been introduced as already discussed (105,110,134,148), but alternative solutions have also been suggested to increase oxygen concentration. For example,

1. Co-immobilization of catalase-glucose oxidase. The enzyme catalase destroys H_2O_2 as shown in Eq. 16 with liberation

$$H_2O_2 \xrightarrow{\text{catalase}} H_2O + \tfrac{1}{2}O_2 \tag{16}$$

   of oxygen, thus reducing total demand of oxygen by 50%. Moreover, under such conditions the enzyme GOD is protected from deactivation.
2. Production of oxygen within the biosensor by inserting a platinum wire around the probe acting as anode and electrolyticlaly generating oxygen from water (115).
3. Limiting glucose diffusion but increasing oxygen diffusion into the enzyme-electrode. The electrode design is of a particular shape, allowing axial diffusion of glucose and oxygen but only radial oxygen diffusion into the enzyme gel (108). Other electrode designs have been suggested where glucose diffusion is restricted but free diffusion of oxygen is maintained (91,113,119,120,151).

Less common than oxidases, glucose dehydrogenase-based electrodes have been developed to monitor potentiometrically (44) or amperometrically (152) glucose concentration. In most instances glucose dehydrogenase systems that are also dependent on the oxygen saturation of the sample and redox mediators (e.g., Meldola blue) have been incorporated (65,71,152). However, recently whole-blood glucose has been determined by using a glucose dehydrogenase possessing a pyrroloquinoline quinone molecule as a prosthetic group (153). The enzyme glucose dehydrogenase isolated from *Acinetobacter calcoaceticus* cannot use oxygen as an electron acceptor and does not require the presence of NAD or NADP. The enzyme catalyzes the oxidation of glucose using a variety of electron acceptors (such as phenazine methosulphate and 2,6-dichlorophenolindophenol).

A serious limitation either in in vivo glucose monitoring or in extracorporeal flow systems is the rapid loss of response due to a coating of the electrode membrane by blood constituents (platelets and fibers). Several attempts have been made to minimize these undesirable effects, e.g., by using repellent membranes, either negatively charged (125) or hydrophobically treated (92,113,147), by using antiplatelet agents for immobilizing the enzyme on cellulose acetate membranes containing heparin (155), or by reducing the flow stream towards the electrode (92), thus diminishing platelet deposition.

Galactose has been determined by immobilizing the enzyme galactose oxidase at an oxygen electrode (157), or at a micro-platinum electrode operating anodically to monitor H_2O_2 formation (see Eq. 15) (145,158). The enzyme was immobilized onto collagen membranes (124) or cellulose acetate membranes and was applied selectively to plasma and whole blood determinations (145,159).

$$\text{D-Galactose} + O_2 \xrightarrow[\text{oxidase}]{\text{galactose}} \text{galactohexodialdose} + H_2O_2 \qquad (17)$$

Maltose determination has been realized by using several chemically activated collagen membranes and by testing three different modes of enzyme co-immobilization: asymmetric coupling, random co-immobilization, and two membranes placed one on the other (124). The two enzymes involved were glucoamylase and glucose oxidase. Glucoamylase was immobilized on the membrane face exposed to the bulk into which the maltose-containing samples were injected. The hydrolysis of maltose occurred according to the reaction:

$$\text{Maltose} + H_2O_2 \xrightarrow{\text{glucoamylase}} \text{2-D-glucose} \qquad (18)$$

The glucose produced migrates through the membrane and is then oxidized on the inner face with immobilized GOD in close contact to the platinum anode. Extended calibration curves were observed, from 2×10^{-7} to 4×10^{-3} M.

The enzymes glucoamylase and glucose oxidase have also been embedded in polyacrylamide gels and maintained over an oxygen electrode to monitor maltose. The resulting device was covered by a polyurethane foil to avoid starch interferences increasing the linear concentration range up to 250 mg/dl (99,160).

A three-enzyme-based electrode has been described for the determination of sucrose (124,161). The enzymes invertase, mutarotase, and glucose oxidase were utilized in the following reaction sequence:

$$\text{Sucrose} + H_2O \xrightarrow{\text{invertase}} \alpha\text{-D-glucose} + \text{fructose} \qquad (19)$$

$$\downarrow \text{ mutarotase}$$

$$\beta\text{-D-Gluconolactone} + H_2O_2 \xleftarrow[\text{GOD}]{O_2} \beta\text{-D-glucose} \qquad (20)$$

Asymmetrical coupling (invertase and mutarotase co-immobilized on one face and glucose oxidase on the other) was found to improve the electrode response. The sensors could be used for hundreds of assays with a linearity of approximately 1×10^{-4} to 2×10^{-3} M. Lactose electrodes have been constructed by asymmetric coupling of β-galactosidase and glucose oxidase (124).

Lactose, sucrose, galactose, and glucose probes have also been constructed by immobilizing the corresponding enzymes to nylon net which was then used to cover an oxygen probe. The resulting single or multienzyme devices exhibited easy handling and high activities (118).

Flow injection systems have been successfully utilized for galactose determination in urine and lactose in milk (162). The enzymes galactose oxidase and peroxidase were immobilized in a reactor, and the H_2O_2 formed was amperometrically detected in the presence of a mediator.

B. Urea Electrodes

Like glucose, urea has been the subject of numerous investigations, mainly with immobilized potentiometric gas selective electrodes which monitor the ammonia formation in the reaction:

$$(NH_2)_2CO + 2H_2O + H^+ \xrightarrow{\text{urease}} HCO_3^- + \begin{array}{c} 2NH_4^+ \\ H^+ \upharpoonleft\downharpoonright OH^- \\ NH_3 \end{array} \tag{21}$$

Urease has been immobilized in a polyacrylamide matrix and placed over a cation-selective electrode which responds to NH_4^+ (73). Because sodium and potassium interfere with the ammonium sensor, improvements have been made by using a silicon rubber-based nonactin ammonium electrode (53) or by working in the differential mode with a combination uncoated ammonium electrode (74,163). Further improvement in the selectivity of urea determination has been achieved by replacing the ammonium sensor by an ammonia gas-sensitive electrode, allowing only gaseous species to diffuse through. Urease has been directly polymerized onto the surface of an ammonia gas electrode probe by means of glutaraldehyde (75). Sufficient NH_3 was produced in the enzyme reaction layer, even at pH values as low as 7–8, to allow direct and selective urea determination. Still better, totally interference-free, direct reading electrodes for urea have been prepared by chemically attaching a thin enzyme layer to polyacrylic acid, and performing measurements at a solution pH of 8.5, where good enzyme activity was still obtained. An air gap NH_3 electrode was used to yield sensitive measurements with linear ranges of 30 mM to 0.5 μM (164). The electrode could be used for blood serum analyses continuously for up to 1 month (500 samples) with a precision and accuracy variation of less than 2%. Improved NH_3 electrodes have been reported by Mascini and Guilbault (37), by immobilizing the enzyme urease with BSA and glutaraldehyde on a new Teflon membrane.

Immobilization of urease by two different techniques has been investigated and the methods compared for urea determination in urine (165). The enzyme has been physically entrapped in a gel (agar-agar) or chemically immobilized onto a cellulose acetate membrane with glutaraldehyde. Chemical immobilization requires more enzyme for optimal response due to some loss of enzyme activity, but the stability is superior to the physically entrapped enzyme. Urea in urine samples has been determined, and results compared favorably with a spectrophotometric method. The sensor of the enzyme probe was a NH_3 gas electrode. Free ammonia interferes with the assay and must first be determined separately. From 200 to 1000 assays can be performed on one electrode with a C.V. of 2.5% over the range of 0.5–10 mM. At least 20 assays/h on serum samples can be performed using the electrode yielding results with excellent correlation to results obtained by the spectrophotometric diacetyl procedure.

Finally, urea has been determined in diluted plasma (31), urine, and serum samples (32) by covering the tip of glass pH electrodes with a thin layer of immobilized urease physically entrapped in polymer gels. More recently antimony pH-sensitive electrodes, covered with a chemically immobilized urease membrane, have been described (171).

C. Alcohol and Acid Electrodes

1. Ethanol

The oxidation of lower primary aliphatic alcohols may be catalyzed by alcohol oxidase:

$$RCH_2OH + O_2 \xrightarrow[\text{oxidase}]{\text{alcohol}} RCHO + H_2O_2 \qquad (22)$$

The hydrogen peroxide production or the O_2 depletion may be detected amperometrically as in the determination of glucose above. Ethanol has been determined using 1 ml of sample over the range of 0–10 mg/100 ml by poising the electrode anodically (81). The technique should be adequate for clinical determination of blood ethanol, but methanol is a serious interference. However, by setting the working potential cathodically at −0.6 V vs. S.C.E., O_2 depletion is measured, and 0.4 to 50 mg% of ethanol can be assayed with little interference (166).

Electrochemical sensors for ethanol, lactate, and malate were designed by constraining a dehydrogenase enzyme and NAD^+ onto the surface of a platinum electrode (167). The NAD^+ is converted to NADH by oxidation of the substrates catalyzed by their respective dehydrogenases, and the NADH formed is detected anodically giving a current dependent on substrate concentration:

$$NAD^+ + \text{substrate} \xrightarrow{\text{substrate dehydrogenases}} NADH + \text{oxidized substrate} + H^+ \tag{23}$$

$$NADH \longrightarrow NAD^+ + 2e^- + H^+ \tag{24}$$

An acetylated dialysis membrane holds both enzyme and NAD^+, exhibiting low permeability to NAD^+. However, the membrane restricted permeability to the substrates also. Ethanol electrodes were applied to the analysis of plasma ethanol, resulting in excellent accuracy and precision in the physiologically significant range of ethanol concentrations. Electrochemically activated glassy carbon electrodes have been found to possess catalytic activity towards NADH oxidation at potentials as low as 0 to + 0.2 V versus Ag/AgCl (pH 7.0) and offers advantages over platinum electrodes that need to be maintained at high positive potentials (168). The electrodes have been successfully applied for lactate, pyruvate, ethanol, and glucose-6-phosphate determinations by using the respective dehydrogenase enzymes (169). The specificity of the enzyme alcohol oxidase towards alcohol is poor, but recently a device has been constructed that is absolutely specific for ethanol (82). This was done by using yeast alcohol dehydrogenase (ADH) in addition to alcohol oxidase (AO). The unique specificity of these two enzymes create a window through which only ethanol can pass.

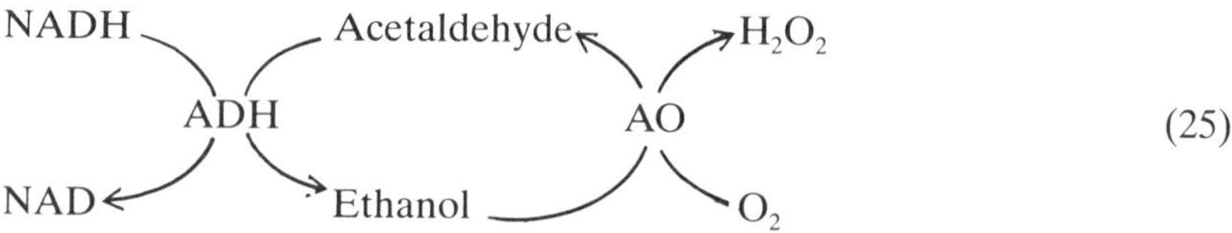

(25)

As illustrated in Eq. 25, the enzyme AO catalyzes the oxidation of ethanol to acetaldehyde; the presence of ADH will quickly reform ethanol from acetaldehyde. The oxygen uptake is measured, and, in the process, methanol or isopropyl alcohol do not recycle. The enzyme ADH has been applied in an interesting example of a bioelectrochemical fuel cell. Indeed, the quinoprotein ADH utilized in the presence of a stable redox mediator, N,N,N′,N′-tetramethyl-4-phenylenediamine, in an appropriate bioelectrochemical cell, allows the conversion of alcohol substrates to power in a fairly efficient way. Moreover, the enzyme has been shown to possess selectivity towards straight chain primary alcohols, like methanol, decan-1-ol, 1,4-butanediol, and 1,3-propanediol. Secondary or tertiary alcohols yield no current. The device allows quantitative measurements of alcohols at low nmol levels when operating in a coulometric mode (173).

2. Lactate

The need for a rapid and sensitive method for lactate determination in biological samples has contributed to the development of efficient lactate selective enzyme-based electrodes (see Refs. 85,174,175 and refs. cited therein).

Actually, several enzymes utilizing lactate as substrate have been immobilized over the tip of electrodes, operating either amperometrically or potentiometrically. In the former mode, depletion of oxygen or production of hydrogen peroxide can be monitored on the basis of the reaction catalyzed by the enzyme lactate oxidase (LOD).

$$\text{Lactate} + O_2 \xrightleftharpoons{\text{LOD}} \text{pyruvate} + H_2O_2 \tag{26}$$

Reagentless meaurements have been performed using oxygen-based electrodes in the presence of the enzyme lactate 2-monooxygenase (LOD), catalyzing the decarboxylation of lactate (38):

$$\text{Lactate} + O_2 \xrightarrow{\text{LOD(EC 1.13.12.4)}} \text{acetate} + CO_2 + H_2O \tag{27}$$

Alternatively, the oxidation of lactate can be carried out by the NAD^+ cofactor of lactate dehydrogenase (LDH):

$$\text{Lactate} + NAD^+ \xrightleftharpoons{\text{LDH(EC.1.1.1.27)}} \text{pyruvate} + NADH \tag{28}$$

the NADH formed being detected anodically (167,168,184). Moreover, cytochrome type LDH can be applied to follow the oxidation of lactate by ferricyanide (176) or ferriferrocytochrome C (175):

$$\text{Lactate} + 2\,Fe(CN)_6^{3-} \xrightleftharpoons{\text{LDH(EC1.1.2.3)}} \text{pyruvate} + 2\,Fe(CN)_6^{4-} + 2H^+ \tag{29}$$

The ferrocyanide formed is detected amperometrically.

The potentiometric mode of detection is also based on Eq. 29 and the change in the concentration ratio of ferricyanide to ferrocyanide in the enzyme layer is monitored (177).

Various immobilization techniques have been utilized for LOD and LDH. The enzymes have been physically entrapped (gelatin (85,177), polyacrylamide (85), poly(vinyl alcohol) (174)), sandwiched between membranes, (175,180) chemically immobilized (glutaraldehyde-BSA (181), nylon nets (38,39), cellulose acetate membranes (174,178,179), and polyamide membranes (183)) or linked directly onto the electrode surface (168,182).

In an attempt to improve biosensor sensitivity, several strategies have been employed: (1) addition of a redox mediator, such as phenazine methosulfate (139,185), (2) use of enzymes with high specific activities (38,179), or 3) attachment of the enzyme (LDH) onto the electrode surface (168,182). The sensitivities generally achieved allow lactate determination in the range of concentrations between 5 μM and 5 mM. However, recently, in an effort to produce improved sensors, high gains in sensitivity have been achieved using the amplification principle (6,174,184):

$$\text{Lactate} + \text{OX} \xrightleftharpoons{\text{LOD or LDH}} \text{pyruvate} + \text{RED} \tag{30}$$

$$\text{Pyruvate} + \text{NADH} + \text{H}^+ \xrightleftharpoons{\text{LDH}} \text{lactate} + \text{NAD}^+ \tag{31}$$

Amplification factors of 10 (6), 60 (184), and even 250 (under optimal conditions) (174) have been reported, and applications have been described for microsamples of biological fluids. A second way of increasing the enzyme electrode performance has been demonstrated recently using LOD immobilized on commercially available preactivated membranes, like polyamide (183). These membranes, covering an amperometric probe detecting hydrogen peroxide, allow fast responses (between 1 and 2 min), and the resulting biosensor exhibits accurate lactate measurements within the linear range of 0.25 μM to 0.25 mM. The concentration of lactate in sera, different kinds of dairy products (cream cheese, plain yogurt, sweet and acid reconstituted whey from powder), red wine, white wine, and sauerkraut was analyzed using both the amperometric procedure and conventional spectrophotometric methods (183).

3. Pyruvate

The determination of pyruvate can be successfully carried out using the LDH enzyme electrode and following electrochemically the NADH consumption or NAD^+ formation (Eq. 27) (69,169). Recently, the construction and characterization of an enzyme-based microelectrode for pyruvate measurements in cerebrospinal fluids (CSF) has been reported (69). Pyrolytic carbon microelectrodes have been subjected to alternating potentials (triangular wave form from 0.0 to 2.6 V vs. Ag/AgCl) in the presence of BSA, followed by surface activation (cycling between −0.8 and + 1.5 V). The resulting electrodes, covered with a protein sheeth, are dipped into a LDH solution in 50% glycerol for irreversible enzyme immobilization. Fast electrochemical responses towards NADH oxidation were observed with these electrodes. Very small volumes of CSF could be analyzed (50 μl) and the decrease of NADH concentration was linearly related to pyruvate in the range from 10 μM to 2 mM.

The enzyme pyruvate oxidase has also been utilized for pyruvate determination (187):

$$\text{Pyruvate} + H_3PO_4 + O_2 \rightarrow CH_3OPO_3H_2 + \text{acetate} + CO_2 + H_2O_2 \tag{32}$$

The assay is based on monitoring the decrease in dissolved oxygen. The physically entrapped enzyme requires thiamine pyrophosphate, calcium ions, and FAD for efficient activity. A linear relationship between pyruvate concentration from 0.06 to 0.8 Mm and the current decrease was observed at pH 7.05 and 7.35.

4. Urate

The enzyme electrodes developed for uric acid determination are based on the monitoring of reactants or products of the uricase catalyzed reaction:

$$\text{Uric acid} + O_2 + 2H_2O \xrightarrow{\text{uricase}} \text{allantoin} + H_2O_2 + CO_2 \tag{33}$$

Detection of hydrogen peroxide from this reaction has been studied using platinum electrodes (76,189) but it was found that at the positive potential required (+ 0.6 V vs. S.C.E.) to oxidize the peroxide, uric acid oxidation occurs at the electrode surface, precluding selective measurements (76). Thus, in order to improve the uricase electrode selectivity, bienzyme amperometric devices in the presence of a redox mediator (ferrocyanide) have been used (110). The enzymes uricase and peroxidase were immobilized together, and ferricyanide produced measured cathodically at 0.0V vs. Ag/AgCl:

$$Fe(CN)_6^{4-} + H_2O_2 + 2H^+ \xrightarrow{\text{peroxidase}} 2H_2O + Fe(CN)_6^{3-} \tag{34}$$

The monitoring of oxygen depletion is preferentially effected, and successful results have been obtained by immobilizing a layer of glutaraldehyde-bound uricase over the tip of a platinum electrode, and covering the biocatalyst with a thin layer of dialysis membrane. The uptake of oxygen in the solution, due to the enzymatic reaction, was measured by applying a potential of −0.6 V vs. SCE. Linear calibration curves were obtained between 10 μM and 0.1 M uric acid. Alternatively, a third mode of uricase-catalyzed reaction monitoring is the detection of liberated carbon dioxide (188,199). The characteristics of a CO_2-uricase immobilized electrode have been established with regard to the mode of enzyme insolubilization (collagen and glutaraldehyde matrices), response sensitivity, and response time. After 10 days of activity, a significant decrease of the response was observed, but the initial performance level could be recovered by dipping the membranes in Cu^{2+} solutions (188).

5. Salicylate

Salicylate has been determined in blood serum using an immobilized salicylate hydroxylase (36,190). The enzyme catalyzes the formation of catechol and carbon dioxide from salicylate and reduced pyridine nucleotide in the presence of flavine adenine dinucleotide as cofactor:

$$\text{C}_6\text{H}_4(\text{COOH})(\text{OH}) + \text{NADH} + \text{H}^+ + \text{O}_2 \xrightarrow[\text{hydroxylase}]{\text{salicylate}} \text{C}_6\text{H}_4(\text{OH})_2 + \text{NAD}^+ + \text{H}_2\text{O} + \text{CO}_2 \qquad (35)$$

Either the CO_2 formation can be followed potentiometrically (190) or the O_2 consumption can be measured amperometrically at an oxygen electrode (191). For the former the enzyme has been physically immobilized with a dialysis membrane. The response is linear in the range of 5–300 μg/ml of salicylate concentration. The second technique consists of chemical immobilization (glutaraldehyde + BSA) of the enzyme onto a pig intestine mounted on the tip of the O_2 electrode. Samples containing 1.0×10^{-5}–1.87×10^{-3} M (1.6–300 μg/ml) salicylate were assayed with a coefficient of variation (C.V.) between 1.3 and 6.0 %, and recoveries were between 98.7 and 103%.

6. Oxalate

Oxalate has been determined either by using oxalate decarboxylase (192–194) or oxalate oxidase (36) immobilized electrodes. In the former, the CO_2 liberated and detected by a CO_2 electrode is proportional to the logarithm of the oxalate concentration. Linearity has been reported in the range from 0.2 to 10 mM and the electrodes were stable longer than one month. Human control urine samples, spiked with oxalate aliquots, have been analyzed with a recovery of 97.7% and a C.V. of 4.5% (192). Oxalate oxidase has been immobilized onto an O_2 electrode (36), an amperometric H_2O_2 sensor (192), or a potentiometric CO_2 probe (192) on the basis of the following reaction:

$$\text{Oxalic acid} + \text{O}_2 \xrightarrow{\text{oxalate oxidase}} 2\,\text{CO}_2 + \text{H}_2\text{O}_2 \qquad (36)$$

Oxalic acid has been directly determined in urine samples with recoveries ranging between 97.6–105.0% (36) or in a 1:40 diluted sample with a recovery of 95.5% and a C.V. of 4.9% (192). The electrodes were stable over periods longer than 3 weeks.

7. Ascorbate

Besides the construction of ascorbate electrodes for ascorbic acid determination in food (196), fruit juices (197), or pharmaceutical preparations (198), the enzyme ascorbate oxidase has been advantageously utilized in combination with in vivo microelectrodes (195). In the latter technique, the enzyme layer is confined over a micro carbon paste electrode and constitutes a barrier consuming ascorbic acid. Indeed, the in vivo device, dedicated to monitoring of brain catechols, must be free of easily oxidizable species (ascorbic acid) which interfere in endogenous catechol determination. Ascorbate oxidase catalyzes the reaction:

$$\text{L-Ascorbic acid} + \tfrac{1}{2}O_2 \longrightarrow \text{dehydroascorbic acid} + H_2O \qquad (37)$$

and the enzyme can be immobilized over a Clark type oxygen electrode. The biocatalyst has been immobilized on polyamide nets using glutaraldehyde (197), cross-linked with collagen-glutaraldehyde (196) or albumin-glutaraldehyde (198), or linked to cellulose acetate membranes (198). A comparative study has been conducted using several modes of biocatalyst immobilization: a soluble form, chemically linked matrices, a peel of cucumber, and living bacterial cells (198). Chemical immobilization offers the greatest stability (16–18 days) with linear ranges from 4μM to 0.7 mM. D-Isoascorbic acid, glucose, and fructose interfere in the determination (196,198).

D. Creatine and Creatinine Electrodes

Creatinine is the degradation product of creatine. The molecule has been determined in serum using the Jaffe reaction with formation of creatinine-picrate, the reaction being monitored potentiometrically with a picrate electrode. A prior separation step was required in order to remove interfering substances (202). An improved creatinine sensor has been realized by immobilizing creatinine deiminase at an ammonia gas sensor. Direct reading of creatinine in the concentration range from 1 to 100 mg% is possible (200). A multienzyme electrode has been constructed by coimmobilizing creatinine amidohydrolase (CA), creatinine amidinohydrolase (CI), and sarcosine oxidase (SO) onto a cellulose acetate membrane (203). The system responds linearly up to 100 mg of creatinine and creatine per liter in human serum, with a detection limit of 1 mg/liter. Only 25 μl of serum sample is required for measurement. More than 500 assays have been carried out, and the enzyme is very stable if stored at 4°C in air. Sensors have been developed which eliminate the interference of endogenous ammonia during the determination of creatinine in blood serum and urine samples (204,205,208). The biocatalyst layer consists of a bienzyme system immobilized over an ammonia electrode catalyzing the following reactions:

$$NH_4^+ + \alpha\text{-ketoglutarate} + NADH \overset{\text{G1DH}}{=\!=\!=} \text{glutamate} + NAD^+ + H_2O \tag{38}$$

$$\text{Creatinine} + H_2O \xrightarrow[\text{deiminase}]{\text{creatinine}} N\text{-methylhydantoin} + NH_3 \tag{39}$$

Endogenous ammonia is converted to a nonresponsive product (Eq. 38); creatinine does not react with G1DH and is converted to ammonia when it reaches the immobilized creatinine deiminase (Eq. 39). A more sophisticated device, an oxygen electrode with immobilized nitrifying bacteria and immobilized deiminase membrane, has been constructed for creatinine detection (206). The ammonia released in the reaction (Eq. 39) is oxidized to nitrite and nitrate by nitrifying bacteria, and the consumption of oxygen is monitored by an oxygen electrode. Interestingly, the device has been optimized with regard to sensitivity by developing a new immobilized creatinine deiminase membrane. The enzyme was attached to poly(trichloroethylglutamate) in the organic solvent DMSO, with little loss of enzyme activity (207).

E. Cholesterol Electrodes

Cholesterol can be determined by immobilizing the enzyme cholesterol oxidase in a layer over an oxygen electrode or hydrogen peroxide electrode. Cholesterol oxidase has been chemically immobilized onto nylon net (118,209) or collagen (210) membranes, and also fixed onto a pO_2 electrode:

$$\text{Cholesterol} + O_2 \xrightarrow[\text{oxidase}]{\text{cholesterol}} \Delta 4\text{-cholestenone-3} + H_2O_2 \tag{40}$$

The oxygen consumption is monitored amperometrically; the current decrease is linearly related to the cholesterol concentration. Several applications have been realized for the microdetermination of human bile cholesterol (209) and serum cholesterol (210–215) using the bienzyme system cholesterol oxidase–cholesterol esterase:

$$\text{Cholesterol esters} \xrightarrow[\text{esterase}]{\text{cholesterol}} \text{cholesterol} \tag{41}$$

Alternatively, the hydrogen peroxide detection mode has permitted measurement of total serum cholesterol in a highly specific and routine manner using the cholesterol bienzyme system. The chemical immobilization of the enzymes onto alkylamine glass beads provided a stable enzyme stirrer to completely convert all the total cholesterol esters into first cholesterol (Eq. 41), then hydrogen peroxide (Eq. 40), which is monitored by the current flow at a Pt electrode (216).

Cholesterol has been determined in small plasma samples by means of a multiple enzyme system utilizing cholesterol ester hydrolase and cholesterol oxidase with anodic detection of H_2O_2 at a Pt sensor (217). The former enzyme is soluble and cholesterol oxidase is chemically bound to a collagen membrane. The recovery of cholesterol was 97% and the C.V. 5%, with a range of 25–300 mg% cholesterol.

Likewise, cholesterol oxidase has been immobilized under mild conditions on collagen films and the H_2O_2 liberated by its activity is detected at a platinum electrode (124,220). The linear range of cholesterol determination is 0.1 μM to 8 mM, and the sensitivity is higher than reported earlier (216). More than 150 experiments have been performed without significant decline in sensitivity. Human serum free cholesterol has been determined, but the use of a nonenzymatic electrode is required to substract out interfering oxidizable species (ascorbate, urate, tyrosine, etc.).

A rapid (one-minute) electrochemical assay for cholesterol in biological materials has also been described (218). Ten microliters of sample are injected into a 50°C thermostated cuvet containing soluble cholesterol oxidase and esterase. The H_2O_2 produced is anodically detected at a platinum anode covered with an acetate/polycarbonate membrane, which prevents ascorbic acid, uric acid, or bilirubin from being detected. The linearity is 100–500 mg/dl. The enzymes cholesterol oxidase and cholesterol ester hydrolase have been immobilized in a modified 2-hydroxyethylmethacrylate gel and maintained over a platinum electrode operating at +0.6 V vs. Ag/AgCl. Free cholesterol in serum samples has been measured as well as total cholesterol using this bienzyme sequence electrode (219).

F. Protein, Amino Acid, and Amine Electrodes

D-Amino acid and L-amino acid selective electrodes, operating on the basis of the biocatalytic oxidation of amino acids, have been described:

$$\text{Amino acid} + H_2O + O_2 \xrightarrow{\text{L-AAO}} \text{R-CO-COOH} + NH_4^+ + H_2O_2 \quad (42)$$

L-amino acid enxymes immobilized onto pO_2 electrodes have been developed for L-methionine, L-leucine, L-phenylalanine, L-tyrosine, L-cysteine,

L-lysine, and L-isoleucine (221). The stability was longer than 4 months, the response time less than one minute, and greater selectivity over the peroxide-based sensors was observed (222). Kinetic measurements of hydrogen peroxide formation (Eq. 42) have resulted in fast electrode responses (less than 12 sec) using chemically bound L-amino acid oxidase (LAAO) covering a platinum electrode (80). The L-amino acids cysteine, leucine, tyrosine, phenylalanine, tryptophane, and methionine were assayed.

Potentiometric selective electrodes have been constructed by immobilizing L-AAO on chemically modified graphite electrodes (43). Interaction of hydrogen peroxide with carbon surface groups appears to be the major contributor to the potentiometric response. Several amino acid electrodes monitoring the NH_4^+ ion have also been reported (112,223–226).

L-Glutamate has been determined using the enzyme glutamate decarboxylase in combination with a pCO_2 electrode (201). But the most promising results with regard to selectivity and sensitivity were obtained with the dehydrogenase enzyme (64,71,185). The systems developed consist of the coupling of the glutamate dehydrogenase to redox mediators 3-β-naphthoyl-Nile-Blue (71) and Meldola Blue (64) adsorbed on graphite electrodes and utilizing an oxygen-independent catalyzed electrochemical NADH oxidation. Considerable sensitivity enhancement was obtained using the glutamate dehydrogenase (G1DH)-mediator-enzyme substrate amplification technique (64).

$$\text{Glutamate} + \text{NAD}^+ + \text{H}_2\text{O} \underset{}{\overset{\text{G1 DH}}{\rightleftharpoons}} \alpha\text{-ketoglutarate} + \text{NH}_4^+ + \text{NADH} \tag{43}$$

$$\alpha\text{-Ketoglutarate} + \text{alanine} \underset{\text{transaminase (GPT)}}{\overset{\text{glutamate pyruvate}}{\rightleftharpoons}} \text{glutamate} + \text{pyruvate} \tag{44}$$

In Eqs. 43 and 44 glutamate is recycled; thus small amounts of glutamate will produce a buildup of NADH in the reaction.

L-Tyrosine determination in biological fluids has been conducted by immobilizing apo-L-tyrosine decarboxylase over a CO_2-sensitive gas electrode (227).

L-Lysine has been determined in grains and foodstuffs with a L-lysine decarboxylase (purified from *E. coli*) mounted onto a CO_2 electrode (228). The enzyme immobilization technique, BSA-glutaraldehyde, has been optimized by varying the amounts of enzyme, albumin, and glutaraldehyde. Electrode stability is longer than 40 days, it exhibits a response time of 5–10 min, and a linear range of L-lysine concentration of 50 μM to 0.1 M.

L-Lysine and L-arginine have been determined in a rapid fashion by using a bienzyme immobilized system of decarboxylase and diamine oxidase:

$$R\text{-}CHNH_2COOH \xrightarrow{\text{decarboxylase}} RCH_2NH_2 + CO_2 \quad (45)$$

$$RCH_2NH_2 + O_2 + H_2O \xrightarrow{\text{amine oxidase}} RCHO + NH_3 + H_2O_2 \quad (46)$$

The uptake of oxygen was linearly related to the substrate concentration in the 10–100 mM region (229). L-lysine has also been monitored in a fermenter by using a L-lysine oxidase electrode operating in a continuous flow system. The enzyme was immobilized directly onto the selective gas membrane (a pO_2) by copolymerization with gelatin using glutaraldehyde (172). L-Arginine, L-phenylalanine, and L-ornithine interfere in the determination. Highly selective L-arginine sensors have been described, where the ammonia liberated in the reaction sequence catalyzed by the enzymes arginase-urease is potentiometrically monitored. The probes exhibit linear responses in the range of L-arginine concentrations of 0.1 mM to 0.01 M (35) and 30 μM to 3 mM (230).

Several sensors for L-alanine have been described by immobilizing L-alanine dehydrogenase over an ammonia gas-sensing electrode (231) or over a pO_2 sensor (71). The enzyme specifically catalyzes the deamination of alanine in the presence of the coenzyme NAD^+:

$$\text{L-Alanine} + NAD^+ + H_2O \rightarrow \text{pyruvate} + NADH + NH_4^+ \quad (47)$$

The co-immobilization of a second enzyme, NADH oxidase (71) or lactate dehydrogenase (231), permits the regeneration of NAD^+. Measurements may be made in Tris-HC1 or carbonate buffers, but borate and glycine buffers inhibit L-alanine dehydrogenase (232).

Highly selective L-histidine electrodes have been constructed by immobilizing either purified histidine ammonia-lyase (from *Pseudomonas* sp.) on an ammonia gas-sensing electrode (233) or histidine decarboxylase (extracted from *Lactobacillus* 30a) on a carbon dioxide sensor (234). The enzymes were either chemically immobilized via the cross-linking glutaraldehyde-BSA (233,234) method or maintained in a soluble form between the gas membrane and a dialysis membrane (233). Optimization of the sensors was realized by utilizing the reduced form of the enzyme, in the presence of glutathione and Mn^{2+} and studying the influence of inhibitors and activators (233). Accurate determinations of histidine in urine samples, with good correlation to a fluorimetric procedure, have been reported (234).

A potentiometric L-threonine selective sensor has been constructed for the determination of L-threonine in biological fluids and foods. The enzyme threonine deaminase (from *B. stearothermophilus*) was utilized in conjunction with an NH_3 gas-sensing electrode. The bioprobe exhibits a linear response to the L-threonine concentration over the 0.1–200 mM range (235).

L-Tryptophan has been determined in a comparative manner by using both crude extract tryptophanase from *Escherichia coli* and the living bacterial cells and holding the extract or cells over an ammonia selective electrode (20). Physical immobilization in agar gel and chemical cross-linking with glutaraldehyde were found not to improve the stability longer than the simple use of the crude enzyme extract. The electrodes operate in the presence of the cofactor pyridoxal phosphate in phosphate buffers. Stabilities were less than 5 days for the enzyme electrode and approximately 3 weeks for the bacterial electrode. Several compounds may interfere with the assays, namely cysteine, serine, tryptamine, and 5-hydroxy tryptophan.

A very specific tyrosinase electrode has recently been reported for total serum protein determination (236). The assay consists first of hydrolysis of the protein in serum with pepsin and the liberated tyrosine is subsequently determined. According to the following:

$$\text{Tyrosine} + O_2 \xrightarrow{\text{tyrosinase}} \text{dopaquinone} + H_2O_2 \qquad (48)$$

the enzyme tyrosinase has been immobilized onto small intestine with BSA-glutaraldehyde, and the resulting enzyme layer cast over a platinum electrode operating cathodically to monitor the O_2 uptake. The electrode useful lifetime was from 15 to 20 days with 10–20 assays/day, the limit of detection was 10 μM, and the linearity extended up to 0.25 mM. The method is accurate and reliable and requires no tedious sample treatment in contrast to the fluorimetric technique.

Total protein measurements in a flow system have also been described (237). The protein is first cleaved in a reactor by protease, and the liberated amino acids (AA) are converted to NH_3 by passage over glass-immobilized LAAO (Eqs. 49–51). Several enzyme immobilization modes on controlled pore glass have been investigated using an ammonia gas-sensing electrode. BSA was determined in the range of 0.1 to 100 μg/m1. The results achieved were reported to be superior to those obtained with determination of liberated sulfide (238).

$$\text{Protein} \xrightarrow{\text{hydrolysis}} \text{LAA} \qquad (49)$$

$$\text{LAA} \xrightarrow{\text{LAAO}} NH_4^+ + H_2O_2 \qquad (50)$$

$$NH_4^+ + OH^- \rightarrow NH_3 + H_2O \qquad (51)$$

Glutamine has been selectively determined using a glutamine selective electrode (239). A comparison was carried out between several biocatalysts (isolated enzymes, bacterial cells, kidney mitochondrial fractions) immobilized at an ammonia gas-sensing electrode. Quantification of glutamine in cerebrospinal fluid controls has also been reported (240).

G. Choline and Choline Esters Electrodes

Phosphatidyl choline can be determined in serum samples using a bienzymic immobilized platinum electrode operating amperometrically at +0.60 V vs. S.C.E. (241). The enzymes phospholipase D and choline oxidase catalyze the following reactions:

$$\text{Phosphatidyl choline} \xrightarrow{\text{phospholipase D}} \text{phosphatydic acid} + \text{choline} \qquad (52)$$

$$\text{Choline} + 2O_2 + H_2O \xrightarrow{\text{choline oxidase}} \text{betaine} + 2H_2O_2 \qquad (53)$$

Various immobilization techniques have been tested in order to optimize the electrode performances.

Acetylcholine and choline have been determined in the concentration range of 1–10 μM in 0.1–1 ml of sample with a peroxide sensor (platinum electrode at +0.6 V vs. Ag/AgCl) (242). The enzymes acetylcholinesterase and choline oxidase were immobilized on nylon net. Stability of the electrodes was longer than one month. Potentiometric determination of choline by immobilizing choline oxidase and peroxidase with BSA-glutaraldehyde on a platinum plate electrode silanized with (γ-aminopropyl) triethoxysilane has also been realized (129). The electrode response time was 10 sec, and its stability was longer than 2 months.

VII. IMMUNOELECTRODES

The use of enzyme labels and enzyme-associated labels in immunoassay systems has already provided specific and sensitive substrate measurements (see Ref. 243 and references cited therein). In fact, the linkage of an enzyme directly to an antigen or antibody, or the direct binding of the antigen or antibody to a carrier, such as glass or collagen, can be effected as easily as the binding of an enzyme (243). The enzyme-antigen or enzyme-antibody linkage used in an enzyme immunoassay (EIA) has many advantages over radioimmunoassays (RIA) such as elimination of expensive counting equipment, no radioactive waste to deal with, and cheap and stable reagents.

A. Linked Antibodies

The activity of creatinine kinase isoenzyme MB (CK-MB) has been determined by immunoinhibition with goat-antihuman CK-M antibodies. The residual activity of CK-B in serum is detected at a platinum electrode by the coupling of an electroactive pair $Fe(CN)_6^{3-}/Fe(CN)_6^{4-}$ to NADH produced from the HK-GPD (hexokinase-glucose phosphate dehydrogenase) system:

CK − MB
↓ Antibody CK−M
CK−B

CrP → Cr; ADP ⇄ ATP; G-6-P → 6-P-Gluconate; Glucose; $NADP + H_2O$ ⇄ $NADPH + H^+$; Diaphorase $Fe(CN)_6^{4-}$, $E = +0.36$ V → $Fe(CN)_6^{3-}$ (54)

The whole assay takes only 10 min, and the linearity of a calibration plot of serum CK-MB enzyme extends up to 875 U/liter (244). Similarly, isoenzyme CK-MB has been determined in blood serum by immobilizing isoenzyme antibodies onto alkylamine glass beads with glutaraldehyde (245). The beads are packed into a rotating porous cell. After incubation with stirring, the CK-M isoenzymes in the serum sample are inhibited and are bound to the antibodies inside the stirrer. The residual CK-B isoenzyme activity is then measured as above (Eq. 54). The binding capacity of the immunostirrers to CK-M isoenzyme was estimated to be 800 U/liter, with an efficiency of 97.8%. Reproducibility of the measurements over a period of 52 days were of the order of a C.V. of 4–5%.

Lactate dehydrogenase isoenzyme (LD-1) has been measured in serum using a flow injection system with electrochemical detection in a thin layer transducer after immunochemical separation (246). LD catalyzes the oxidation of lactate to pyruvate:

$$\text{Lactate} + \text{NAD}^+ \xrightarrow{\text{LD}} \text{Pyruvate} + \text{NADH} + \text{H}^+ \qquad (55)$$

After immunochemical separation of LD-1, amperometric detection of NAD^+/NADH was achieved with a redox mediator, ferricyanide/ferrocyanide, at a glassy carbon electrode poised at +0.07 V. vs Ag/AgCl.

An enzyme immunoelectrode suitable for the assay of human serum albumin and insulin was described (247) using an oxygen electrode covered with an antibody-containing nylon net kept in place with an O-ring. From 1 to 25 μg/ml of albumin and 5–100 μg/ml of insulin can be assayed.

A specific sensor for the tumor antigen (AFP) prepared from a membrane of immobilized antibody and an O_2 probe has been developed (248). Anti-AFP antibody is covalently immobilized onto a membrane prepared from cellulose triacetate, 1,8-diamino-4-aminomethyl octane, and glutaraldehyde. The sensor was applied to an EIA based on the competitive antigen–antibody reaction with catalase-labeled antigen. After competitive binding of free and catalase-labeled AFP, the sensor is examined for catalase activity, by amperometric measurement after addition of H_2O_2. AFP can be assayed in the range 10^{-8}–10^{-11} g/ml.

Several assays for human chorionic gonadotropin (HCG) have been developed on the basis of the antigen–antibody reaction in conjunction with selective electrodes (249–251). The assay has been monitored using a cyanogen bromide-treated electrode coated with the corresponding antibody (249). The potential of the modified electrode shifts in the positive direction after contact with a solution of the antigen. The change in potential is approximately proportional to the HCG concentration. The technique, applied to human urine samples, has shown a specific response to only HCG. Alternatively, acetylcholinesterase (AChE) as the label enzyme and a pH electrode can be used for specific HCG determination (250). The required separation of enzyme conjugate (AChE-antigen) from the assay mixture was accomplished with an immobilized antibody. The antibody to HCG was fixed in a membrane form with a polyethylene net. The AChE antigen and the free antigen competitively react with the membrane. After completion of reaction, the activity of the enzyme-antigen fixed on the membrane is measured through the use of a pH electrode (glass) placed against the membrane in an acetylcholine solution. The pH variation is dependent on fixed enzyme activity and, consequently, it is a measurement of free antigen in the test solution. Sensitivities of 1 nM in HCG have been observed.

A "sandwich" assay for HCG using glucose oxidase (GOD) as label has been recently described (251). The procedure is based on an amperometric enzyme immunoassay with an electrode-immobilized antibody. The antibody-electrode (activated glassy carbon) is used both to separate the active complex formed and to monitor the activity of the bound enzyme label. Two monoclonal antibodies linking different antigenic sites were used:

$$\text{E-anti HCG} + \text{antigen} \rightarrow \text{E-anti HCG} - \text{antigen} \tag{56}$$

$$\text{E-anti HCG} - \text{antigen} + \text{anti HCG} - |\!\!\!\text{electrode} \rightarrow \text{E-anti HCG} - \text{antigen} - \text{anti HCG-}|\!\!\!\text{electrode} \tag{57}$$

First, the immobilized GOD-anti HCG was allowed to react with the antigen (Eq. 56). In a second step the antibody-electrode is utilized to capture the GOD-anti HCG complex (Eq. 57). After electrode capture, the catalytic activity of the bound enzyme label was measured by cyclic voltammetry in the presence of a judicious redox mediator. The electrode can be reused up to 40 times just by soaking in urea for 5 min to break the antibody–antigen bond. Sensitivity of the assay has been reported to be 9 mIU HCG/ml in serum.

Several antibody electrodes have been developed for hepatitis-B surface antigen (HBs Ag) determination in biological fluids (252–254). Horseradish peroxidase was linked to anti-HBs Ag gamma-globulins and the resulting labeled antibody immobilized on gelatin membranes. The antigen concentration was measured potentiometrically with an iodide selective electrode modified by fixing the active membrane onto the iodide sensor (252,253). The electrode was dipped in HBs Ag solution to extract the antigen, then placed in a solution of Anti-HBs Ag labeled with HRP according to the "sandwich" principle (243). The amount of HRP fixed was measured in a peroxide iodide solution according to the HRP catalyzed reaction:

$$H_2O_2 + 2I^- + 2\,H^+ \xrightarrow{\text{HRP}} I_2 + 2H_2O \tag{58}$$

One-tenth of a microgram of antigen per liter has been accurately detected (253). An oxygen electrode-based enzyme immunoassay has also been developed for the amperometric determination of HBs Ag. Glucose oxidase as the enzyme label was conjugated to specific antibodies and subsequently immobilized onto a pig skin membrane. Consumption of oxygen was measured by the oxygen sensor. After immersion in HBs Ag positive serum samples, the response was proportional to the enzyme activity retained on the membrane and therefore to the HBs Ag concentration. The calibration graph of signal vs. log HBs Ag concentration was linear over the range of 0.1–100 μg/liter (255).

Bovine serum albumin (BSA) and cyclic AMP (cAMp) have been determined by a competitive-binding EIA (255). With urease as the label, an ammonia gas-sensing electrode has been used to directly measure the amount of urease-labeled antigen bound to a double-antibody solid phase by continuously measuring the rate of ammonia produced from urea as substrate. Examples were given by using urease-BSA conjugate to carry out a competitive binding EIA for BSA and urease-cyclic nucleotide (cGMP) conjugate for cAMP. The method yields accurate and sensitive assays for proteins (BSA $\langle$ 10 ng/ml) and antigens (cAMP $\langle$ 10 nM) with fairly good selectivity over cGMP, AMP, and GMP.

The potentiometric determination of estradiol-17 in solution has also been described as an application of a solid-phase competitive enzyme immunoassay (256). The anti-estradiol-17 antibodies were immobilized on a pig skin membrane. After incubation with HRP-labeled steroid and estradiol, the membrane was mounted over an iodide selective electrode in order to measure the enzymatic activity. Measurements were realized in the presence of H_2O_2 and iodides. The electrode potential was a function of antigen concentration at levels ranging from 57 pmol/liter to 9.2 nmol/liter.

Recently, very low levels of digoxin have been accurately determined in human plasma using a competitive heterogeneous enzymic immunoassay with electrochemical detection (detection limit of 50 pg/ml) (257). The enzyme label (alkaline phosphatase) was conjugated to digoxin. Digoxin in the sample and labeled digoxin compete for the solid-phase antibody coated on the walls of reagent tubes. The labeled digoxin bound to the solid phase antibody was determined by incubation with the enzyme substrate (phenylphosphate). The phenol produced was quantitated by oxidation in a thin layer cell under flow through conditions. The potentiometric CO_2 gas-sensing electrode in conjunction with HRP-antidigoxin labeled antibody and digoxin-BSA polystyrene beads has also been utilized for digoxin determination (263). After competitive reaction of the free digoxin, with coated digoxin-antibody, the beads were centrifuged and resuspended. The amount of anti-digoxin-HRP bound to the beads is then assayed by measuring the rate of CO_2 production from the peroxide-pyrogallol reaction. Detection of picomolar concentrations of digoxin has been reported using this method.

B. Bound Antigens

An enzyme immunoassay method using adenosine deaminase as the enzyme label has been described (130). Potentiometric rate measurements were made with an ammonia gas-sensing electrode to determine the activity of the enzyme label bound to agarose bead-immobilized second antibody. The activity is related to the concentration of either a model haptenic dinitrophenyl, DNP, or anti-DNP antibody. The detection limit is 50 ng antibody.

An immunosensor for determining specific protein has been described (259,260). A liquid antigen containing cardiolipin, phosphatidylcholine, and cholesterol was immobilized to an acetyl cellulose membrane. The membrane-bound antigen retained immunological reactivity to Wasserman antibody. The asymmetrical potential developed was dependent on the concentration of the antibody. This approach was used to develop a method for assay of syphilis antibody in blood serum (261). The contact potential between bound antigen and antibody was measured in the assay and very low mVs (1–3 mV) were observed.

Potentiometric digoxin antibody measurements with antigen-ionophore based membrane electrodes have been successfully applied for serum samples (262). The antigen (digoxin) corresponding to the antibody to be measured is chemically coupled to an ionophore (*cis*-diaminobenzo-18-crown-16) to form an antigen-carrier conjugate. The conjugate is incorporated into a plastic support membrane and that membrane is mounted in the sensing tip of a conventional potentiometric membrane electrode (K^+ selective sensor). The potential change which occurs after addition of the appropriate antibody is proportional to the antibody concentration. Interferences are negligible and sensitivity is high due to the high affinity of digoxin antibodies for the drug digoxin.

Highly sensitive detection of immunoglobulin G (IgG) has been described by operating under flow-through conditions (264,265). HRP and GOD have been utilized as enzyme label for the assays. In the former technique the IgG was precipitated by mixing the label antibody and the precipitate was further dissolved and fed through the flow system. The analyses were based on the continuous flow detection of liberated fluoride after the enzymatic reaction between HRP, H_2O_2 and *p*-fluoroaniline (Eq. 59). The sensitivity of the technique allows detection of as little as 1 μg/ml of IgG in human serum (264).

$$2NH_2-C_6H_4-F + H_2O_2 \xrightarrow{HRP} F-C_6H_4-N{=}C_6H_4{=}NH + H^+ + F^- + H_2O \tag{59}$$

In the latter technique a "sandwich" ELISA was developed. The process involves an incubation time in a microreactor preceding the thin layer amperometric detection system. The labeled second antibody (anti-IgG-glucose oxidase conjugate) converts the injected glucose to hydrogen peroxide which is detected anodically. Sensitivities in the femtomole to picomole range have been reported using this method (265).

VIII. REDOX-MEDIATED AMPEROMETRIC BIOSENSORS

As mentioned previously, there is a great deal of interest in developing sensors that couple small electron carriers with redox proteins in order to facilitate the electron transfer between the enzyme and the electrode material. Actually, ferrocene [bis(n^5-cyclopentadienyl)iron] and its derivatives have been extensively investigated as possible mediators for enzymatic reactions and successfully applied in several bioelectrochemical devices suitable for real sample analysis (266,267). The first electrode developed used 1,1′-dimethylferrocene to shuttle electrons from a flavoprotein to the electrode and was characterized versus glucose measurements (134) and successfully commercialized for blood sample analysis (268). In this glucose assay, the ferricinium ion ($Fe(cp)^+$) competes favorably with the natural mediator oxygen for the oxidation of the reduced glucose oxidase:

$$\begin{aligned} \text{glucose} + \text{GOD}_{(ox)} &\longrightarrow \text{gluconolactone} + \text{GOD}_{(red)} \\ \text{GOD}_{(red)} + 2\text{Fe(cp)}^+ &\longrightarrow \text{GOD}_{(ox)} + 2\text{Fe(cp)} + 2\text{H}^+ \\ 2\text{Fe(cp)} &\longleftrightarrow 2\text{Fe(cp)}^+ \qquad + 2\,e^- \end{aligned} \tag{60}$$

The reoxidation of reduced ferricinium(Fe(cp)) is monitored at the electrode surface poised at sufficiently low positive potential (+160 mV vs. S.C.E.) to minimize interferences. Due to its ingenious microsized and disposable conception, any loss of enzyme and/or mediator has been elegantly circumvented in this radically new type of device (269). For continuous in-line monitoring, namely, in fermentation controls (270), as well as for in vivo sensing devices (271), this configuration suffers from poor enzyme immobilization, and this constitutes a major drawback in long-term monitoring. Several modifications have been proposed in order to improve enzyme attachment on the graphite surface via chemical linking to hexadecylamine-pretreated surfaces (271) or via covalent linking to agarose beads (271).

The organometallic-mediated conception has also been applied to several other flavoproteins and quinoproteins, for example, for the detection of D-galactose, glycolate, and L-amino acids (272), for ATP and creatine kinase(273), and for cholesterol (272). In an effort to construct implantable glucose probes that would not suffer from the inherent drawback of leaching of "soluble" mediating species, several recent strategies have been presented which consist of the chemical incorporation of the organometallic mediator into the thin polymeric backbone covering the electrode surface (275) or mixed with the electrode matrix (276).

By taking advantage of the possibility to entrap enzymes into electrodeposited polymers, ferrocene-containing polypyrrole films with entrapped glucose oxidase have been prepared and characterized versus glucose measurement (275). Considering the fact that the latter configuration suffered from poor stability (2 days), a recent new conception of electrode construction has been presented to be adequate for long-term glucose monitoring. By attaching ferrocene to a polysiloxane backbone and mixing it into a carbon paste matrix (graphite particles + Nujol) in the presence of glucose oxidase, the authors were able to observe effective electron-transfer mediation without loss of the incorporated mediator (276). In further attempts to improve redox mediation between electrode and enzyme, the covalent binding of several ferrocene units to the protein part of the enzyme has been reported and the resulting amperometric probes characterized versus glucose measurements (277,278).

Several amperometric immunoassays which take advantage of electron transfer mediation by ferrocene derivatives have also been successfully proposed. The assays involve generally electrochemical mediation in conjunction with glucose oxidase, for example, for the measurement of human chorionic gonadotrophin (HCG) (279) or thyroxine (T4) (280,281). More recently, homogeneous immunoassays for HCG with a monoclonal anti-HCG antibody and glucose oxidase co-immobilized onto the surface of a glassy carbon electrode have been successfully carried out (282). Beside ferrocene and its derivatives, other mediators have been incorporated into various bioelectrochemical systems. Modified electrodes prepared by adsorbing redox polymers, containing quinone groups, on glassy carbon electrodes have been shown to be suitable for the electrooxidation of glucose oxidase and 1-lactate oxidase (283). On the other hand, the carbon paste matrix has been shown to be an appropriate reservoir of mediators like *p*-benzoquinones (284) and ubiquinones (285). The two latter allow rapid electron transfer between the electrode interface and the deposited enzymatic layer of oligosaccharide dehydrogenase (maltose assay) and D-gluconate dehydrogenase (D-gluconate assay), respectively. Other mediated electrochemical devices involve the use of systems which catalyze the regeneration of the soluble nicotinamide coenzymes of dehydrogenases. There is a variety of redox mediators which have been proposed for the selective electrocatalytic oxidation of nicotinamide coenzymes (66); among these, the phenoxazine derivatives exhibit the most promising features (66, 286). For further well-documented and detailed informations regarding bioelectrochemical devices, the reader is referred to recent relevant books on biosensors (287–289).

IX. COMMERCIAL AVAILABILITY

Single, self-contained enzyme electrode probes are available commercially from Universal Sensors (New Orleans, LA), including probes for glucose, urea, uric acid, lactate, alcohols, amino acids, aspirin, and sugars. A glucose probe based on the technology of Pierre Coulet et al. is available from Tacussel, Inc. (Lyon, France), and alcohol oxidase-based probes for alcohols and sugars are available from Provesta Corporation (Bartlesville, OK).

Instruments and systems utilizing enzyme probes are available from a number of companies including Yellow Springs Instrument Co., Universal Sensors, and Provesta in the United States; Tacussel, Seres, and MediSense in Europe; and Fuji Electric and Omron Toyobo in Japan.

REFERENCES

1. G. G. Guilbault, *Analytical Uses of Immobilized Enzymes,* Marcel Dekker, New York, 1984.
2. J. Savory, R. L. Bertholf, J. C. Boyd, D. E. Bruns, R. A. Felder, M. Lovell, J. R. Shipe, M. R. Willis, J. D. Czaban, K. F. Coffey, and K. M. O'Connell, *Anal. Chim. Acta 180:*99 (1986).
3. L. D. Bowers, *Anal. Chem. 58:*513 A (1986).
4. G. A. Rechnitz, *Anal. Chim. Acta 180:*281 (1986).
5. E. A. H. Hall, *Enzyme Microb. Technol. 8:*651 (1986).
6. F. Scheller, F. Schubert, R. Renneberg, H-G. Muller, M. Janchen and H. Weise, *Biosensors 1:*135 (1985).
7. J. D. Czaban, *Anal. Chem. 57:*345A (1985).
8. J. Koryta, *Electrochim. Acta 31:*515 (1986).
9. D. Monroe, *Am. Clin. Prod. Rev. 4(1):*24 (1985).
10. D. Monroe, *Am. Clin. Prod. Rev. 4(3):*46 (1985).
11. P. W. Carr, L. D. Bowers, *Immobilized Enzymes in Analytical and Clinical Chemistry,* Wiley-Interscience, New York, 1980.
12. G. G. Guilbault, *Enzyme. Eng. 6:*395 (1982).
13. G. G. Guilbault, in *Immobilized Enzymes, Antigens, Antibodies and Peptides,* Vol. 1, (H. H. Weetall, Ed.), Marcel Dekker, New York, 1975, p. 293.
14. G. A. Rechnitz, *Science 214:*287 (1981).
15. A. P. F. Turner, I. Karube, G. S. Wilson, in *Biosensors: Fundamentals and Applications,* Oxford University Press, 1987.
16. D. N. Gray, M. H. Keyes, B. Watson, *Anal. Chem. 49:*1067A (1977).
17. S. Suzuki, I. Karube, *Appl. Biochem. Bioeng. 3:*154 (1981).
18. M. A. Arnold, *Ion-Selective Electrode Rev. 8:*85 (1986).
19. I. Karube, in *Fundamentals and Applications of Chemical Sensors,* (D. Schuetzle, R. Hammerle, J. W. Butler, Eds.), A. C. S. Symposium Series, American Chemical Society, Washington, D.C., 1986, p. 331.
20. B. J. Vincke, J-C. Vire, G. J. Patriarche, in *Electrochemistry, Sensors and Analysis* (M. R. Smyth, J. G. Vos, Eds.), 1986, p. 147.
21. L. J. Blum, C. Bertrand, P. R. Coulet, *Anal. Letters 16:*541 (1983).
22. J. W. Ross, J. H. Riseman, J. A. Krueger, *Pure Appl. Chem. 36:*473 (1973).
23. M. Riley, in *Ion-Selective Electrode Methodology,* Vol. 2, (A. K. Covington, Ed.), CRC Press Inc., Boca Raton, FL, 1979, p. 23.
24. W. J. Albery, P. N. Bartlett, A. E. Cass, D. H. Craston, and B. G. D. Haggett, *J. Chem. Soc. Faraday Trans. 82:* 1033 (1986).
25. M. J. Green and H. A. O. Hill, *J. Chem. Soc. Faraday Trans. 82:*1237 (1986).
26. G. Davis, *Biosensors 1:*161 (1985).
27. M. R. Tarasevich, in *Comprehensive Treatise of Electrochemistry,* Vol. 10 (S. Srinivasan, Y. A. Chizmadzhev, J. O'M. Bockris, B. E. Conway, and E. Yeager, Eds.), Plenum Press, New York, 1985, p. 231.
28. A. S. Attiyat and G. D. Christian, *Am. Biotechnol. Lab. 2:*8 (1984).
29. H. H. Weetall and W. H. Pitcher, Jr., *Science 232:*1396 (1986).

30. G. G. Guilbault, *Handbook of Enzymatic Analysis,* Marcel Dekker, New York, 1977.
31. P. Vadgama, *The Analyst 111:*875 (1986).
32. R. Tor and A. Freeman, *Anal. Chem. 58:*1042 (1986).
33. D. Kirstein, F. Scheller, B. Olsson, and G. Johansson, *Anal. Chim. Acta 171:*345 (1985).
34. K. Mosbach, in *Immobilized Enzymes,* Vol. 44 (S. P. Kolowick, N. O. Kaplan, Eds.), Academic Press, New York, 1978.
35. P. Valle-Vega, C. T. Young, and H. E. Swaisgood, *J. Food Science 45:*1026 (1980).
36. M. A. Nabi-Rahni, G. G. Guilbault, and G. Neto De Olivera, *Anal. Chem. 58:*523 (1986).
37. M. Mascini and G. G. Guilbault, *Anal. Chem. 49:*795 (1977).
38. M. Mascini, D. Moscone, and G. Palleschi, *Anal. Chim. Acta 157:*45 (1984).
39. M. Mascini, S. Fortunati, D. Moscone, G. Palleschi, M. Massi-Benedetti, and P. Fabietti, *Clin. Chem. 31:*451 (1985).
40. P. R. Coulet, J. H. Julliard, and D. C. Gautheron, *Biotechnol. Bioeng. 16:*1055 (1974).
41. C. H. Assolant-Vinet and P. R. Coulet, *Anal. Lett. 19:*875 (1986).
42. V. J. Razumas, J. J. Jasaitis, and J. J. Kulys, *Bioelectrochem. Bioenerg. 12:*297 (1984).
43. R. M. Ianniello and A. Yacynych, *Anal. Chim. Acta 131:*123 (1981).
44. C. C. Liu, J. P. Weaver, and A. X. Chen, *Bioelectrochem. Bioenerg. 8:*379 (1981).
45. M. F. Suaud-Chagny and J. F. Pujol, *Analusis 13:*25 (1985).
46. T. Yao, *Anal. Chim. Acta 148:*27 (1983).
47. U. Oesch, D. Ammann, and W. Simon, *Clin. Chem 32:*1448 (1986).
48. G. G. Guilbault and T. J. Rohm, *Intern. J. Environ. Anal. Chem. 4:*51 (1974).
49. G. G. Guilbault, *Bull. Soc. Chim. Belg. 84:*679 (1975).
50. G. J. Moody and J. D. R. Thomas, *The Analyst 100:*609 (1975).
51. R. L. Solsk, CRC, *Crit. Rev. Anal. Chem. 14:*1 (1982).
52. G. G. Guilbault and F. R. Shu, *Anal. Chem. 44:*2161 (1972).
53. G. G. Guilbault and G. Nagy, *Anal. Chem. 45:*417 (1973).
54. N. J. Szuminsky, A. K. Chen, and C. C. Liu, *Biotechnol. Bioeng. 26:*642 (1984).
55. J. P. Joseph, *Anal. Chim. Acta 169:*249 (1985).
56. R. M. Ianniello and A. Yacynych, *Anal. Chim. Acta 146:*249 (1983).
57. M. E. Lopez and G. A. Rechnitz, *Anal. Chem. 54:*2085 (1982).
58. M. E. Lopez, *Anal. Chem. 56:*2360 (1984).
59. G. G. Guilbault, J. P. Czarnecki, and M. A. Nabi Rahni, *Anal. Chem. 57:*2110 (1985).
60. C. R. Bradley and G. A. Rechnitz, *Anal. Chem. 57:*1401 (1985).
61. R. W. Murray, in *Electroanalytical Chemistry,* Vol. 13 (A. J. Bard, Ed.), Marcel Dekker, New York, 1984, p. 191.
62. A. J. Bard and L. R. Faulkner, *Electrochemical Methods,* Wiley, New York, 1980.

63. R. A. Durst and E. A. Blubaugh in *Fundamentals and Applications of Chemical Sensors* (D. Schuezle, R. Hammerle, and J. W. Butler, Eds.), A. C. S. Symposium Series, American Chemical Society, Washington, D.C., 1986, p. 245.
64. F. Schubert, D. Kirstein, F. Scheller, R. Appelqvist, L. Gorton, and G. Johansson, *Anal. Lett. 19:*1273 (1986).
65. R. Appleqvist, G. Marko-Varga, L. Gorton, A. Torstensson, and G. Johansson, *Anal. Chim. Acta 169:*237 (1985).
66. L. Gorton, *J. Chem. Soc. Faraday. Trans. 82:*1245 (1986).
67. M. R. Tarasevich, *Bioelectrochem. Bioenerg. 6:*587 (1979).
68. P. J. Elving, W. T. Bresnahan, J. Moiroux, and Z. Samec, *Bioelectrochem. Bioenerg. 9:*365 (1982).
69. M. F. Suaud-Chagny and F. G. Gonon, *Anal. Chem. 58:*412 (1986).
70. M. L. Fultz and R. A. Durst, *Anal. Chim. Acta 140:*1 (1982).
71. H. Huck, A. Schelter-Graf, J. Danzer, P. Kirch, and H-L. Schmidt, *The Analyst 109:*147 (1984).
72. G. G. Guilbault and J. G. Montalvo, *J. Am. Chem. Soc. 92:*2533 (1970).
73. G. G. Guilbault and J. G. Montalvo, *J. Am. Chem. Soc. 91:*2164 (1969).
74. G. G. Guilbault and E. Hrabankova, *Anal. Chim. Acta 52:*287 (1970).
75. T. Anfalt, A. Granelli, and D. Jagner, *Anal. Lett. 6:*969 (1973).
76. G. G. Guilbault and M. Nanjo, *Anal. Chem. 46:*1769 (1974).
77. H. Nilsson, A. Akerlund, and K. Mosbach, *Biochim. Biophys. Acta 320:*529 (1973).
78. G. J. Papariello, A. K. Mukerji, and G. M. Shearer, *Anal. Chem. 45:*790 (1973).
79. G. G. Guilbault and E. Hrabankova, *Anal. Lett. 3:*53 (1970).
80. G. G. Guilbault and G. J. Lubrano, *Anal. Chim. Acta 69:*183 (1974).
81. G. G. Guilbault and G. J. Lubrano, *Anal. Chim. Acta 69:*189 (1974).
82. T. R. Hopkins, *Am. Biotechnol. Lab. 3:*32 (1985).
83. G. Nagy, L. H. von Storp, and G. G. Guilbault, *Anal. Chim. Acta 66:*443 (1973).
84. C. Tran-Minh and J. Beaux, *Anal. Chem. 51:*91 (1979).
85. M. R. Weaver and P. M. Vadgama, *Clin. Chim. Acta 155:*295 (1986).
86. H. H. Weetall, *Anal. Chem. 46:*602A (1974).
87. L. C. Clark, Jr. and C. Lyons, *Ann. N. Y. Acad. Sci. 102:*29 (1962).
88. L. C. Clark, Jr. and M. Sachs, *Ann. N.Y. Acad. Sci. 148:*133 (1968).
89. G. G. Guilbault and G. J. Lubrano, *Anal. Chim. Acta 64:*439 (1973).
90. G. Sittampalam and G. S. Wilson, *J. Chem. Education 59:*71 (1982).
91. M. Shichiri, R. Kawamori, Y. Yamasaki, and N. Hakvi, *Lancet 20:*1129 (1982).
92. I. Hanning, P. Vadgama, A. K. Covington, K. G. M. M. Alberti, *Anal. Lett. 19:*461 (1986).
93. J.-L. Romette, B. Froment, and D. Thomas, *Clin. Chim. Acta 95:*249 (1979).
94. B. J. Vincke, J-M. Kauffmann, M. Devleeschouwer, and G. J. Patriarche, *Analusis 12:*141 (1984).

95. M. T. Flanagan and N. J. Carroll, *Biotechnol. Bioeng. 28:*1093 (1986).
96. G. P. Hicks and S. J. Updike, *Anal. Chem. 38:*726 (1966).
97. S. J. Updike and G. P. Hicks, *Nature 214:*986 (1967).
98. L. D. Mell and J. T. Maloy, *Anal. Chem. 47:*299 (1975).
99. D. Pfeiffer, F. Scheller, M. Janchen, and K. Betermann, *Biochimie 62:*587 (1980).
100. C. C. Liu, L. B. Wingard, S. K. Wolfson, S. J. Yao, A. L. Drash, and J. G. Schiller, *Bioelectrochem. Bioenerg. 6:*19 (1979).
101. C. C. Liu, F. M. Fryburg, and A. K. Chen, *Bioelectrochem. Bioenerg. 8:*703 (1981).
102. L. B. Wingard, *Bioelectrochem. Bioenerg 9:*307 (1982).
103. I. K. Al-Hitti, G. J. Moody, and J. D. R. Thomas, *The Analyst 109:*1205 (1984).
104. N. C. Foulds and C. R. Lowe, *J. Chem. Soc. Faraday. Trans. 82:*1259 (1986).
105. M. A. Lange and J. Q. Chambers, *Anal. Chim. Acta 175:*89 (1985).
106. T. Ikeda, H. Hamada, and M. Senda, *Agric. Biol. Chem. 50:*883 (1986).
107. O. Migawaki and L. B. Wingard, *Biotechnol. Bioeng. 26:*1364 (1984).
108. D. A. Gough, J. Y. Lucisano, and P. H. S. Tse, *Anal. Chem. 57:*2351 (1985).
109. J. F. Castner and L. B. Wingard, *Anal. Chem. 56:*2891 (1984).
110. J. J. Kulys, V. S. A. Laurinavicius, M. V. Pesliakiene, and V. V. Gureviciene, *Anal. Chim. Acta 148:*13 (1983).
111. J. J. Kulys, M. V. Pesliakiene, and A. S. Samaluis, *Bioelectrochem. Bioenerg. 8:*81 (1981).
112. C. Tran-Minh and G. Broun, *Anal. Chem. 47:*1359 (1975).
113. W. H. Mullen, F. H. Keedy, J. Churchouse, and P. M. Vadgama, *Anal. Chim. Acta. 183:*59 (1986).
114. G. G. Guilbault and G. J. Lubrano, *Anal. Chim. Acta 97:*229 (1978).
115. S. Enfors, *Enzyme Microb. Technol. 3:*29 (1981).
116. F. Schubert, D. Kirstein, K. L. Schroder, and F. W. Scheller, *Anal. Chim. Acta 169:*391 (1985).
117. D. Kirstein, F. Schubert, and F. Scheller, *Anal. Lett. 15:*1479 (1982).
118. M. Mascini, R. M. Ianniello, and G. Palleschi, *Anal. Chim. Acta 146:*135 (1983).
119. K. Ito, S. Ikeda, K. Asai, H. Naruse, K. Ohkwa, H. Ichihashi, H. Kamei, and T. Kondo, in *Fundamentals and Applications of Chemical Sensors* (D. Schuetzle, R. Hammerle, and J. W. Butler, Eds.), A. C. S. Symposium Series, American Chemical Society, Washington, D.C., 1986, p. 373.
120. G. J. Moody, G. S. Sanghera, and J. D. R. Thomas, *The Analyst 111:*605 (1986).
121. M. Mascini, M. A. Mateescu, and R. Pilloton, *Bioelectrochem. Bioenerg. 16:*149 (1986).
122. D. R. Thevenot, P. R. Coulet, R. Sternberg, and D. C. Gautheron, *Bioelectrochem. Bioeng. 5:*548 (1978).
123. D. R. Thevenot, R. Sternberg, P. R. Coulet, J. Laurent, and D. C. Gautheron, *Anal. Chem. 51:*96 (1979).

124. C. Bertrand, P. R. Coulet, D. C. Gautheron, *Anal. Chim. Acta 126:*23 (1981).
125. E. Lobel and J. Rishpon, *Anal. Chem. 53:*51 (1981).
126. S. J. Updike, M. C. Shults, and M. Busby, *J. Lab. Clin. Med. 93:*519 (1979).
127. L. Sokol, C. Garber, M. C. Shults, and S. Updike, *Clin. Chem. 26:*89 (1980).
128. J. Havas, E. Porjesz, G. Nagy, and E. Pungor, *Hung. Sci. Instr. 49:*53 (1980).
129. T. Yao, *Nippon Kagaku Kaishi 8:*1335 (1984).
130. F. R. Shu and G. S. Wilson, *Anal. Chem. 48:*1679 (1976).
131. C. Bourdillon, J-P. Bourgeois, and D. Thomas, *J. Am. Chem. Soc. 102:*4231 (1980).
132. R. M. Ianniello and A. M. Yacynych, *Anal. Chem. 53:*2090 (1981).
133. L. B. Wingard, Jr., D. Ellis, S. J. Yao, J. G. Schiller, C. C. Lui, S. K. Wolfson, and A. L. Drash, *J. Solid Phase Biochem. 4:*253 (1979).
134. A. E. G. Cass, G. Davis, G. D. Francis, H. A. O. Hill, W. J. Aston, I. J. Higgins, E. V. Plotkin, L. D. L. Scott, and A. P. F. Turner, *Anal. Chem. 56:*667 (1984).
135. R. A. Kamin and G. S. Wilson, *Anal. Chem. 52:*1198 (1980).
136. G. G. Guilbault and G. J. Lubrano, *Anal. Chim. Acta 60:*254 (1972).
137. R. Kobos and G. A. Rechnitz, *Anal. Lett. 10:*751 (1977).
138. J. J. Kulys, M. V. Pesliakiene, and A. S. Samalius, *Bioelectrochem. Bioenerg. 8:*81 (1981).
139. J. J. Kulys, *Enzyme Microb. Technol. 3:*344 (1981).
140. N. K. Cenas and J. J. Kulys, *Bioelectrochem. Bioenerg. 8:*103 (1981).
141. T. T. Ngo and H. M. Lenhoff, *J. Appl. Biochem. 25:*373 (1980).
142. J. Havas, E. Porjesz, G. Nagy and E. Pungor, *Anal. Chem. Symp. Ser. 8:*241 (1981).
143. K. Bertermann, P. Elze, F. Scheller, D. Pfeiffer, and M. Janchen, *Anal. Lett. 15:*397 (1982).
144. M. Lindh, K. Lindgren, A. Carlstroem, and P. Masson, *Clin. Chem. 28:*726 (1982).
145. G. J. Buffone, J. M. Johnson, S. A. Lewis, and J. W. Sparks, *Clin. Chem. 26:*339 (1980).
146. R. Sternberg, A. Apoteker, and D. Thevenot, *Anal. Chem. Symp. Ser. 2:*461 (1980).
147. T. Tsuchida and K. Yoda, *Enzyme Microb. Technol. 3:*326 (1981).
148. A. P. F. Turner and J. C. Pickup, *Biosensors 1:*85 (1985).
149. E. J. Fogt, L. M. Dodd, E. M. Jenning, and A. H. Clemens, *Clin. Chem. 24:*1366 (1978).
150. K. S. Chuan and I. K. Tan, *Clin. Chem. 24:*150 (1978).
151. O. Garcia, J. Ceylen, W. Sansen, and F. Colin, *Extended Abstract at the Journees d'Electrochimie* (Paris, France) *8:*10 (1983).
152. G. Marko-Varga, R. Appelqvist, and L. Gorton, *Anal. Chim. Acta 179:*371 (1986).
153. W. H. Mullen, S. Churchouse, and P. M. Vadgama, *The Analyst 110:*925 (1985).

154. U. Wollenberger, F. Scheller, D. Pfeiffer, Y. A. Bog Danovskaya, and M. R. Tarasevich, *Anal. Chim. Acta. 187:*39 (1986).
155. M. Shichiri, R. Kawamori, R. Goriya, Y. Yamasaki, M. Nomura, N. Hakui, and H. Abe, *Diabetologia 24:*179 (1983).
156. J. F. Castner and L. B. Wingard, Jr., *Biochemistry 23:*2203 (1984).
157. F. Cheng and G. Christian, *Anal. Chim. Acta 104:*47 (1979).
158. F. Lang, E. Gstrein, J. Geibel, W. Rehwald, H. Volkl, and H. Oberleithner, *Bioelectrochem. Bioenerg. 11:*365 (1983).
159. P. Taylor, E. Kmetec, and J. Johnson, *Anal. Chem. 49:*789 (1977).
160. D. Pfeiffer, F. Scheller, M. Jänchen, K. Bertermann, and H. Weise, *Anal. Lett. 13:*1179 (1980).
161. I. Sato, I. Karube, and S. Suzuki, *Biotechnol. Bioeng. 18:*269 (1976).
162. H. Lundbaeck and B. Olsson, *Anal. Lett. 18:*871 (1985).
163. K. Yasuda, H. Miyagi, Y. Hamada, and Y. Takata, *The Analyst 109:*61 (1984).
164. G. G. Guilbault and M. Tarp, *Anal. Chim. Acta 73:*355 (1974).
165. B. J. Vincke, M. J. Devleeschouwer, and G. J. Patriarche, *Anal. Lett. 16:*673 (1973).
166. G. G. Guilbault and M. Nanjo, *Anal. Chim. Acta. 75:*169 (1975).
167. W. J. Blaedel and R. C. Engstrom, *Anal. Chem. 52:*1691 (1980).
168. J-M. Laval, C. Bourdillon, and J. Moiroux, *J. Am. Chem. Soc. 106:*4701 (1984).
169. N. Cenas, J. Rozgaite, and J. Kulys, *Biotechnol. Bioeng. 16:*551 (1984).
170. J. J. Kulys, *Biosensors 2:*3 (1986).
171. J. J. Kulys, V. V. Gureviciene, V. A. Laurinavicius, and A. V. Jonuska, *Biosensors 2:*35 (1986).
172. J. L. Romette, J. S. Yang, H. Kusakabe, and D. Thomas, *Biotechnol. Bioeng. 25:*2557 (1983).
173. G. Davis, A. H. O. Hill, W. J. Aston, I. J. Higgins, and A. P. F. Turner, *Enzyme Microb. Technol. 5:*383 (1983).
174. F. Mizutani, T. Yamanaka, Y. Tanabe, and K. Tsuda, *Anal. Chim. Acta 177:*153 (1985).
175. H. Durliat and M. Comtat, *Anal. Chem. 52:*2109 (1980).
176. W. P. Soutter, F. Sharp, and D. M. Clark, *Br. J. Anaesth. 50:*445 (1978).
177. T. Shinbo, M. Sugiura, and N. Kamo, *Anal. Chem. 51:*100 (1979).
178. F. Scheller, F. Schubert, B. Olsson, L. Gorton, and G. Johansson, *Anal. Lett. 19:*1691 (1986).
179. L. C. Clark, L. K. Noyes, T. A. Grooms, and C. A. Gleason, *Clin. Biochem. 17:*288 (1984).
180. F. Mizutani, K. Sasaki, and Y. Shimura, *Anal. Chem. 55:*35 (1983).
181. T. Tsuchida, H. Takasugi, K. Yoda, and K. S. Takizawa, *Biotechnol. Bioeng. 27:*837 (1985).
182. J-M. Laval and C. Bourdillon, *J. Electroanal. Chem. 152:*125 (1983).
183. G. Bardeletti, F. Sechaud, and P. Coulet, *Anal. Chim. Acta 187:*47 (1986).
184. F. Mizutani, Y. Shimura, and K. Tsuda, *Chem. Lett. 199* (1984).

185. A. Malinauskas and J. J. Kulys, *Anal. Chim. Acta 98:*31 (1978).
186. C. Bourdillon, J. Thomas, and D. Thomas, *Enzym. Microb. Technol. 4:*175 (1982).
187. F. Mizutani, K. Tsuda, I. Karube, S. Suzuki, and K. Matsumoto, *Anal. Chim. Acta 118:*65 (1980).
188. T. Kawashima, A. Arima, N. Hatakeyama, N. Tominaga, and M. Ando, *Nippon Kagaku Kaishi 10:*1542 (1980).
189. M. Janchen, G. Grunig, and K. Bertermann, *Anal. Lett. 18:*1799 (1985).
190. T. Fonong and G. A. Rechnitz, *Anal. Chim. Acta 158:*357 (1984).
191. M. A. Nabi Rahni, G. G. Guilbault, and G. Olivera de Neto, *Anal. Chim. Acta. 181:*219 (1986).
192. C. R. Bradley and G. A. Rechnitz, *Anal. Lett. 19:*151 (1986).
193. S. Yao, J. Wolfson, and J. Takarsky, *J. Bioelectrochem. Bioenerg. 2:*348 (1975).
194. R. K. Kobos and T. A. Ramsey, *Anal. Chim. Acta 121:*111 (1980).
195. G. Nagy, M. E. Rice, and R. N. Adams, *Life Sci. 31:*2611 (1982).
196. K. Matsumoto, K. Yamada, and Y. Osajima, *Anal. Chem. 53:*1974 (1981).
197. P. Posadka and L. Macholan, *Collect. Czech. Chem. Commun. 44:*3395 (1979).
198. B. J. Vincke, M. J. Devleeschouwer, G. J. Patriarche, *Anal. Lett. 18:*1593 (1985).
199. T. Kawashima and G. A. Rechnitz, *Anal. Chim. Acta 83:*9 (1976).
200. G. G. Guilbault and P. R. Coulet, *Anal. Chim. Acta 152:*223 (1983).
201. G. G. Guilbault and F. Shu, *Anal. Chim. Acta. 56:*333 (1971).
202. E. Diamandis and T. Handjiioannou, *Clin. Chem. 27:*455 (1981).
203. T. Tsuchida and K. Yoda, *Clin. Chem. 29:*51 (1983).
204. K. Kihara and E. Yasukawa, *Anal. Chim. Acta 183:*75 (1987).
205. M. Mascini, S. Fortunati, D. Moscone, and G. Palleschi, *Anal. Chim. Acta 171:*175 (1985).
206. I. Kubo, I. Karube, and S. Suzuki, *Anal. Chim. Acta 151:*371 (1983).
207. I. Kubo and I. Karube, *Anal. Chim. Acta 187:*31 (1986).
208. G. G. Guilbault, S. P. Chen, and S. S. Kuan, *Anal. Lett. 13:*1607 (1980).
209. M. Mascini, M. Tomassetti, and M. Iannello, *Clin. Chim. Acta 132:*7 (1983).
210. I. Satoh, I. Karube, and S. Suzuki, *Biotechnol. Bioeng. 19:*1095 (1977).
211. J. M. Dietschy, L. E. Weeks, and J. J. Delente, *Clin. Chim. Acta 73:*407 (1976).
212. U. Wollenberger, F. Scheller, and P. Atrat, *Anal. Lett. 13:*825 (1980).
213. Y. Kameno, N. Nakano, and S. Baba, *Clin. Chim. Acta 77:*245 (1977).
214. A. Noma and K. Nakayama, *Clin. Chem 22:*236 (1976).
215. A. Kumar and G. Christian, *Clin. Chim. Acta 74:*101 (1977).
216. N. Huang, S. S. Kuan, and G. G. Guilbault, *Clin. Chem. 23:*671 (1977).
217. L. C. Clark, C. Emory, C. Glueck, and M. Campbell, in *Enzyme Engineering,* Vol. 3 (E. K. Pye and H. H. Weetall, Eds.), Plenum Press, New York, 1978.

218. L. C. Clark, G. Duggan, T. Grooms, L. M. Hart, and M. E. Moore, *Clin. Chem. 27:*1978 (1981).
219. U. Wollenberger, M. Kuhn, F. Scheller, H. Deppmeyer, and M. Janchen, *Bioelectrochem. Bioenerg. 11:*307 (1983).
220. C. Bertrand, P. R. Coulet, and D. C. Gautheron, *Anal. Lett. 12:*1477 (1979).
221. A. Kumar and G. Christian, *Clin. Chem. 21:*325 (1975).
222. G. G. Guilbault and M. Nanjo, *Anal. Chim. Acta 73:*367 (1974).
223. G. G. Guilbault and E. Hrabankova, *Anal. Chim. Acta 56:*285 (1971).
224. G. G. Guilbault and E. Hrabankova, *Anal. Chem. 42:*1779 (1970).
225. G. G. Guilbault and G. Nagy, *Anal. Lett, 6:*301 (1973).
226. R. Llenado and G. Rechnitz, *Anal. Chem. 46:*1109 (1974).
227. J. Havas and G. G. Guilbault, *Anal. Chem. 54:*1991 (1982).
228. W. C. White and G. G. Guilbault, *Anal. Chem. 50:*1481 (1978).
229. L. Macholan, *Collect. Czech. Chem. Commun. 43:*1811 (1978).
230. D. P. Nikolelis and T. P. Hadjuoannou, *Anal. Chim. Acta. 147:*33 (1983).
231. C. P. Pau and G. A. Rechnitz, *Anal. Chim. Acta 160:*141 (1984).
232. D. P. Nikolelis, *Anal. Chim. Act. 167:*381 (1985).
233. R. R. Walters, P. A. Johnson, and R. P. Buck, *Anal. Chem. 52:*1684 (1980).
234. P. M. Kovach and M. E. Meyerhoff, *Anal. Chem. 54:*217 (1982).
235. T. Iida, S. Machida, N. Iijima, and T. Mitamura, *Anal. Chem., Symp. Ser. 17:*631 (1983).
236. T. Toyoda, S. S. Kuan, and G. G. Guilbault, *Anal. Chem. 57:*1925 (1985).
237. M. Mascini and R. Giardini, *Anal. Chim. Acta 114:*329 (1980).
238. P. W. Alexander and G. A. Rechnitz, *Anal. Chem. 46:*250 (1974).
239. M. A. Arnold and G. A. Rechnitz, *Anal. Chem. 52:*1170 (1980).
240. S. Kuriyama and G. A. Rechnitz, *Anal. Chim. Acta 131:*91 (1981).
241. I. Karube, K. Hara, I. Satoh, and S. Suzuki, *Anal. Chim. Acta 106:*243 (1979).
242. M. Mascini and D. Moscone, *Anal. Chim. Acta 179:*439 (1986).
243. C. Blake and B. J. Gould, *The Analyst 109:*533 (1984).
244. C. Yuan, S. Kuan, and G. G. Guilbault, *Anal. Chem. 53:*190 (1981).
245. C. Yuan, S. Kuan, and G. G. Guilbault, *Anal. Chim. Acta 124:*169 (1981).
246. T. Toyoda, S. S. Kuan, and G. G. Guilbault, *Anal. Lett. 18:*345 (1985).
247. B. Mattiasson and H. Nilsson, *FEBS Lett. 78:*251 (1977).
248. M. Aizawa, A. Morioka, and S. Suzuki, *Anal. Chim. Acta. 115:*61 (1980).
249. N. Yamamoto, Y. Nagasawa, S. Shuto, H. Tsubomuro, M. Saval, and H. Okumara, *Clin. Chem. 26:*1569 (1980).
250. M. Mascini, F. Zolesi, and G. Palleschi, *Anal. Lett. 15:*101 (1982).
251. G. A. Robinson, V. M. Cole, S. J. Rattle, and G. C. Forrest, *Biosensors 2:*45 (1986).
252. J. L. Boitieux, G. Desmet, and D. Thomas, *Clin. Chim. Acta 88:*329 (1978).
253. J. L. Boitieux, G. Desmet, and D. Thomas, *Clin. Chem. 25:*318 (1979).
254. J. L. Boitieux, D. Thomas, and G. Desmet, *Anal. Chem. Acta 163:*309 (1984).
255. M. E. Meyerhoff and G. A. Rechnitz, *Anal. Biochem. 95:*483 (1979).

256. J. L. Boitieux, C. Lemay, G. Desmet, and D. Thomas, *Clin. Chim. Acta 113:*175 (1981).
257. K. R. Wehmeyer, H. B. Halsall, W. R. Heineman, C. R. Volle, and I., Wen-Chen, *Anal. Chem. 58:*135 (1986).
258. C. Gebauer and G. Rechnitz, *Anal. Lett. 14:*97 (1981).
259. M. Aizawa, S. Suzuki, Y. Nagamara, and R. Shinobara, *Chem. Lett.:*779 (1977).
260. M. Aizawa, S. Kato, and S. Suzuki, *J. Membr. Sci. 2:*125 (1977).
261. S. Suzuki, M. Aizawa, Y. Nagamura, R. Shinohara, and I. Ishiguro, *J. Solid Phase Biochem. 4:*25 (1979).
262. M. Y. Keating and G. A. Rechnitz, *Anal. Chem. 56:*801 (1984).
263. M. Y. Keating and G. A. Rechnitz, *Anal. Lett. 18:*1 (1985).
264. P. W. Alexander and C. Maltra, *Anal. Chem. 54:*68 (1982).
265. W. Uditha de Alvis and G. S. Wilson, *Anal. Chem. 57:*2754 (1985).
266. A. E. G. Cass, G. Davis, M. J. Green, and H. A. O. Hill, *J. Electroanal. Chem. 190:*117 (1985).
267. A. E. G. Cass, G. Davis, H. A. O. Hill, and D. Nancarrow, *Biochim. Biophys. Acta 828:*51 (1985).
268. D. R. Matthews, R. R. Holman, E. Bown, J. Steemson, A. Watson, S. Hughes, and D. Scott, *Lancet 4:*778 (1987).
269. A. P. F. Turner and A. Swain, *Am. Biotechnol. Lab. 11:*10 (1988).
270. S. L. Brooks, R. E. Ashby, A. P. F. Turner, M. R. Calder, and D. J. Clarke, *Biosensors 3:*45 (1987/88).
271. J. C. Pickup, G. W. Shaw, and D. J. Claremont, *Biosensors 3:*335 (1987/88).
272. J. Dicks, W. J. Aston, G. Davis, and A. P. F. Turner, *Anal. Chim. Acta 182:*103 (1986).
273. G. Davis, M. J. Green, and H. A. O. Hill, *Enzyme Microbiol. Tech. 8:*349 (1986).
274. M. Ball, J. Frew, M. J. Green, and H. A. O. Hill in *Electrochemical Sensors for Biomedical Applications* (C. K. N. Li, Ed.) The Electrochemical Soc., 1986, p. 16.
275. N. C. Foulds and C. R. Lowe, *Anal. Chem. 60:*2473 (1988).
276. P. D. Hale, T. Inagaki, H. I. Karan, Y. Okamoto, and T. A. Skotheim, *J. Am. Chem. Soc. 111:*3482 (1989).
277. Y. Degani and A. Heller, *J. Am. Chem. Soc. 110:*2615 (1988).
278. P. N. Bartlett and R. G. Whitaker, *J. Chem. Soc. Chem. Commun.:*1603 (1987).
279. G. A. Robinson, V. M. Cole, S. J. Rattle, and G. C. Forrest, *Biosensors 2:*45 (1986).
280. G. A. Robinson, G. Martinazzo, and G. C. Forrest, *J. Immunoassay 7:*1 (1986).
281. G. A. Robinson, V. M. Cole, and G. C. Forrest, *Biosensors 3:*147 (1987/88).
282. T. Yao and G. A. Rechnitz, *Biosensors 3:*307 (1988).
283. N. K. Cenas, A. K. Pocius, and J. J. Kulys, *Bioelectrochem. Bioenerg. 12:*583 (1984).

284. T. Ikeda, T. Shibata, and M. Senda, *J. Electroanal. Chem. 261:*351 (1989).
285. T. Ikeda, K. Miki, F. Fushimi, and M. Senda, *M. Agric. Biol. Chem. 51:*747 (1987).
286. L. Gorton and A. Hedlund, *Anal. Chim. Acta. 213:*91 (1988).
287. A. P. F. Turner, I. Karube, and G. S. Wilson (Eds.), *Biosensors Fundamentals and Applications,* Oxford University Press, 1987.
288. R. Schmid, (Ed.), Biosensors International Workshop 1987. GBF Monograph Series Vol. 10, VCH Verlagsgesellschaft, Weinheim, 1987.
289. T. T. Ngo (Ed.),. *Electrochemical Sensors in Immunological Analysis,* Plenum Press, New York, 1987.

8

Immobilized Antibody- and Receptor-Based Biosensors

Richard F. Taylor

Arthur D. Little, Inc.
Cambridge, Massachusetts

I. INTRODUCTION

Advances in biotechnology during the past decade have significantly affected science and our daily lives. Genetically engineered drugs, synthetic vaccines, transgenic plants and animals, and synthetic enzymes are but a few of the emerging products of this technological revolution.

Diagnostics and detection have also been transformed by biotechnology. Hybridoma technology, for example, has strongly influenced new immunoassay products. Advances in DNA analysis and synthesis have stimulated the development of DNA probe assays, which are proving a rapid means to detect and diagnose genetic diseases and pathogenic microorganisms.

Diagnostics and detection are also being affected by the emergence of new bioelectronic products. Bioelectronics, a subdiscipline of biotechnology, represents the merging of molecular biology and electronics to result in new technologies and products. Such products are applicable to a wide range of industries including medicine (diagnostics, therapeutics, and prosthesis devices), processing (process control, quality control/quality assurance, and waste stream monitoring), environmental monitoring (toxic and hazardous substances), computing (memory intensive systems, image processing, and 3D arrays), and artificial intelligence (language processing and robotics).

Biosensors are bioelectronic products. Biosensors are real-time (or near real-time) measuring and detection devices which utilize immobilized biological molecules (including enzymes, antibodies, and receptors) as their detection component and electronic components to amplify and report the detection event. As detection devices, biosensors have high specificity, high sensitivity, rapid data output, and the near-infinite variety available from the numbers and types of biological molecules which can form their basis. In addition, biosensor development is aiming for products which are portable, easily used by nontechnical personnel, reagentless, and low cost. The economics and markets for biosensors have been reviewed elsewhere (1,2).

The purpose of this chapter is to review the use of immobilized proteins, specifically antibodies and receptors, in biosensor applications. In order to meet this purpose, an overview of biosensor technology will be presented prior to discussion of specific biosensors developed to date.

II. BIOSENSOR TECHNOLOGY

A. Historical Overview

Biosensor technology is grounded in protein and cell immobilization. The first prototype biosensor was an enzyme electrode reported in 1962 which utilized immobilized glucose oxidase on a Clark pO_2 electrode for measuring the concentration of glucose in solution (3). This prototype enzyme electrode later served as the basis for the development of the first commercialized enzyme electrode and glucose analyzer. Advances followed with the development of potentiometric enzyme electrodes, microbial biosensors, and immunosensors (Table 1). Most recently, biological receptors have been successfully immobilized onto electronic transducers for measurement of classes of substances binding to each receptor.

Chapter 7 of this text presented a detailed discussion of immobilized enzyme electrode biosensors. The focus here will be on the use of antibodies, binding proteins, and receptors for biosensor fabrication.

B. Definition of Biosensor Technology

Figure 1 represents the basic component parts common to all biosensors. These are a bioactive surface which interacts with the substance, or analyte, to be detected; an electronic transducer which detects the biochemical event occurring between the bioactive surface and the analyte; and the support electronics which amplify and report the transducer output signal. The integration of these three components is basic to and required for a biosensor. Of these three biosensor components, development during the past 10 years has focused on transduction and bioactive surface (immo-

Table 1 Historical Development of Biosensors

Date	Biosensor	Ref.
1962	First prototype amperometric biosensor: glucose oxidase-based enzyme electrode for glucose determinations	3
1969	First prototype potentiometric biosensor: acrylamide-immobilized urease on an ammonia electrode for urea determinations	4
1972–74	First commercial enzyme electrode (for glucose) and glucose analyzer based on the electrode (Yellow Springs Instruments)	
1975	First microbe-based biosensor: immobilized *Acetobacter xylinium* in cellulose on an oxygen electrode for ethanol determination	5
	First binding protein sensor: immobilized concanavalin A in a polyvinyl chloride membrane on a platinum wire electrode for measuring yeast mannan	6
1975–76	First immunosensors: immobilized ovalbumin on a platinum wire electrode for ovalbumin antibody determination; antibody to human immunoglobulin G (hIgG) immobilized in an acetylcellulose membrane on a platinum electrode for measuring hIgG	6,7
1986	First tissue-based biosensor: antennules from blue crabs mounted in a chamber with a platinum electrode for measuring amino acids	8
1987	First receptor-based biosensor: acetylcholine receptor immobilized onto a capacitance transducer for measuring cholinergic agents	9

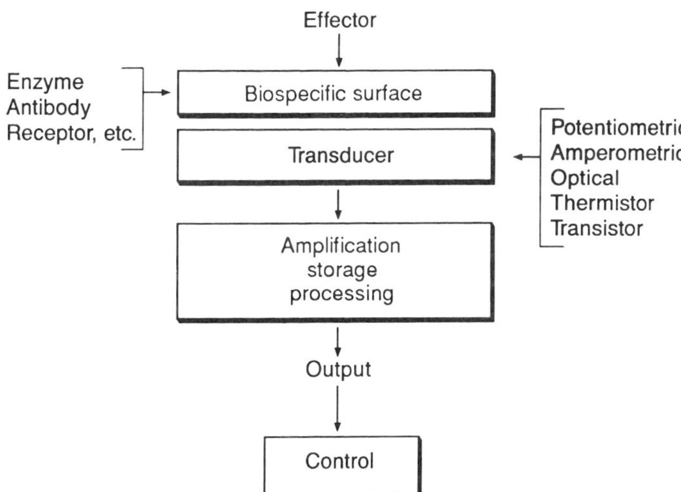

Figure 1 Basic biosensor component parts.

bilization) technology since electronic amplification and signal reporting in biosensor systems can utilize already developed technologies.

Table 2 summarizes common biosensor transducer technologies. The basis for these technologies and details on transducer fabrication have been reported in detail elsewhere (2,10–13). Many variations of the transducers listed in Table 2 are possible. Because of the numbers and types of transducers available, however, transducer technology is no longer a major limitation to biosensor development. Rather, the integration of the transducer with the bioactive surface presents a greater challenge.

The major technological focus in current biosensor research and development is bioactive surface preparation and stabilization on an appropriate

Table 2 Biosensor Transducer Technologies

Base technology/type	Parameter(s) measured	Examples[a]
Photometric		
Light absorption, scattering, or refractive index	Changes in light intensity, color or emission	Ellipsometry, Internal reflectometry, Laser light scattering,
Fluorescence or luminescence activation, quenching or polarization	Changes in fluorescence or luminescence	Surface plasmon resonance, Optrodes, Fiber optic wave guides, Fluorescence polarization
Electronic		
Amperometric	Change in applied current	Enzyme, antibody and whole cell electrodes
Potentiometric	Changes in voltage	Enzyme electrodes, FETs, ENFETs, some immunosensors
Capacitance/Impedance	Changes in impedance	Interdigitated electrode capacitors, Conductimeters
Acoustical/Mechanical		
Mass, density	Changes in weight	Piezoelectric devices (e.g., QCM devices)
Acoustical	Changes in amplitude, phase, or frequency of an acoustic wave	SAW devices
Calorimetric		
Thermistor	Change in temperature	Enzyme and immunoenzyme reactors

[a]Abbreviations: FET, field effect transistor; ENFET, enzyme-FET; QCM, quartz crystal microbalance; SAW, surface acoustic wave.

transducer. This requires protein immobilization with retention of activity and stabilization of activity under storage and use conditions. Such immobilization has used primarily entrapment, adsorption, and covalent immobilization methods. The specific types of immobilization being used for biosensor fabrication and specific biosensor examples will be discussed below.

C. Biosensor Types

Combinations of immobilization and transducer technologies have been used to produce two major types of biosensors: enzyme/metabolic biosensors and bioaffinity sensors (Fig. 2) (14,15). Enzyme/metabolic biosensors are exemplified by enzyme and cell electrodes and have been discussed in detail in Chapter 7 of this text. They function by enzymatic action on the analyte to be detected. Some product of that reaction is then detected by their ion- or gas-specific electrode transducer. These sensors are available commercially and represent the first generation of biosensor technology.

Bioaffinity sensors represent the second generation of biosensor technology. These sensors are based on binding interactions between the immobilized biomolecule and the analyte of interest. In their purest form, bioaffinity sensors detect the binding event itself and readout is proportional to analyte–biomolecule binding (Fig. 3). These have been termed

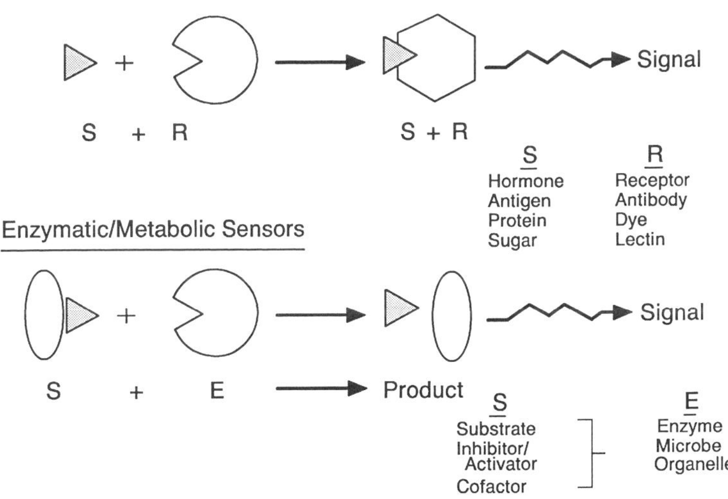

Figure 2 Major biosensor types.

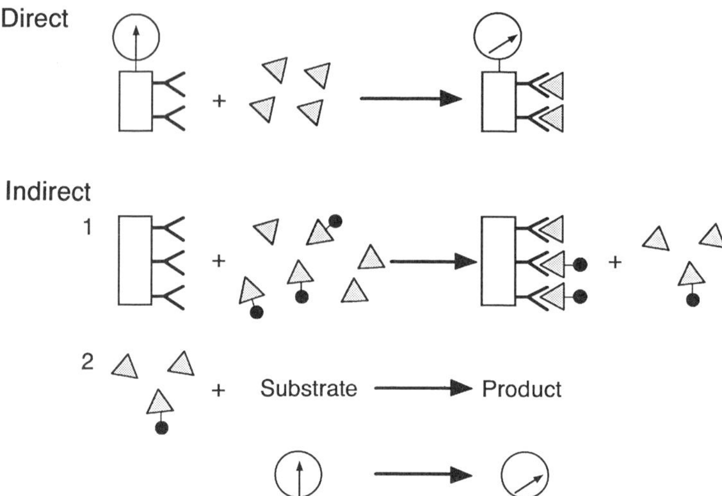

Figure 3 Direct and indirect biosensor formats. Indirect biosensors utilize multiple steps and/or reagents.

direct bioaffinity sensors (15). Other bioaffinity sensors may utilize detection of other events which come about at or near the bioactive surface as the result of the binding event. For example, as shown in Figure 3, the sample analyte may compete with added analyte conjugated with an indicator (such as an enzyme or fluorescent indicator). The amount of analyte conjugate left in solution (or, alternatively, bound to the bioactive surface) is then quantitated, e.g., by enzyme activity or fluorescence. In many ways, this type of indirect bioaffinity sensor is a modification of enzyme or fluorimetric immunoassays commonly used in clinical diagnostics but in a biosensor format.

As discussed here, only integrated detection and reporting devices are considered as biosensors. Many commercial companies are developing laboratory instruments for application to diagnostic needs in clinical, analytical, and pharmaceutical laboratories. While the detection basis for these instruments may be drawn from biosensor technology (e.g., immobilized enzymes, antibodies or receptors on light addressable capacitance, surface plasmon resonance, fiber optic wave guide, and other transducers), their application purpose, assay and support equipment complexity, and cost are not consistent with the definition of a biosensor as a portable, automated, and cost-effective detection device for use in nonlaboratory environments.

III. BIOMOLECULE IMMOBILIZATION IN BIOSENSOR FABRICATION

A. Overview of Immobilization Requirements

Immobilization technology is key to biosensor fabrication. As will be discussed below, a variety of immobilization methods have been used in the development of successful biosensors. It is important to note, however, that with very few exceptions, bioaffinity sensor development to date has not progressed past laboratory or (at best) manufacturing prototypes. While enzyme electrode-based biosensors have been commercially available for over 15 years, no bioaffinity sensor has been commercialized to date.

Biomolecular immobilization remains the primary challenge to commercialization of bioaffinity sensors. The transfer of a prototype biosensor from the laboratory to manufacturing with retention of functionality, reliability, and quality (accuracy and precision) requires a simple method which reproducibly immobilizes and stabilizes the detector biomolecule onto the transducer. For this reason, elegant multistep immobilization methods, which are simple to perform in the laboratory by an experienced technician, are difficult to transfer to manufacturing. Likewise, an immobilization method which requires extraordinary handling conditions (such as preparation/storage in the cold or the use of complex buffers and reagents) will also encounter problems in manufacturing. The aim of any biosensor immobilization method should be a one-step process in which immobilization media or mediator and the biomolecule are applied directly to the transducer under normal environmental conditions. The immobilization technology should also result in a stable product, able to be stored under reasonable conditions (e.g., 4–40°C at varying humidities) for at least one year with retention of >90% of the original activity of the biomolecule. If possible, the immobilization method should also be inert to conditions required for dissociation of the analyte from the immobilized biomolecule, thus allowing for biosensor recycling.

Many of these requirements have been met for enzyme electrode-based biosensors and, as a result, these biosensors have been successfully commercialized. Bioaffinity sensors appeared approximately 7 years after the first enzyme electrode (Table 1). This time lag, together with additional problems associated with immobilizing molecules as labile as biological receptors, may explain the problems which have been encountered in commercialization of bioaffinity sensors. New immobilization methods, which will be discussed below, now promise to advance bioaffinity sensor technology and lead to viable products in the 1990s.

B. Immobilization Methods

Biomolecule immobilization onto transducers utilizes a variety of methods. These have been reviewed elsewhere in this text (see Chapters 2,3,4, and 7) and in various monographs (15–19). Table 3 summarizes the major immobilization approaches applied to biosensors. All methods have both advantages and disadvantages for biosensor applications, and thus choice of method and optimization are critical to biosensor development.

1. Entrapment and Adsorption

The oldest immobilization methods for biosensors are simple entrapment of the biomolecule into a polymerized film and adsorption onto a film or

Table 3 Biomolecule Immobilization Methods for Biosensors

Method	*Advantages*	*Disadvantages*
Entrapment into/by membranes and films coating or associated with a transducer	Nonchemical treatment; mild reaction conditions	Leakage possible; membrane/film may limit analyte diffusion
Entrapment into a lipid bilayer or a liposome coating or associated with a transducer	Nonchemical treatment; mild reaction conditions; mimics natural membranes	Leakage possible; very susceptible to changes in pH, temperature, and ionic conditions
Adsorption directly to the transducer or a film/membrane on the transducer	Nonchemical treatment; mild reaction conditions	Leakage possible; very susceptible to changes in pH, temperature, and ionic conditions
Biological binding to a specific molecule already immobilized on the transducer	Strong, reversible, noncovalent binding; mild reaction conditions	Leakage possible; binding site for matrix must be different from the analyte site
Cross-linking into membranes/films	Mild reaction conditions; high analyte permeability; can mimic natural membranes	Possible loss of activity
Covalent binding directly to the transducer	Strong bond resistant to large changes in pH, ionic strength, and temperature	Possible loss of activity; transducer recycling unlikely
Covalent binding to a membrane/film on the transducer	Strong bond resistant to changes in pH, ionic strength and temperature; transducer may be recycled	Possible loss of activity; stability may depend on membrane/film-transducer bond

directly to the transducer. Many of the enzyme electrodes utilize these types of immobilization (16–18) as do the early bioaffinity sensors (see below). Entrapment and adsorption are the most gentle immobilization methods, and their use can result in maximum retention of biological activity after immobilization. However, the weak bonds between the biomolecule and the matrix or transducer can be easily disrupted to result in leakage of the biomolecule off the transducer and resulting biosensor degradation.

Highly hydrophobic bilayer lipid membranes and liposomes have also been used for entrapment of labile molecules such as biological receptors (see below and Refs. 15,19). Again, the resulting films or layers, while functional in the laboratory for short periods of time, are inherently unstable with time especially with changes in pH and temperature. It is thus unlikely that such lipid-based membranes will be viable for commercial mass production.

In spite of their limitations, entrapment and adsorption immobiliization methods may be necessary in cases where a very labile biomolecule is used and/or recycling of the transducer is required.

2. Biological Immobilization

Immobilization utilizing biological binding differs from adsorption due to its specificity, the strength of the bonds formed and the opportunity for directional immobilization. For example, protein A, which strongly binds to the Fc region of antibodies, has association constants with many antibodies, in excess of 10^6 M^{-1} (20). The strength of the protein A–antibody complex formed together with the directional orientation of the antibody which results can be used for biosensor immobilization (see below). Another example of biological immobilization is the avidin–biotin complex with a K_a of approximately 10^{15} M^{-1} (21). Since avidins only require the bicyclic ring system of biotin for recognition and binding, the carboxylic group of the biotin side chain may be modified for reaction with amino, imidazole, hydroxyl, sulfhydryl, and aldehyde functional groups on proteins. This versatility again allows directional immobilization of the biomolecule onto the transducer surface.

Biological immobilization can encounter problems if the binding site for the immobilization reaction is the same as or near the analyte binding site. Leakage can also be a problem, although the high binding constants minimize this possibility. Finally, binding molecules such as protein A and avidin are expensive, and cost can become a limiting factor in their use for biosensor products.

3. Covalent Binding

Covalent binding of biomolecules either directly onto the transducer or onto a film/membrane coating the transducer has been utilized for a variety of biosensors. Covalent binding provides stable bioactive surfaces which are resistant to wide ranges of pH, temperature, and ions. Typical agents used for covalent binding include cynanogen bromide, carbodiimides, carbonyldiimidazole, glutaraldehyde, and various succinimide derivatives.

Covalent binding can also result in loss of biomolecule activity by reaction of the agent with the biomolecule active site and/or denaturation and destruction of the biomolecule. In addition, most covalent immobilization agents require narrow pH ranges for optimal coupling, often at low (pH 4–6) or high (pH 9–11) pH, which may change or destroy the biomolecule. It is thus necessary to evaluate the effect of the covalent coupling method on the biomolecule to be immobilized prior to developing and optimizing the method.

4. Cross-Linking

Cross-linking provides a means of biomolecule immobilization with the positive aspects found with entrapment and covalent immobilization. Cross-linking involves the use of an agent which induces intermolecular linkages between proteins or proteins and films/membranes. Cross-linking agents are usually bifunctional, i.e., they contain two reactive groups and can thus act as a bridge between two functional groups on different proteins, on the a protein and a matrix, or on the same protein (17,22). Such agents can be homobifunctional (i.e., both reactive groups are the same) or heterobifunctional (i.e., the reactive groups are different). The most common cross-linking agent is glutaraldehyde, although many others are available such as hexamethylene diisocyanate, 1,5-difluoro-2,4-dinitrobenzene, dimethyl suberimidate, disuccinyl suberate, and N-gamma-maleimidobutyrloxy succinimide ester.

The primary use for cross-linking agents in biosensor fabrication is to stabilize physically adsorbed proteins in a film or membrane. Glutaraldehyde has also been used to form membranes from proteins such as albumin and collagen which entrap small concentrations of biomolecules (23) and act as the bioactive surface in some biosensors. In these and other cases, cross-linked proteins retain their activity while being more resistant to extremes of pH, temperature, and ion concentrations than adsorbed proteins.

Cross-linking can cause some loss in biomolecule activity. This is probably due to intramolecular cross-linking and modification or blockade of the biomolecule active site. Such losses may be minimized by utilizing low concentrations of the biomolecule and cross-linking agent in the immobilization reaction.

IV. ANTIBODY-BASED BIOAFFINITY SENSORS

A. Overview

The specificity, sensitivity, and availability of antibodies has resulted in hundreds of immunodiagnostic products in the past 15 years. These same attributes are being exploited for development of antibody-based biosensors (immunosensors). While still limited by complexity (multireagent and/or multistep formats), time of detection (minutes to hours), costly support equipment, and difficulties in mass production, these limitations are being met and overcome by advances in immunosensor technology. For example, recently developed immunosensors are responding to low parts-per-billion (ng/ml, ppb) or parts-per-trillion (pg/ml, pptr) analyte concentrations in seconds utilizing simplified (including homogeneous) formats and portable hardware.

Table 4 lists a number of immunosensors developed to at least the laboratory prototype stage, their detection basis, sensitivity, assay time, and mode of biomolecule immobilization. A further distinction is made between direct and indirect immunosensors, as discussed above (Figure 3). Immunosensors have also been the subject of a number of reviews (15,24–28).

The first, prototypical immunosensors utilized calomel electrode cells as the transducer separated by a membrane containing the immobilized detection biomolecule (29,44,52). Interaction of the analyte with the membrane resulted in voltage or current changes. While not true biosensors, these cells were the direct predecessors of immunosensors.

The first immunosensor laboratory prototypes were simply electrodes coated with antibodies or antigens (7,30–32,40). Changes in potential or current were related to the amount of analyte present. Unfortunately, these first immunosensors were very susceptible to environmental effects (pH, temperature, and ion content of the medium) and to nonspecific binding. In addition, problems were encountered in reproducing their fabrication in different laboratories. These problems led to conclusions by some authors that steady-state potentiometric immunosensors were little more than experimental artifacts (67,68).

These conclusions appear unfounded in light of the rapid advances being made in immunosensor development. As reported in Table 4, a significant number of direct immunosensors (both potentiometric and amperometric) have been successfully developed for analytes such as IgG, hIgG, HSA, AFP, IgE, and DNP (36,38,40,46). These sensors utilize more sophisticated transducers and immobilization methods to overcome problems with background and nonspecific binding (see below).

Table 4 Examples of Antibody-Based Biosensors

Biosensor	Detection basis	Immobilized biomolecule	Immobilization method	Sensitivity	Assay time	Ref.
Direct Measurement						
Wasserman antibody	Change in potential, calomel electrode cell	DPG + cholesterol + PC	Entrapment into a cellulose triacetate membrane	1:800 titer	10–30 min	29
Wasserman antibody	Change in potential, FET	DPG + cholesterol + PC	Entrapment into a PVC membrane	NR	5–10 min	30
hCG α hCG	Change in potential, titanium electrode	α hCG hCG	Direct binding to cyanogen bromide-activated titanium wire	65 IU/ml, ppm	20–50 min	31,32
hCG	Change in current glassy carbon electrode, ferrocene mediator	α hCG and GOD	Covalent carbodiimide coupling of GOD to the electrode, SMCC coupling of hCG to the GOD	10 IU/ml	15 min	33
α DNP α BSA	Change in potential, Ag-AgCl electrode	DNP- and BSA-dibenzo-18-crown-6 ion carriers	Direct binding into PVC and triacetyl-cellulose membranes	ppm	5–20 min	34,35
DNP α DNP	Change in potential, K^+ ISE	α DNP DNP	Entrapment of DNP in a collagen membrane; incorporation of DNP into a PVC membrane	ppb	15–20 min	36
HSA α HSA α BSA	Change in reflectance, surface plasmon resonance	α HSA HSA BSA	Adsorption to silver- or gold-coated glass cover slips (wave guides)	ppb	1–2 h	37

HSA	Change in gate voltage, ISFET	α HSA	Adsorption onto a GA-cross-linked PVB membrane coating the gate insulator	ppt	1–5 min	38
hIgG	Change in frequency, SAW piezo-electric transducer	α hIgG	Direct linkage to active aldehyde groups on silylated piezoelectric crystals	ppt	NR	39
IgG α IgG	Change in potential, titanium wire	α IgG IgG	Adsorption onto polypyrrole or polythiophene electrode films	ppb	minutes	40
hIgG	Change in scattered light, optical fiber	α hIgG	CDI coupling to carboxyl-functional, 1-μm latex beads	ppb	15 min	41
hIgG	Change in optical signal, ellipsometry with amplification	α hIgG	Adsorption onto epoxysllanized silicon wafers, and hydrophobic 12-μm silica beads (amplifier reagent)	ppm	1–2 h	42
α hIgG HSA	Change in reflectance, surface plasmon resonance	hIgG α HSA	Adsorption to gold-coated glass diffraction gratings	ppm	5–15 min	43
α IgG	Change in potential, iridium oxide capacitance electrode	IgG	Entrapment into a stearic acid LB film and adsorption of the film onto the electrode	ppm	2–5 min	43a
Human blood group types	Change in potential, calomel electrode cell	Blood group A and B erythrocyte lipids	Entrapment in cellulose triacetate membranes	ID of types A and B	30 min	44

Table 4 (Continued)

Biosensor	Detection basis	Immobilized biomolecule	Immobilization method	Sensitivity	Assay time	Ref.
AFP	EIA, oxygen electrode	PAPTG	CDI coupling to GA-activated pig skin/polypropylene film	ppb	1–2 h	45
AFP IgE	Change in capacitance, gold/silica semiconductor	α AFP α IgE	GA coupling to capacitor surface silane groups	ppb	15–60 min	46
Thyroxine	Change in current, platinum electrode, ferrocene mediator	GOD, Catalase and α Thyroxine	Adsorption onto electrode followed by entrapment with a collagen membrane	ppb	15–30 min	47
MTX	Change in optical waveguide evanescent wave	α MTX	Direct coupling to GA-activated optical wave guides	ppm	5–10 min	48
Digoxin	Change in potential, KCl-agar electrode	Digoxin-benzo-15-crown-5 conjugate	Incorporation into a PVC membrane	ppm	5–30 min	49
Benzo[a]pyrene	Change in fluorescence, optical fiber	α Benzo[a]pyrene	Periodate coupling to silanized quartz fibers	ppb	10–60 min	50
Benzopyrene tetraol	Change in fluorescence, optical fiber	α Benzopyrene tetraol	Adsorption onto protein A coated, 7-μm silica beads	ppb	45–60 min	51
Candida albicans	Change in potential, Ag-AgCl electrode cell	α *C. albicans*	GA-catalyzed incorporation into cellulose triacetate membrane	10^4 cells/ml	45–60 min	52
C. albicans	Change in frequency, piezoelectric crystals	α *C. albicans*	Direct linkage to active aldehyde groups on silanized piezoelectric crystals	10^7 cells/ml	30–60 min	52,53

Indirect Measurements						
hIgG	EIA-linked oxygen electrode	α hIgG	Direct binding to epoxy-activated cellulose membrane	ppt	60 min	7
hIgG	Competitive FIA, optical fiber	α hIgG	Direct coupling to silanized, GA-activated optical wave guide	ppm	10–15 min	48
IgG	EIA-linked, magnetic antibodies, oxygen electrode	α IgG	Covalent coupling to magnetic particles	ppb	60 min	54
IgG	Competitive FIA, optical fiber	α IgG	GMBS-catalyzed coupling to silanized optical fibers	ppb	1–2 h	55
hCG	EIA-linked, oxygen electrode	α hCG	Direct binding to epoxy-activated cellulose membrane	0.02 IU/ml	>60 min	56
hCG	Competitive EIA, pH electrode	α hCG	GA cross-linkage into albumin membrane	5 IU/ml	>2 h	57
hCG	EIA-linked, magnetic antibodies, oxygen electride	α hCG	Direct coupling to cyanogen bromide activated, magnetizable cellulose	0.2 IU/ml	20 min	58
HSA	EIA-linked, oxygen electrode	α HSA	GA Coupling to silanized molecular sieve	ppm	60 min	52
HBsAg	Competitive EIA, iodide ISE	α HBsAg	GA cross-linkage into protein membranes	ppb	2–3 h	59
AFP	EIA-linked, oxygen electrode	α AFP	GA cross-linkage into a cellulose membrane	ppm	>2 h	60
α DNP-cap-PE	Complement-mediated TPA^+ release, ISE	DNP-cap-PE	Entrapment into liposomes containing TPA^+	1:4000 titer	30–45 min	61
LDH5 isoenzyme	EIA-linked, ISE	α LDH5	CDI coupling to carboxyl-functional electrode	ppb	1–2 h	62

Table 4 (Continued)

Biosensor	Detection basis	Immobilized biomolecule	Immobilization method	Sensitivity	Assay time	Ref.
Theophylline	Complement-mediated liposome lysis, EIA-linked, oxygen electrode	Theophylline-PE conjugate	Entrapment into liposomes	ppb-pptr	30–60 min	63
Theophylline	Competitive EIA, oxygen electrode	α Theophylline	Direct coupling to GA-activated nylon net pretreated with hexanediamine	ppb	1–24 h	64
17-β-Estradiol	Competitive EIA, oxygen electrode	α 17-β-Estradiol	GA entrapment into a gelatin membrane	ppm	minutes	65
α Gangliosides	Complement-mediated TPA^+ release ISE	Gangliosides	Entrapment into TPA^+-containing liposomes	1:2000 titer	5–10 min	66

Abbreviations: NR, not reported; h, human; α, antibody to the indicated material; ppt, parts-per-thousand (mg/ml); ppm, parts-per-million (μg/ml); ppb, parts-per-billion (ng/ml); pptr, parts-per-trillion (pg/ml); ISE, ion-selective electrode; FET, field effect transistor; ISFET, ion-specific FET; SAW, surface acoustic wave; EIA, enzyme-linked immunoassay; FIA, fluorescence immunoassay; PC, phosphatidylcholine; PE, phosphatidylethanolamine; DPG, cardiolipin; CG, chorionic gonadotropin; GOD, glucose oxidase; DNP, dinitrophenol; DNP-cap-PE, dinitrophenylaminocaproyl PE; BSA, bovine serum albumin; HSA, human serum albumin; IgG, immunoglobulin G; IgE, immunoglobulin E; AFP, α-fetoprotein; MTX, methotrexate; PAPTG, p-aminophenylthio-β-D-galactopyranoside; TPA^+, tetrapentylammonium ion; HBsAg, hepatitis B surface antigen; LDH, lactate dehydrogenase; CDI, carbodiimide; GA, glutaraldehyde; SMCC, succinimidyl 4-(N-maleimidomethyl)cyclohexane-1-carboxylic acid; GMBS, N-gamma-maleimidobutyryloxy succinimide ester; PVC, polyvinyl chloride; PVB, polyvinylbutyral; LB, Langmuir-Blodgett.

Newer immunosensors are not limited to simple electrode-based transducers but rather are utilizing a variety of optical and electrochemical transducers including surface plasmon resonance, optical fiber wave guides, ellipsometry, interdigitated electrode capacitors, and field effect transistors (FETs). Given these advances and the state of the technology, immunosensors promise to have a significant impact on antibody-based detection and diagnostic products in the next 10 years. More detailed reviews on immunosensor development and applications have been presented elsewhere (2,13,15,69).

B. Immunosensor Immobilization Methods

As with all biosensors, immobilization of an antibody (or antigen) for an immunosensor should utilize a simple method to result in retention of biological activity, high analyte sensitivity, and sensor stability during storage and use. Few immunosensors to date have accomplished all of these requirements.

The immobilization methods used for immunosensors include a variety of adsorption, entrapment, cross-linking, and covalent methods (Table 4). A key consideration in antibody coupling to the transducer is to maintain, reproducibly, the highest possible binding activity after immobilization while conserving the amount of antibody used. This is especially important in cases where a monolayer of antibody is desired on the transducer such as with optical fiber-based immunosensors. In such cases, efforts are being made to immobilize the antibody to the transducer through the antibody Fc (or "tail") region, thus orientating the antibody Fab (antigen-binding) regions outward and away from the transducer surface. Lack of antibody orientation in immunosensors which depend on small changes in parameters such as reflectance or fluorescence at the surface of the transducer can result in low analyte sensitivity.

One method for directing antibody orientation is to oxidize the sugar residues in the antibody Fc region (e.g., by periodate oxidation) and then to react the oxidized antibody with the transducer surface which has been activated to react with the sugar aldehydes (e.g., by hydrazine activation) (see Chapter 4). This method has yet to be applied to an immunosensor.

A second method for antibody orientation is to utilize specific binding proteins such as protin A which, again, binds to the antibody Fc region. For example, in the fabrication of an immunosensor for a benzo[a]pyrene metabolite (51), the antibody for the metabolite was adsorbed to a protein A-coated support to result in orientation of the immobilized antibody.

In a nonbiosensor study directed at monoclonal antibody immobilization on agarose and polymeric supports, a comparison between nondirectional (cyanogen bromide- and N-hydroxysuccinimide-activated) and directed (hydrazine-activated) immobilization showed that directional immobilization increased specific antigen binding by approximately threefold (70). In addition, this study confirmed earlier studies on immobilized antibody density and binding activity, i.e., an antibody density of 0.34 mg/ml of polymeric beads bound approximately two times more antigen than beads containing an antibody density of 5.6 mg/ml. These results are consistent with studies on immobilized enzymes, which have shown that high amounts of immobilized protein do not necessarily correlate with increased support activity (see Chapter 4). Rather, in such cases, the high amounts of protein immobilized may be due to the formation of large multiprotein complexes which mask or block active and binding sites. Thus, optimization studies are required in the development of a bioactive surface, especially in cases such as optical biosensors where thin or monolayers of biomolecules are required.

High retention of antibody binding activity after immobilization onto optical fibers is also possible using nondirectional immobilization methods. For example, a comparison has been made between the use of various silylation reagents for initial preparation of the optical surface for immobilization, and various heterobifunctional cross-linking reagents for attachment of antibodies to the silylated fiber (71). It was found that the best retention of antibody activity (up to 55% binding activity retention or 0.55 mol antigen bound/mol antibody immobilized) resulted when the fibers (or glass cover slips) were first silylated with mercaptomethyldimethylethoxysilane or 3-mercaptopropyltrimethoxysilane, and the thiol-terminal silane was then reacted with N-gamma-maleimidobutyryloxy succinimide ester to result in a succinimide-activated surface. Antibody terminal amino groups react with the succinimide groups to form a stable covalent linkage. In addition, a number of blocking agents were evaluated for elimination of nonspecific antigen binding to the immobilized antibody surfaces including BSA, gelatin, lysozyme, Triton X-100, and Tween 20. It was found that treatment of fibers with 2 mg/ml BSA reduced nonspecific antigen binding by approximately 90%.

In the case of many transducers, and especially electrochemical transducers, antibody monolayers are not required and random orientation of antibodies can result in a sensitive and accurate immunosensor. For example, potentiometric, amperometric, and capacitance immunosensors tend to utilize relatively thick membranes and films (from 1 to 50 μm thick) for entrapment or attachment of the antibody or antigen (7,30,47,52,54,56,57, 59,60,65). Such membranes/films, if highly permeable, can present the antibody to the incoming analyte throughout its structure, thus minimizing antibody orientation as a limiting factor to sensitivity. For example, we have developed a generic technology for the rapid immobilization of antibodies and receptors into stabilized films (approx. 10μm thick as measured by ellipsometry), which are applied to interdigitated electrode capacitance transducers to result in a functional biosensor (9,15). The method for preparation of the films and the support electronics package for the biosensor are described below. We used this technology to prepare a biosensor for serum antibody (raised in goat) to human IgG (hIgG). In one formulation, hIgG was immobilized into a bovine serum albumin (BSA) membrane containing stabilizers such as phosphatidyl choline and polymerized with glutaraldehyde. The BSA, hIgG, stabilizers, and glutaraldehyde were mixed and applied directly to the interdigitated chip. Within 15 minutes, polymerization was complete and after aging at room temperature overnight, the hIgG chip and a parallel non-hIgG membrane-coated control chip were simultaneously challenged with serum containing anti-hIgG. As shown in Figure 4, we found a direct dose-response (in this case increased output impedance with increased serum antibody concentration) output as the biosensor was challenged with dilutions of serum. The sensitivity of this (nonoptimized) antibody-based biosensor is in the mid- to low-ppb (ng/ml) range. The biosensor reaches output signal equilibrium in 5–10 seconds, and the sensor may be recycled by washing with NaCl or other salt solutions. The

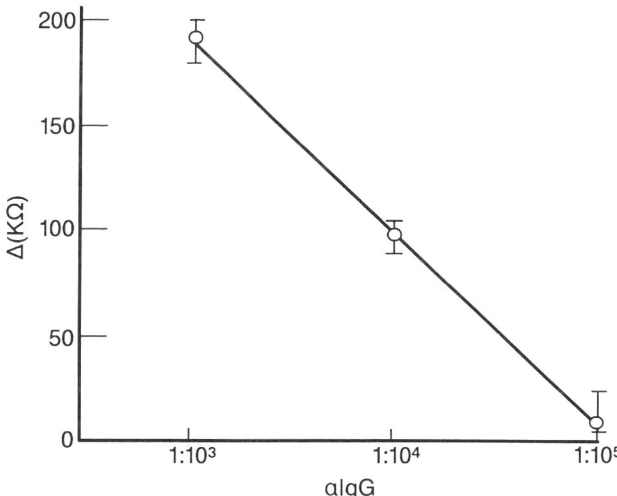

Figure 4 Response of a capacitance anti-human IgG (α IgG) biosensor to challenge by serum containing α IgG. $\Delta K\Omega$ represents the difference between the ouput impedance from the test chip (containing hIgG) minus the output impedance from the control chip (no hIgG). Measurements were made on the electronics units described in Figures 9 and 10.

stability of the biosensor is at least 9 months stored dry at room temperature and at least 2–3 days in use.

C. Immobilization and Immunosensor Performance

The latter biosensor example illustrates three basic challenges common to all biosensors including immunosensors: background interference, regeneration, and storage/use stability. All are directly linked to the immobilization technology used for the biosensor. Background interference includes environmental effects (such as changes in temperature, pH, and ionic content of the test medium) and nonspecific binding to the bioactive surface. As described above, a variety of compounds including proteins (BSA and gelatin) and detergents (Tritons and Tweens) can be used to minimize background effects. In the case of the anti-hIgG sensor just described, the base BSA membrane itself has low nonspecific binding characteristics (<2% binding of nonantigen protein). In addition, the use of control and test sensor chips allows subtraction of the control chip output (which represents background) from the test chip output to automatically correct for background (see below). The use of low nonspecific binding surfaces and automatic background correction using a control surface can successfully minimize background problems. As these methods advance in effectiveness, immunosensor selectivity should routinely achieve pptr analyte levels.

Recycling of immunosensors involves dissociation of the antibody–antigen complex. A variety of chaotrops have been used to accomplish this at biosensor surfaces including 0.5–1 M NaCl or KCl, ethylene glycol, 8 M urea, 1–100 mM citric acid, 2–4 M $MgCl_2$, 0.05–0.2 mM Tris-HCl with/without 1 mM KCl, glycine-HCl, 0.1 M borate buffer, and 4 M phosphate buffer (33,34,36,38,45–47). While antibody-antigen dissociation using such agents can lead to complete regeneration of sensor activity, they may also cause degeneration of sensitivity and response time. This latter problem can be especially acute in the case of indirect immunsensors which utilize immobilized enzymes (33).

Immunosensor storage and use stability has not received intensive study. In those cases where it has, the immobilized antibody detection component of the sensors may be used from as few as 1–2 cycles to as many as 50 cycles with no loss in sensitivity or response time (34,36,38,45,46). Storage stability (wet or dry at 4–25°C) ranges from a few days to over 18 months. Each sensor differs in its stability characteristics, which are a function of the sensor immobilization technology and architecture. Until immunosensor fabrication is achieved in significant unit numbers and statistically significant storage studies are carried out, the real storage and use stability of the sensors will not be known.

V. RECEPTOR-BASED BIOSENSORS

A. Overview

Biological receptors are proteins or complexes of protein with carbohydrate and/or lipid which interact specifically with biological and chemical substances (also known as messengers, transmitters, or ligands). The ligands may be low molecular weight compounds such as neurotransmitters, amino acids, steroid hormones, or metal ions; larger molecules such as peptide hormones and toxins, proteins and complexed proteins; or total life forms such as viruses and bacteria. In all cases, the binding of the ligand to its specific receptor results in activation or blockade of a series of biochemical reactions leading to a physiological event in the receptor-bearing organism. Such events include neural transmission, hormonal regulation, and viral infection. Biological receptors, then, are critical mediators of both intra- and intercellular communication.

A recent review listed over 30 defined receptor classes able to interact with over 300 individual and classes of chemical and biological materials (15). Table 5 presents a summary of these receptor classes. All of these receptors and subclasses of these receptors are candidate detection biomolecules for biosensors.

Unlike antibodies, receptors do not react with one specific substance. Rather, receptors interact with classes of substances in which each substance has a common structural component able to bind to the receptor binding site(s). Because of this binding site commonality, many seemingly unrelated substances can interact with the same receptor. In the case of naturally occurring ligands, receptors interact with messangers normally found in the organism but also bind many bioactive substances and toxins which have evolved to specifically interact with the receptor either in predator or prey. The many plant and invertebrate alkaloid and peptide toxins and psychoactive substances acting on vertebrate neural receptors exemplify this evolution. Man himself utilizes the specificity of receptor binding to direct the synthesis and use of specific drugs and therapeutic agents such as anesthetics, stimulants, tranquilizers, synthetic hormones, and antimicrobial agents. There is an extensive literature base on receptors and their function, and this area is the subject of many excellent reviews (72–77).

Receptor-based biosensors provide a unique capability for diagnostics and detection, enabling the detection of an unknown analyte based on its biological activity. Antibody-based biosensors are limited to detection of (usually) one analyte and thus are best suited to detection and quantification of that analyte in samples known to contain it. Receptor-based biosensors cannot be specific for a single analyte but can be used to screen samples for the class of analytes which bind to the receptor. Thus receptor-

Table 5 Examples of Receptor Classes and Their Ligands

Receptor	Ligand examples	
	Naturally occurring (source)	Synthetic
Cholinergic (nicotinic, muscarinic)	Acetylcholine (Ub)[a]	Carbachol
	α-Conotoxins (snail)	Organophosphates
	Viscotoxins (mistletoe)	Succinyldicholine
	Cobra/Krait toxins	Decamethonium
	Muscarine (mushroom)	Hexamethonium
Cholinergic (presynaptic)	Black widow spider toxin	Hemicholiniums
	Botulinum toxins	
	Tetrahydrocannabinol (plant)	
Adrenergic (alpha-, beta-)	Epinephrine (animal, plant)	Dibenamine
	Ergot alkaloids (fungi)	Phentolamine
	Yohimbine (plant)	Propranolol
Adrenergic (presynaptic)	Cocaine (plant)	Amphetamines
	Mescaline (plant)	Imipramine
Dopaminergic	Dopamine (animal)	Apomorphine
	Bromocriptine (fungi)	Spiperone
Serotonergic	Serotonin (animal, plant)	Imipramine
	LSD (plant, fungi)	Mianserin
	Psilocybin (mushroom)	Trazodone
Opiate	Codeine (plant)	Heroin
	Morphine (plant)	Methadone
	Enkephalins (animal)	Naloxone
Histamine	Histamine (Ub)	Cimetidine
		Mepyramine
γ-Amino butyric acid (GABA)	GABA (Ub)	Barbiturates
	Muscimol (mushroom)	Lindane
	Picrotoxin (plant)	Benzodiazepine
Glycine	Glycine (Ub)	Benzodiazepines
	Strychnine (plant)	Chlormethiazole
Glutamate	Glutamic acid (Ub)	4-Fluoroglutamate
	Kanic acid (algae)	Methyl ibotenate
Adenosine	Adenosine (Ub)	Alloxazine
	Caffeine (plant)	Etazolate
	Theophylline (plant)	Phenyltheophylline
Adenosine triphosphate (ATP)	ATP (Ub)	Antazoline
	Quinidine (plant)	Dipyridylisatogen
Sodium (potassium) channels	Scorpion toxin	Phencyclidines
	Palytoxin (coral)	DDT
	Tetrodotoxin (invertebrates)	Ketamine
	Veratridine (plant)	Amantadine

Table 5 (Continued)

Receptor	Ligand examples: Naturally occurring (source)	Ligand examples: Synthetic
Calcium channel	Batrachotoxin (frog)	Phencyclidines
	Maitotoxin (algae)	Verapamil
	Taicatoxin (snake)	Dibenamine
Peptide hormone (various)	Angiotensins (animal)	Peptide analogs (various)
	Insulin (animal)	
	Glucagon (animal)	
	Vasopressin (animal)	
Eicosanoid	Leukotrienes (animal)	Various analogs
	Prostaglandins (animal)	
Immune system cell	Anaphylatoxins (animal)	Imidodisulfamides
	Interleukins (animal)	Various analogs
Estrogen	Estradiol (animal)	Diethylstilbestrol
	Estriol (animal)	Tamoxifen
	Estrone (animal)	
Androgen	Testosterone (animal)	Flutamide
	Androsterone (animal)	Methyltrienolone
Progesterone	Progesterone (animal)	Norethindrone
	Pregnanediol (animal)	Dihydrogestrone
Glucocorticoid	Cortisol (animal)	Dexamethasone
	Corticosterone (animal)	Phenylbutazone
Retinol (cellular)	Retinol (plant)	
	Citral (plant)	
Retinoic acid	Retinoic acid (Ub)	Various analogs
	13-*cis*-Retinoic acid (Ub)	
Sweet taste	Sucrose (plant)	Aspartame
	Glucose (Ub)	Saccharin
	Fructose (Ub)	Sodium cyclamate
Choleragen	Gangliosides (animal)	Thiogalactopyranosides
	Raffinose (plant)	
Auxin	Indolacetic acid (plant)	2,4-D; 2,4,5-T[a]
	Phenylacetic acid (plant)	Napthylacetate
Cytokinin	Kinetin (plant)	Azidopurines
	Benzyladenine (plant)	Aminobenzylpurines
E Colicins	E Colicins (bacteria)	
	Bacteriophage BF23	
	Vitamin B_{12} (bacteria)	
Viral (glycoproteins, glycolipids)	HIV-1	
	Polyoma virus	
	Reovirus	

[a]Ub = Ubiquitous, found in plants, animals and microbes; 2,4-D = 2,4-dichlorophenoxyacetic acid; 2,4,5-T = 2,4,5-trichlorophenoxyacetic acid.

and antibody-based biosensors provide complementary functions: the former as a first-line screening detection device, the latter as a confirmatory detection device. For example, in the case of detection of abused drugs such as opiate narcotics, an opiate receptor-based biosensor could detect the presence of one of these substances in the field (e.g., opium, heroin, morphine, codeine, etc., and their immediate metabolites), and then antibody-based biosensor panels could identify the actual compound. Working in concert, antibody- and receptor-based biosensors may provide significant advances in real-time detection and identification for medical, industrial, environmental, and defensive applications.

B. Receptor-Based Biosensor Immobilization Methods

Similar to antibody-based biosensors, immobilization technology is key to receptor-based biosensor development. Table 6 lists examples of receptor-based biosensors developed to at least the laboratory prototype stage. Again, similar to immunosensors, both direct and indirect detection modes have been used for these sensors.

1. Binding Molecules

Many of the sensors listed in Table 6 utilize biomolecules which are not, by strict definition, biological receptors. For example, binding pairs such as yeast mannan (or sugar)-concanavalin A, aprotinin-trypsin, riboflavin-aporiboflavin binding protein, and biotin-avidin represent binding protein–ligand interactions rather than receptor–ligand interactions. Additionally, binding proteins such as avidins, concanavalin A, and aporiboflavin binding protein are not complexed proteins as most membrane-associated receptors are. Thus, these binding proteins tend to be more robust toward immobilization, similar to antibodies, than true receptors with their mixed solubility and labile protein–lipid–carbohydrate complexes.

The binding protein- (or peptide-)based biosensors are, however, the direct predecessors of receptor-based biosensors. Because of their stability, the first binding protein biosensors were modeled after potentiometric and amperometric immunosensors. For example, these first binding sensors utilized either covalent linkage or adsorption of the biomolecule directly onto platinum and ion-selective electrodes and then monitored changes in potential related to analyte concentration (31,78,80,83,87,97). Advances in transducer and immobilization technology resulted in more sophisticated binding protein sensors based on piezoelectric crystals, field effect transducers, ellipsometry, and optical fibers (42,79,86,95–97), and used immobilization methods better directed at maintaining the binding activity and stability of the binding molecule.

Table 6 Examples of Receptor-Based Biosensors

Biosensor	Detection basis	Immobilized biomolecule	Immobilization method	Sensitivity	Assay time	Ref.
Direct Measurement						
Yeast mannan	Change in potential, platinum electrode	Con A	Epichlorohydrin coupling to a PVC membrane coating the electrode	ppm	30–45 min	78
Saccharomyces cerevisiae	Change in optical signal, ellipsometry with amplification	Con A	Covalent binding to tresyl-activated, epoxysilanized silicon wafers; and to GA-activated, silanized, 12-μm silica beads (amplifier reagent)	ND	1–4 h	42
hIgG, IgG	Shift in resonant frequency, piezoelectric crystals	Protein A	Covalent binding to silanized, GA-activated crystal electrodes	ppb	10–20 min	79
Aprotinin, Trypsin	Change in potential, titanium electrode	Trypsin, Aprotinin	Covalent binding to cyanogen bromide-activated, acid oxidized electrode	ppm	20–60 min	31
HSA	Change in potential, titanium electrode	Cibacron Blue F3G-A	Adsorption to acid-oxidized electrodes	ppm	minutes	80
Potassium	Change in film photopotential	BRd and valinomycin	Entrapment into lecithin monolayers and layer deposition onto cellulose nitrate and acetate filters	ppm	seconds	82
Potassium, Calcium, Perchlorate	Cyclic voltammetry, glassy carbon electrode	Lipids and valinomycin	Entrapment into lipid monolayers deposited on the electrode	ppm	seconds	83
Light-activated membrane	Change in gate output voltage, ISFET with illumination	BRd membranes	Adsorption into acetylcellulose-soybean PC vesicles covering the ISFET	$1\text{–}3 \times 10^4$ lux	msec	84

Table 6 (Continued)

Biosensor	Detection basis	Immobilized biomolecule	Immobilization method	Sensitivity	Assay time	Ref.
Light-activated membrane	Change in current, patch electrode with illumination	BRd patches	Incorporation into liposomes coating the electrode	ND	msec	85
ATP	Change in potential, liquid membrane ISE with TPA^+ counterion	Macrocyclic polyamine	Entrapment into a PVC film on a PTFE membrane filter	ppb	1–3 min	86
Polylysine, Cytochrome c	Change in potential, chromium electrode	DMPA, PS	Adsorption into asymmetric lipid bilayers coating the electrodes	ppm	10–12 min	87
Amino acids, Purines	Change in potential, platinum electrode	Antennule of blue crabs	Direct contact with the electrode	ppm to ppb	minutes	88,89
Cholinergics	Change in impedance, interdigitated electrodes	nAChR	Entrapment into GA cross-linked albumin membranes with added stabilizers	ppb	1–5 sec	9,90
Acetylcholine	Change in potential, ISFET	nAChR	Covalent linkage to a GA-activated, PVB membrane	ppb	2–5 min	91
Cholinergics	Change in capacitance, interdigitated electrodes	nAChR	Entrapment into asolectin liposomes and adsorption to the electrodes	ppm	5–7 min	92
Opiate narcotics	Change in impedance, interdigitated electrodes	Opiate receptor	Entrapment into GA cross-linked membranes with added stabilizers	ppb	2–5 sec	94

Indirect Measurement						
Glucose	Change in fluorescence, competitive displacement of dextran-FITC, optical fiber	Con A	GA Linkage to periodate oxidized cellulose hollow fibers	ppt	5–7 min	95,96
Biotin	Change in current, competitive displacement of avidin-catalase, oxygen electrode	Biotin analogs	CDI linkage to a cellulose triacetate membrane	ppm	1–2 h	97
Riboflavin	Change in potential, competitive displacement of ARBP, pH electrode	FAD, Acriflavin	Covalent binding to GA- or cyanogen bromide-activated acetylcellulose-octyldecylamine membranes	ppb	10–15 min	81
Cholinergics	Change in fluorescence, competitive displacement of FITC-BTX, optical fiber	nAChR	Adsorption onto quartz optic fibers	ppb	2–5 min	93

Abbreviations: See Table 4. Additional abbreviations: Con A, concanavalin A; BRd, bacteriorhodopson; ATP, adenosine triphosphate: DMPA, dimyristoyl-phosphatidic acid; nAChR, nicotinic acetylcholine receptor; FAD, flavin adenine dinucleotide; ARPB, aporiboflavin binding protein; BTX, bungarotoxin; msec, millisecond.

2. Light-Activated Receptors

A number of prototype biosensors have also used light-activated biomolecules as receptors ("light" receptors) to illustrate light-mediated ion transport across films and membranes (82,84,85). In these cases, bacteriorhodopsin (or purple membranes containing bacteriorhodopsin), a light-activated membrane proton pump in purple bacteria, is incorporated into liposomes or lipid monolayers for stabilization and immobilization onto the transducer. Light induces changes in voltage or current across such membranes or, if an ion-carrier such as valinomycin is also incorporated in the membrane, light can induce transport (and thus detection) of ions such as potassium, sodium, and calcium.

Lipid vesicle and film-based sensors are the result of previous research on the incorporation of biomolecules (such as antibodies, binding proteins, and receptors) into highly hydrophobic BLMs (bilayer lipid membranes), LB (Langmuir-Blodgett) films, and phospholipid vesicles and liposomes (98–108). These studies have provided the basic technology for lipid layer-based sensors being pursued today. In general, the lipid layers or vesicles/liposomes are phospholipid-based (e.g., phosphatidylcholine and phosphatidylserine) with added stabilizers such as sterols, sterol esters, tocopherol, detergents, fatty acids, fatty alcohols, and sphingomyelin. The membrane components are usually dissolved in an organic solvent (such as hexane, decane, or chloroform) and the membrane is cast on a suitable surface (such as a transducer). In the case of biosensor fabrication, the biomolecule can be incorporated into the lipid layer either in the organic solvent or, if unstable in organic solvent, after the film/vesicle has been formed and the organic solvent taken off.

While such lipid layers/vesicles have proved critical to our understanding of membrane/biomolecule interactions, they are hindered in a practical sense by the laborious methods required to produce them and their inherent instability, especially to elevated temperature and drying conditions. It is thus unlikely that a commercially successful biosensor will utilize lipid bilayer/vesicle immobilization technology.

3. Cell Membrane Receptors

The incorporation of true membrane receptors into biosensors also depends on an understanding of lipid layers and the natural stabilization of receptor proteins by such layers. It has been known for many years that the reconstitution of cell membrane receptors requires the use of lipid bilayers or phospholipid vesicles, and such systems are used routinely to provide the functionality of isolated receptors (15). For example, receptors (cholinergic, adrenergic, gamma-aminobutyrate, insulin, and sodium ion channel) isolated as either detergent-solubilized proteins or organic solvent extracted proteolipids can be reconstituted into phospholipid vesicles containing added stabilizers, membrane structural components, and detergents such as tocopherols, cholesteryl esters, and octyl glycoside/Triton-X 100, respectively. Other studies have shown that cell membrane activity is directly influenced by its lipid environment. For example, phospholipids such as phosphatidylcholine and phosphatidylserine are know to be critical for the functionality of the adrenergic, cholinergic, and opiate receptors, and cerebroside sulfate appears to contribute to the binding specificity of opiate receptors (15).

Based on lipid layer and receptor studies, attempts have been made to develop receptor-based biosensors using lipid films/vesicles. This approach was first suggested from studies which incorporated the acetylcholine receptor (AChR) into polymerizable diacetylene containing phospholipids such as di-(10,12-tricosadiyuoyl)- and di-(10,12-uncosadiyuoyl)phosphatidylcholine (109). While it was suggested from this study that the interaction of analyte (such as acetylcholine, ACh) with the membrane-entrapped AChR could be detected by electrodes positioned on either side of the membrane, no data has been presented indicating that this theory was tested by actual experimentation. In another study, attempts have been made to immobilize the AChR from the electric ray, *Torpedo* sp., in a complex multistep process into asolectin liposomes which are then transferred to a capacitance transducer (92). Upon addition of ACh, low (nF) changes in capacitance occur, which appear related to ACh concentration. However, the membrane of this prototype biosensor is very unstable, requiring extensive preparation just prior to use and rapidly (within minutes to a few hours) degrades after preparation. Thus, again, it appears that a practical biosensor based on lipid layers/vesicles is unlikely unless a major breakthrough in lipid layer stabilization on transducers occurs.

Whole organ receptor-based biosensor prototypes have been developed in attempts to overcome the difficulties of receptor stabilization. For example, the antennule of blue crabs have been used to fabricate sensors for amino acids and purines (88,89). The system developed, however, uses a laboratory perfusion device which has not been further developed toward a practical biosensor.

The AChR has been used for the development of three prototype biosensors utilizing nonlipid layer/vesicle immobilization technology (90, 91,93). The first successful development of an AChR-based biosensor utilized an albumin-based membrane stabilized with detergents, lipids, and antioxidants to entrap the receptor (9,90). The development of this biosensor followed a research and development plan as illustrated in Figure 5. A standard methodology was developed for isolation of the receptor from

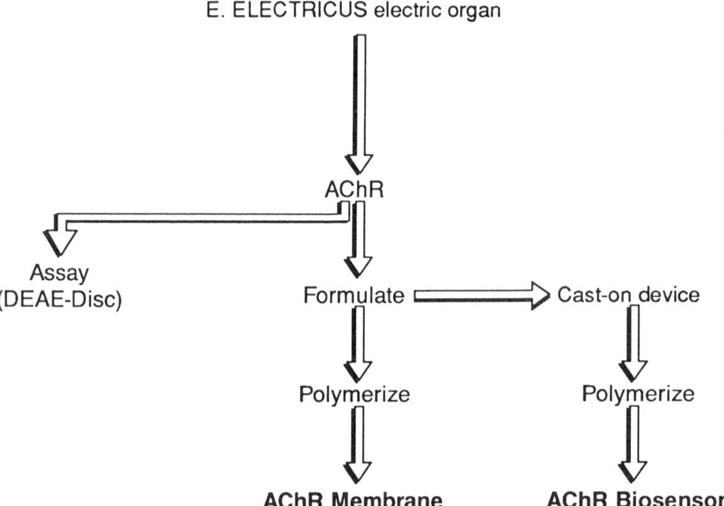

Figure 5 Isolation and immobilization of the acetylcholine receptor (AChR) for initial membrane stability studies and for development of the AChR-based biosensor.

the electric organ of the electric fish, *Electrophorous electricus* based on standard AChR detergent-based extraction methods and a filter disk method for assaying receptor activity (90). The isolated receptor was first used to assess a wide range of immobilization and stabilization methods (9). For example, as shown in Figure 6, membranes based on glutaraldehyde cross-linked albumin containing various detergents, lipids, and stabilizers were found to immobilize the receptor with complete retention of (binding) activity and specificity and to stabilize the binding activity and specificity for at least 6 months. In some cases, certain formulations led to increased binding activity over the control. This appears due to unfolding of the receptor in these formulations leading to increased accessibility to the receptor binding sites and the apparent increase in binding. A similar effect on receptor binding has been observed under high pressure (110). As shown in Figure 6, the nonimmobilized receptor lost all activity after 15 days' storage.

These initial studies were used to develop a one-step fabrication method for the preparation of the AChR-transducer element as depicted in Figure 7. The method utilizes one reaction solution, at room temperature, which contains copolymer (e.g., serum albumin, collagen, gelatin, etc.), stabilizers, lipid, detergent, and the receptor. Polymerizing agent is added (e.g., glutaraldehyde or other bifunctional polymerizing agents) and the mixture

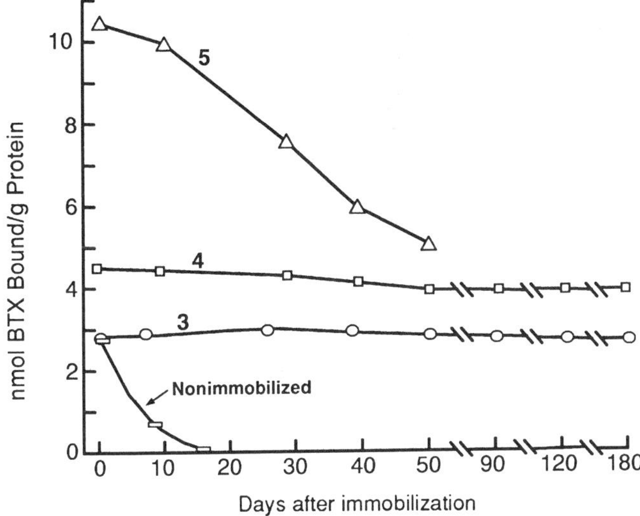

Figure 6 Stability of the immobilized AChR at 25°C as measured by BTX (α-bungarotoxin) binding. Formulations 3, 4, and 5 are albumin-based, glutaraldehyde cross-linked membranes containing the AChR, phosphatidylcholine, detergents, and other stabilizers (9). The nonimmobilized receptor was stored in extraction buffer.

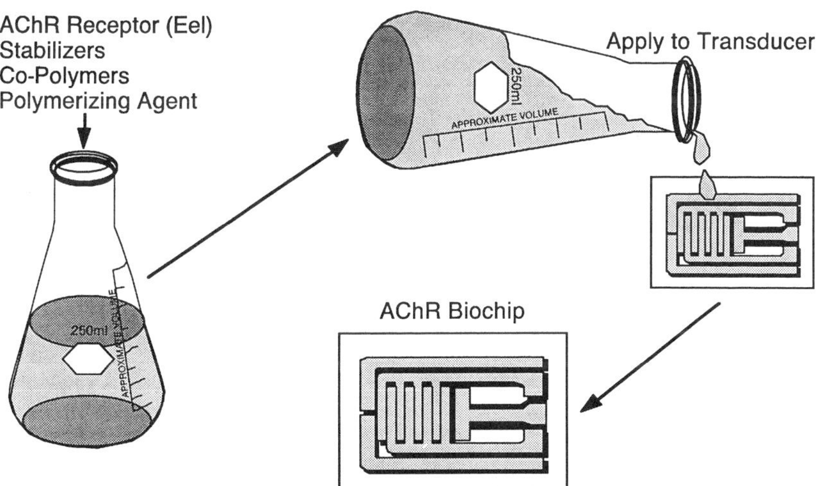

Figure 7 One-step preparation method for the AChR biochip. The immobilization solution is spread directly onto the interdigitated electrodes. Polymerization is complete in 15–30 minutes.

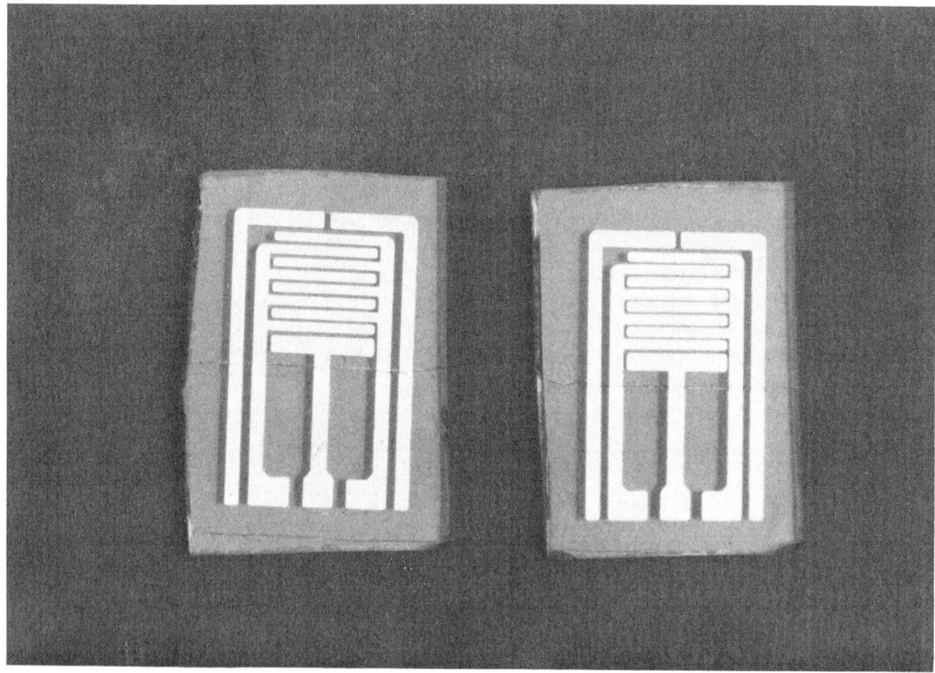

Figure 8 Prototype receptor-based biochip. The gold interdigitated electrodes are prepared by vacuum deposition onto acid-washed fused quartz (90). Current prototypes utilize a printing method to deposit the electrodes on epoxy board.

is spread onto the transducer. Our transducer of choice was an interdigitated (gold) electrode chemiresist deposited initially on quartz (Fig. 8) (and later epoxy board) and used successfully by other workers for chemical sensors (101,111). Polymerization is complete in approximately 30–45 min. The resulting "biochips" have a polymeric membrane layer thickness of approximately 11 ± 2 μm (measured by ellipsometry and micrometry), which adheres tightly to the electrodes and does not peel from the electrodes or crack upon drying. The chips are stored dry (room temperature or 4°C) until used.

The latest version of the chips employs two electrode arrays on an epoxy board able to be plugged into a portable electronics unit (Fig. 9). One of the arrays is coated with membrane containing the receptor while the other, control, array is coated with membrane containing no receptor. During analyte measurements, a differential circuit in the electronics unit (Fig. 10) then automatically compensates for background and nonspecific

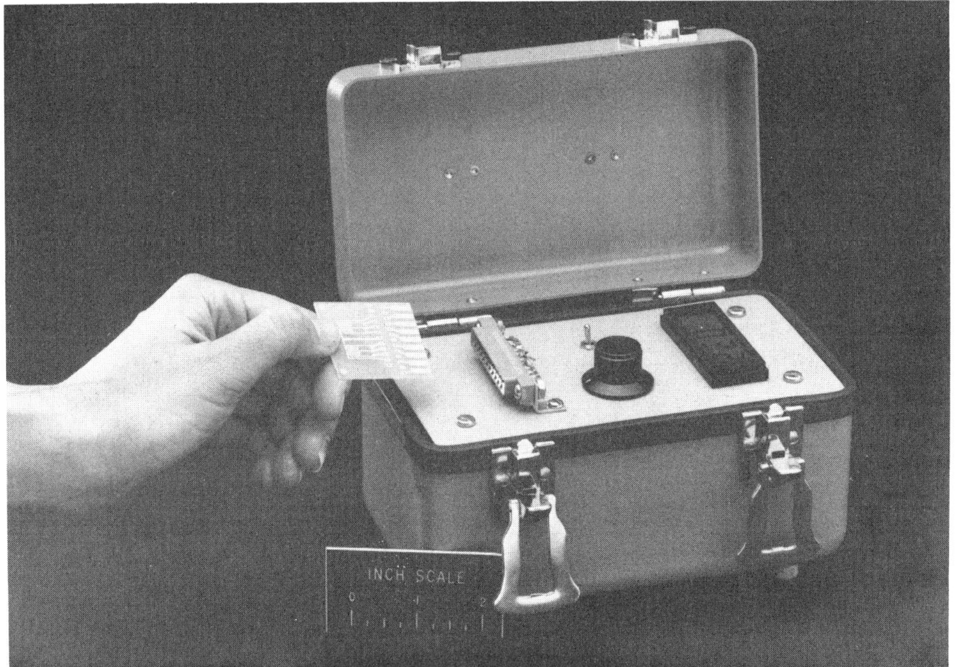

Figure 9 Prototype, portable receptor- and antibody-based biosensor field electronics unit.

binding to the chips by subtracting the control chip response from the receptor chip response to result in corrected (specific binding) output.

To use the AChR chips, they are first equilibrated from dry storage for 10–20 minutes in aqueous buffer solution and then challenged with analyte solution(s). Output signal, e.g., change in impedance, voltage, current, or capacitance, is stabilized in 2–10 seconds and is directly proportional to the concentration of cholinergic agent (Fig. 11). Analysis of the saturation kinetics from experiments such as those illustrated in Figure 11 have shown that the immobilized receptor retains the same binding kinetics as the receptor in solution (15,90). The receptor biochips may also be washed free of cholinergic agents (except irreversible binding agents such as bungarotoxin) and recycled at least 20 times (Fig. 12).

To date, the AChR chips have achieved a sensitivity of 100–300 ppb ACh (1.5–4 μM), or 1–3 ng ACh in the 10–25 μl samples used, with no optimization of receptor density or electronic amplification. Optimization of the sensor is expected to result in sensitivities for cholinergic agents in the pptr (pg/ml) range.

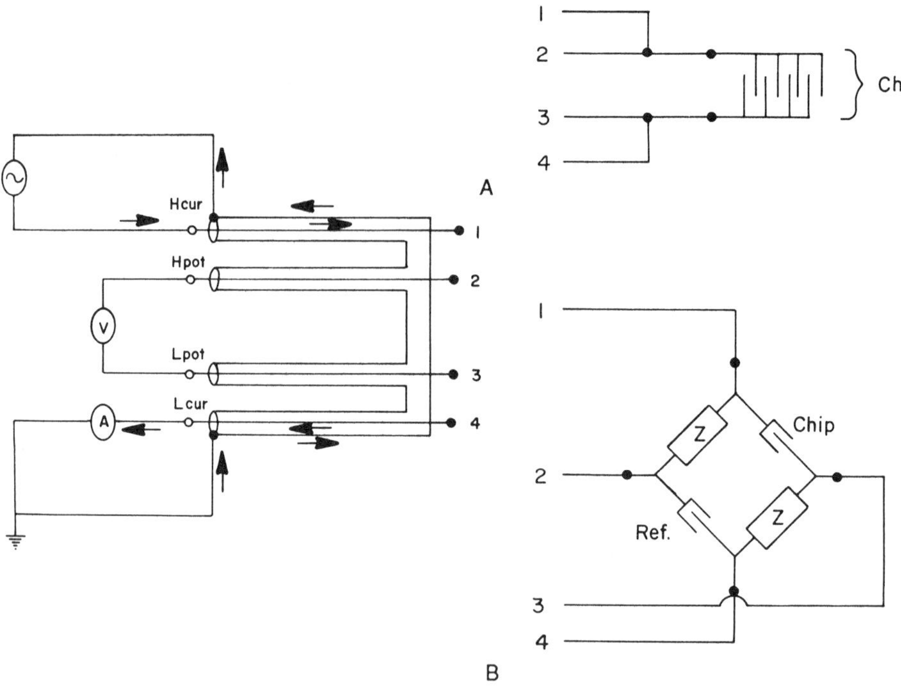

Figure 10 Circuit diagrams for portable biosensor electronics units. A) single biochip configuration; B) differential biochip configuration. Configuration B is used in the unit shown in Figure 9.

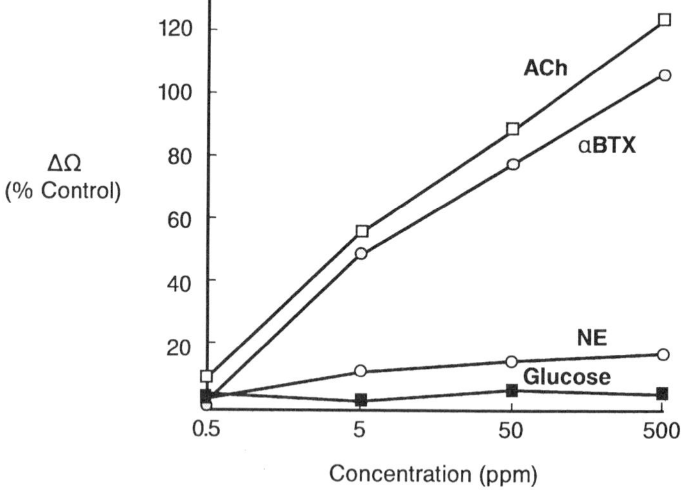

Figure 11 Response of the AChR-based biosensor to increasing concentrations of two cholinergic ligands (ACh and BTX), a noncholinergic ligand (norepinephrine, NE), and glucose. The % control represents specific binding and was corrected for nonspecific binding with a non–receptor containing control chip.

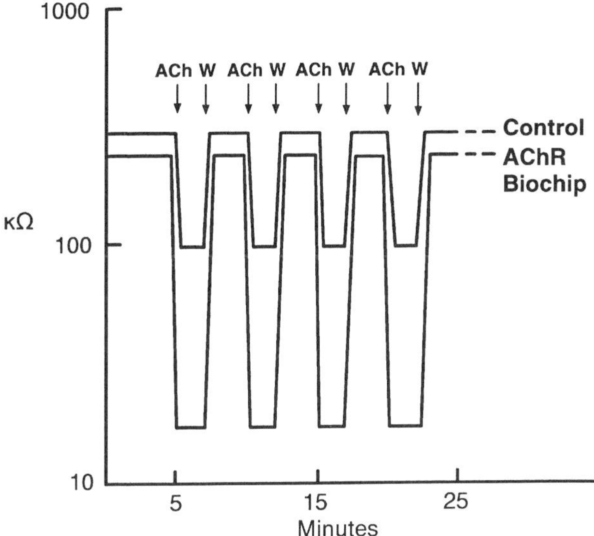

Figure 12 Response of the AChR-based biosensor to repeated applications of ACh. Control (containing no receptor) and test (receptor-containing) chips were challenged with 50 μl of 2.7 μM (5 ppm) ACh, allowed to equilibrate, washed (w) with saline and then rechallenged again after reaching starting equilibrium. In this manner, chips could be challenged at least 15–20 times. The difference in impedance between the control and AChR chips represents specific binding to the receptor.

This same approach has been used to develop a prototype human IgG biosensor (see above) and an opiate receptor (OpR)-based biosensor (94). The OpR was isolated from rat brain by detergent extraction and partially purified by standard methods. In these initial studies, no attempts were made to further separate the OpR into its defined subclasses, and thus a generic opiate biosensor resulted, able to detect all opiate narcotics tested. The OpR was immobilized onto the interdigitated electrode chips in an albumin or gelatin cross-linked membrane stabilized with detergent, lipids, and antioxidants.

As for the AChR-based biosensor, the resulting OpR biochips are storage stable and can be used repeatedly once activated by wetting. The OpR-biochips respond specifically to opiates such as naloxone, levorphanol, morphine, nalorphine, codeine, and leu enkephalin, reaching output signal equilibrium in 2–10 seconds at ppb levels. Output signal (change in circuit impedance, capacitance, voltage, or current) is proportional to analyte concentration. For example, Figure 13 illustrates response of OpR biochips to challenge with naloxone and levorphanol. The biphasic curve is most

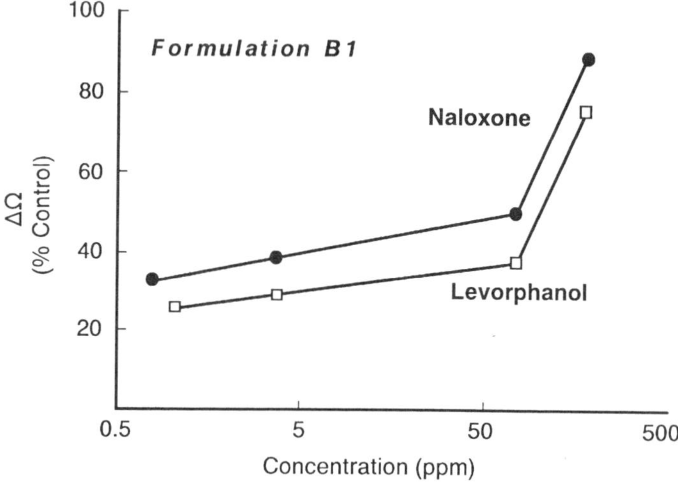

Figure 13 Response of the opiate receptor-based biosensor to two opiates known to bind to the receptor. The % control represents specific binding and was corrected for nonspecific binding using a non–receptor containing control chip.

likely due to the mixed population of opiate receptor subtypes present and their varying degrees of binding activity toward these opiates. Further studies have established a detection range for the (nonoptimized) biochips of 0.1–25 μg/ml of opiate analyte. Again, optimization of the chips is expected to lead to pptr sensitivities for opiates.

Our successful use of an electrochemical transducer for membrane receptor-based biosensors was confirmed in a later report on studies which utilized a glutaraldehyde-activated poly(vinylbutryl) (PVB) membrane covering an ion-sensitive field effect transistor (ISFET) transducer. The AChR from the electric ray, *Torpedo californica,* was immobilized onto the PVB membrane with or without phosphatidyl choline present as a stabilizer. A control membrane was placed on the reference gate of the ISFET and the differential voltage was measured between the control and test gates upon challenge with acetylcholine. These studies reported similar sensitivities to the AChR-biochip discussed above. No studies were carried out, however, to establish the specificity of the ISFET-AChR biosensor for cholinergic agents. In later studies, it was found that when compared to cellulose triacetate, PVC, and agar membranes, the PVB membrane was the most adhesive to the silicon nitride ISFET gates under acidic, alkaline, and neutral conditions (112).

The three independent reports of prototype electrochemical AChR-based biosensors establish the validity of this approach. The molecular basis for the function of these biosensors remains to be established but appears due to the alteration of the electrical field around/through the receptor when it binds with ligand, and/or opening of ion channels (and resulting sodium ion fluxes) in the receptor upon ligand binding. Either mechanism would mimic the function of the receptor in its natural, membrane environment and would lead to changes in the transducer electric field and resulting signal output. This mechanism also defines the sensitivity of such receptor-based biosensors, i.e., a finite number of receptors must interact with analyte in order to produce a cumulative change in the transducer electric field significant enough for detection above background. Thus, immobilization becomes critical in such biosensors: maximal retention of functionality and stability will result in optimal sensitivity.

Another prototype AChR-based biosensor utilizing an optical transducer system has been recently reported as well (93). In this case, the AChR is noncovalently adsorbed onto quartz fibers and change in fluorescence is monitored as fluorescein-labeled bungarotoxin competes with the cholinergic analyte to be measured. While the sensitivity and specificity of this AChR-based biosensor is comparable to that of the electrochemical sensors discussed above, this optical sensor requires the presence of an expensive reagent (fluorescein-bungarotoxin), depends on a competitive reaction, and is nonreversible (cannot be recycled). In addition, the adsorption immobilization method used for the receptor has not been assessed for storage or in-use stability. The authors suggest that because of its surface chemistry, this optical sensor can operate in high ionic strength fluids (such as plasma and urine) while capacitance sensors, such as that discussed above, cannot. As shown by the results above for the determination of anti-hIgG in serum using the capacitance transducer format, this is a premature and unfounded conclusion.

VI. CONCLUSIONS

Antibody- and receptor-based biosensors represent integrated products of both old and new technologies. They promise to revolutionize detection and diagnostics as we know them today and to significantly contribute to the quality of our lives.

Immobilization technology is key to biosensor fabrication. The application of immobilization technology to antibody- and receptor-based biosensors has, however, provided new challenges. The use of labile and functional biomolecules in biosensors demands that immobilization methods be both simple and chemically sensitive to the needs of these molecules. As a result, biosensor technology today is being paced by new developments in biomolecule immobilization aimed not only at providing stable biomolecule surfaces, but the ability to produce such surfaces on a mass- and cost-effective basis. As these requirements are met, biosensors will be applied to the needs of medicine, industry, environmental monitoring, automation, and robotics as both detection and control devices.

REFERENCES

1. R. F. Taylor, Biosensors '89: Technology, applications and markets. *Proc. Biotech '89 USA,* Conference Management Corp., Norwalk, CT, 1989, p. 275.
2. R. F. Taylor, *Biosensors: Technology, Applications and Markets,* Decision Resources Corp., Cambridge, MA, 1990.
3. L. C. Clark, Jr. and C. Lyons, *Ann. N.Y. Acad. Sci. 148:*133 (1962).
4. G. Guilbault, and J. Montalvo, *J. Am. Chem. Soc. 91:*2164 (1969).
5. C. Davies, *Ann. Microbiol.* (Paris) *126A:*175 (1975).
6. J. Janata, *J. Am. Chem. Soc. 97:*2914 (1975).
7. M. Aizawa, A. Morioka, H. Matsuoka, S. Suzuki, Y. Nagamura, R. Shinohara, and I. Ishiguro, *J. Solid-Phase Biochem. 1:*319 (1976).
8. S. L. Belli, and G. A. Rechnitz, *Analyt. Lett. 19:*403 (1986).
9. R. F. Taylor, I. G. Marenchic, and E. J. Cook, U.S. Pat. Appl. 058,389 (1987).
10. C. R. Lowe, *Biosensors 1:*3 (1985).
11. A. P. F. Turner, I. Karube, and G. S. Wilson (eds.), *Biosensors, Fundamentals and Applications,* Oxford University Press, Oxford, 1987.
12. S. M. Angel, *Spectroscopy 2:*38 (1988).
13. A. P. F. Turner, *Sensors Actuators 17:*433 (1989).
14. R. F. Taylor, in *The World Biotech Report,* Online International Inc., New York, 1986, p. 7.
15. R. F. Taylor, in *Bioinstrumentation: Research, Developments and Applications* (Wise, D. L., ed), Butterworths, New York, 1990, p. 355.
16. G. G. Guilbault, *Analytical Uses of Immobilized Enzymes,* Marcel Dekker, New York, 1984.
17. S. A. Barker, in *Biosensors: Fundamentals and Applications,* (A. P. F. Turner, I. Karube, and G. S. Wilson, eds.), Oxford University Press, Oxford, 1987, p. 85.
18. K. Mosbach, (ed.), *Immobilized Enzymes and Cells, Part D, Methods in Enzymology,* Vol. 137, Academic Press, New York, 1988.
19. H. T. Tien, *J. Clin. Lab. Anal. 2:*256 (1988).
20. D. D. Richman, P. H. Cleveland, M. N. Oxman, and K. M. Johnson, *J. Immunol. 128:*2300 (1982).
21. M. Wilchek, and E. A. Bayer, *Anal. Biochem. 171:*1 (1988).
22. J. F. Kennedy, in *Handbook of Enzyme Biotechnology* (Wiseman, A., ed.), Ellis Horwood, Chichester, 1985, p. 380.
23. D. Thomas, and G. Broun, *Met. Enzymol. 44:*901 (1976).
24. J. R. North, *Trends Biotech. 3:*180 (1985).

25. W. R. Heineman and H. B. Halsall, *Anal. Chem. 57:*1321A (1985).
26. J. F. Place, R. M. Sutherland, and C. Dahne, *Biosensors 1:*321 (1985).
27. D. Monroe, *Am. Biotech. Lab. 4*(#6):8 (1986).
28. I. Karube, and M. Gotoh, in *Analytical Uses of Immobilized Biological Compounds for Detection, Medical and Industrial Uses* (G. G. Guilbault and M. Mascini, eds.), D. Reidel Publishing Co., Dordrecht, 1988, p. 267.
29. M. Aizawa, S. Kato, and S. Suzuki, *J. Memb. Sci. 2:*125 (1977).
30. S. Collins, and J. Janata, *Anal. Chim. Acta 136:*93 (1982).
31. N. Yamamoto, Y. Nagasawa, M. Sawai, T. Sudo, and H. Tsubomura, *J. Immunol. Met. 22:*309 (1978).
32. N. Yamamoto, Y. Nagasawa, S. Shuto, H. Tsubomura, M. Sawai, and H. Okumura, *Clin. Chem. 26:*1569 (1980).
33. G. A. Robinson, V. M. Cole, and G. C. Forrest, *Biosensors 3:*147 (1988).
34. R. L. Solsky, and G. A. Rechnitz, *Science 294:*1308 (1979).
35. R. L. Solsky, and G. A. Rechnitz, *Anal. Chim. Acta 123:*135 (1981).
36. D. L. Bush and G. A. Rechnitz, *Anal. Lett. 20:*1781 (1987).
37. R. P. H. Kooyman, H. Kolkman, J. Van Gent, and J. Greve, *Anal. Chim. Acta 213:*35 (1988).
38. M. Gotoh, M. Suzuki, I. Kubo, E. Tamiya, and I. Karube, *J. Mol. Catalysis 53:*285 (1989).
39. J. E. Roderer and G. J. Bastiaans, *Anal. Chem. 55:*2333 (1983).
40. I. Taniguchi, K. Yasukouchi, and I. Tsuji, *Potential-causing element for immunosensor,* Eur. Patent. Appl. 193,154 (1986).
41. H. Matsuoka, S. Tanioka, and I. Karube, *Anal. Lett. 20:*63 (1987).
42. C. F. Mandenius and K. Mosbach, *Anal. Biochem. 170:*68 (1988).
43. D. C. Cullen, R. G. W. Brown, and C. R. Lowe, *Biosensors 3:*211 (1988).
43a. T. Katsube, T. Yaji, K. Suzuki, S. Kobayashi, T. Kawaguchi, and T. Shiro, *Appl. Surf. Sci. 33/34:*1332 (1988).
44. M. Aizawa, S. Kato, and S. Suzuki, *J. Memb. Sci. 7:*1 (1980).
45. J. L. Boitieux, M. P. Biron, and D. Thomas, *Anal. Chim. Acta 222:*235 (1989).
46. F. Gardies, C. Martelet, B. Colin, and B. Mandrand, *Sens. Actuator. 17:*461 (1989).
47. T. Yao, and G. A. Rechnitz, *Biosensors 3:*307 (1988).
48. R. M. Sutherland, C. Dahne, J. F. Place, and A. S. Ringrose, *Clin. Chem. 30:*1533 (1984).
49. M. Y. Keating, and G. A. Rechnitz, *Anal. Chem. 56:*801 (1984).
50. T. Vo-Dinh, B. J. Thomberg, G. D. Griffin, K. R. Ambrose, M. J. Sepaniak, and E. M. Gardenhire, *Appl. Spectroscop. 41:*735 (1987).
51. B. J. Tromberg, M. J. Sepaniak, J. P. Alarie, T. Vo-Dinh, and R. M. Santella, *Anal. Chem. 60:*1901 (1988).
52. I. Karube, and M. Suzuki, *Biosensors 2:*343 (1986).
53. H. Muramatsu, E. Tamiya, and I. Karube, *J. Memb. Sci. 41:*281 (1989).
54. H. H. Weetall and T. Hotalin, *Biosensors 3:*57 (1987).
55. S. K. Bahtia, R. B. Thompson, L. C. Shriver-Lake, M. Levine, and F. S. Ligler, *Soc. Photo-Optic Instru. Eng. 1054:*184 (1989).

56. M. Aizawa, A. Morioka, S. Suzuki, and Y. Nagamura, *Anal. Biochem. 94:*22 (1979).
57. M. Mascini, F. Zolesi, and G. Palleschi, *Anal. Lett. 15:*101 (1982).
58. G. A. Robinson, H. A. O. Hill, R. D. Philo, J. M. Gear, S. J. Rattle, and G. C. Forrest, *Clin. Chem. 31:*1449 (1985).
59. J. L. Boitieux, G. Desmet, and D. Thomas, *Clin. Chem. 25:*318 (1979).
60. M. Aizawa, A. Morioka, and S. Suzuki, *Anal. Chim. Acta 115:*61 (1980).
61. K. Shiba, Y. Umezawa, T. Watanabe, S. Ogawa, and S. Fujiwara, *Anal. Chem. 52:*1610 (1980).
62. J. Rishpon and I. Rosen, *Biosensors 4:*61 (1989).
63. M. Haga, S. Sugawara, and H. Itagaki, *Anal. Biochem. 118:*286 (1981).
64. H. Itagaki, Y. Hakoda, Y. Suzuki, and M. Haga, *Chem. Phar. Bull. 31:*1283 (1983).
65. J. L. Boitieux, C. Lemay, G. Desmet, and D. Thomas, *Clin. Chim. Acta 113:*175 (1981).
66. K. Shiba, T. Watanabe, Y. Umezawa, and S. Fujiwara, *Chem. Lett.:*155 (1980)
67. J. Janata and G. F. Blackburn, *Ann. N.Y. Acad. Sci. 428:*286 (1984).
68. M. Thompson, in *The World Biotech Report,* Vol. 2(5), Online International, Inc., New York, 1986, p. 19.
69. J. McCann, in *Biosensors: Fundamentals and Applications* (A. P. F. Turner, I. Karube, and G. S. Wilson, eds.), Oxford University Press, Oxford, 1987, p. 737.
70. R. S. Matson, and M. C. Little, *J. Chromatogr. 458:*67 (1988).
71. S. K. Bhatia, L. C. Shriver-Lake, K. J. Prior, J. H. Greorger, J. M. Calvert, R. Bredehorst, and F. S. Ligler, *Anal. Biochem. 178:*408 (1989).
72. S. Ehrenpreis, J. H. Fleisch, and T. W. Miittage, *Pharmacol. Rev. 21:*131 (1969).
73. T. Narahashi, *Physiol. Rev. 54:*813 (1974).
74. R. W. Straub, and L. Bolis (eds.), *Cell Membrane Receptors for Drugs and Hormones,* Raven Press, New York, 1978.
75. *Receptors in Classical Pharmacology and Cell Biology,* Conference Proceedings, *Biochem. Pharmacol. 33:*833 (1983).
76. W. A. Catterall, *Science 233:*653 (1984).
77. T. P. Kenakin, *Pharmacol. Rev. 36:*165 (1984).
78. J. Janata, *J. Am. Chem. Soc. 97:*2914 (1975).
79. H. Muramatsu, J. M. Dicks, E. Tamiya, and I. Karube, *Anal. Chem. 59:*2760 (1987).
80. C. R. Lowe, *FEBS Lett. 106:*405 (1979).
81. T. Yao and G. A. Rechnitz, *Anal. Chem. 59:*2115 (1987).
82. M. T. Flanagan, *Thin Solid Films 99:*133 (1983).
83. M. Sugawara, K. Kojima, H. Sazawa, and Y. Umezawa, *Anal. Chem. 59:*2842 (1987).
84. K. Tanabe, M. Hikuma, L. SooMi, Y. Iwaski, E. Tamiya, and I. Karube, *J. Biotech. 10:*127 (1989).
85. P. Yager, *Adv. Exp. Med. Biol. 238:*257 (1988).

86. Y. Umezawa, M. Kataoka, W. Takami, E. Kimura, T. Koike, and H. Nada, *Anal. Chem. 60:*2392 (1988).
87. M. Steizle and E. Sackmann, *Biochim. Biophys. Acta 981:*135 (1989).
88. S. L. Belli, and G. A. Rechnitz, *Anal. Lett. 19:*403 (1986)
89. R. M. Buch and G. A. Rechnitz, *Biosensors 4:*215 (1989).
90. R. F. Taylor, I. G. Marenchic, and E. J. Cook, *Anal. Chim. Acta 213:*131 (1988).
91. M. Gotoh, E. Tamiya, M. Momol, Y. Kagawa, and I. Karube, *Anal. Lett. 20:*857 (1987).
92. M. E. Eldefrawi, S. M. Sherby, A. G. Andreou, N. A. Mansouri, Z. Annau, N. A. Blum, and J. J. Valdes, *Anal. Lett. 21:*1665 (1988).
93. K. R. Rogers, J. J. Valdes, and M. E. Eldefrawi, *Anal. Biochem. 182:*353 (1989).
94. R. F. Taylor, in *Proceedings of the International Biosensors '89 Symposium,* Institute for International Research, Inc., New York, 1988.
95. J. S. Schultz, *Optical Sensor of Plasma Constituents,* U.S. Patent 4,344,438 (1982).
96. S. Mansouri and J. S. Schultz, *Bio/Technology 2:*885 (1984).
97. Y. Ikariyama, M. Furuki, and M. Aizawa, *Anal. Chem. 57:*496 (1985).
98. J. del Castillo, A. Rodriguez, C. A. Romero, and V. Sanchez, *Science 153:*185 (1966).
99. P. Barfort, E. R. Arquilla, and P. O. Volgelhut, *Science 160:*1119 (1968).
100. M. Thompson, U. J. Krull, and P. J. Worsfold, *Anal. Chim. Acta 117:*133 (1980).
101. H. Wohltjen, W. R. Barger, A. W. Snow, and L. Jarvis, *IEEE Trans. Elec. Dev. ED-32:*1170 (1985).
102. W. M. Reicher, C. J. Bruckner, and J. Joseph, *Thin Solid Films 152:*345 (1987).
103. U. J. Krull, *Anal. Chim. Acta 192:*321 (1987).
104. E. Sada, S. Katoh, M. Terashima, and Y. Kikuchi, *Biotech. Bioeng. 30:*117 (1987).
105. U. J. Krull, R. S. Brown, R. F. DeBono, and D. Hougham, *Talanta 35:*129 (1988).
106. J. Kotowski, T. Janas, and H. T. Tien, *Bioelectrochem. Bioenerg. 19:*277 (1988).
107. H. T. Tien, Z. Salamon, J. Kutnik, P. Krysinski, J. Kotowski, D. Liderman, and T. Janas, *J. Mol. Electron. 4:*S1 (1988).
108. U. J. Krull, R. S. Brown, R. N. Koilpillai, R. Nespolo, A. Safarzadeh-Amiri, and E. T. Vandenberg, *Analyst 114:*33 (1989)
109. P. Yager, *Biosensors from Membrane Proteins Reconstituted in Polymerized Lipid Bilayers,* U.S. Stat. Inv. Reg. H201, 1987.
110. R. F. Taylor, in *Undersea Physiology VII* (A. J. Bachrach and M. M. Matzen, eds.), Undersea Medical Society, Bethesda, 1984, p. 607.
111. H. Wohltjen, *Anal. Chem. 56:*87A (1984).
112. M. Gotoh, E. Tamiya, and I. Karube, *J. Memb. Sci. 41:*291 (1989).

9

Therapeutic Applications of Immobilized Proteins and Cells

Thomas M. S. Chang

McGill University
Montreal, Quebec, Canada

I. INTRODUCTION

The feasibility of therapeutic applications for immobilized proteins was demonstrated over 20 years ago (1–7). Recent rapid progress and interests in biotechnology may lead to actual large-scale, routine applications of these methods. Immobilized enzymes, cells, and immunosorbents have important potential uses. However, the most immediate uses for large-scale clinical applications of immobilized systems are in blood substitutes based on immobilized hemoglobin, for therapeutic applications, for microencapsulation of cells, and for hemoperfusion.

II. BLOOD SUBSTITUTES BASED ON IMMOBILIZED HEMOGLOBIN

A. Introduction

At present, there is a worldwide shortage of donar blood for transfusion. Furthermore, donar blood can only be stored for short periods of time using routine procedures. In emergency situations and mass disasters, there may not be time for the required cross-matching and typing of blood group antigens. Possible blood contamination with viruses such as HIV and hepatitis can further hinder rapid blood utilization.

Many of these problems with blood supplies could be avoided by using preparations of the oxygen-carrying hemoglobin of red blood cells. Unfortunately, there are a number of other problems related to the use of free hemoglobin. These include conversion of the hemoglobin from the tetrameric to dimeric form, high oxygen affinity, and high oncotic pressure.

These problems with the use of homoglobin have been solved using immobilization technology. At present, there are two major approaches to hemoglobin immobilization: covalent cross-linking and microencapsulation.

B. Microencapsulated Hemoglobin

In 1957, we began studies on the microencapsulation of red blood cell hemolysates (8). The thickness of the microencapsulating membrane was 100 Å, comparable to that of red blood cells. The diameter of the resulting microcapsules could be smaller or comparable to red blood cells, e.g., 1 to 8 μm. Further studies (1,9–12) focused on other membranes including those based on synthetic polymers, crossed-linked proteins, lipid–protein complexes, lipid–polymer complexes, etc. Typically, such artificial blood cells contained hemoglobin, enzymes, and other materials normally present in red blood cells.

Hemoglobin entrapped within these artificial blood cells does not leak out and is therefore not converted into dimers. The synthetic membrane has no blood group antigens and thus the artificial blood cells do not agglutinate in the presence of sera with different blood group antibodies.

The composition of the encapsulating membrane proved critical to the function of the artificial red blood cells. For example, in early studies on the membranes (8), butyl benzoate was not removed completely from the membrane, resulting in membranes with low permeability. As a result, the membrane was not permeable to molecules with molecular weight of cofactor 2,3-DPG, which was retained inside the artificial cell with hemoglobin. In this case the oxygen affinity of the artificial blood cells was low and the oxygen release characteristics good, most likely due to the better oxygen release characteristics of hemoglobin in the presence of DPG. In those artificial cells where the membrane was very permeable, the oxygen affinity was high.

The major problem with the artificial red blood cells is their rapid removal from the circulation after intravenous injection. Decreasing their mean diameter improved their survival rate only marginally. Studies using different membrane compositions, surface charges, and other properties did improve their survival rates (1,9–12). However, since the smallest artificial blood cells used at this time were in the order of 1 μm in size, the survival time was not sufficient to provide useful blood substitutes.

An increasing number of workers have continued the development of artificial red blood cells (13–22). The most significant improvement from these studies is the use of lipid membrane artificial blood cells of submicrometer size. Of these, the most promising ones are those using single bilayer membrane liposomes rather than lamellar liposomes. Lamellar liposomes cannot micoencapsulate significant amounts of hemoglobin. A number of methods are available to produce single bilayer membrane liposomes including sonication (13), reverse phase evaporation (14), and extrusion (15).

In addition to the use of lipid bilayer membrane liposomes for artificial blood cells, other approaches are available. These include coacervation of lipids, hemoglobin, and albumin (16), and incorporation of heme into the lipid bilayer of very small artificial cells (17).

Submicrometer lipid membrane artificial blood cells have two important characteristics. First, their affinity and capacity are comparable to that of hemoglobin in red blood cells (18). This is because 2,3-DPG or other similar cofactors can be microencapsulated with the hemoglobin in the artificial cells (19). Second, the survival time of the artificial cells in circulation has improved from earlier artificial blood cells to the point where they can now be used in acute blood volume replacement. Thus, rats with 90% of their red blood cells exchanged with lipid membrane artificial blood cells still survived (20–22). The control group with no exchange transfusion died. Before clinical trials of these artificial cells can begin, however, further detailed studies on the toxicity of these artificial blood cells will be required, including studies of their effects on the reticuloendothelial (RE) system, complement activitation, and on formed elements of the blood and coagulation systems.

C. Cross-Linked Hemoglobin

In 1964, we reported the use of a bifunctional agent (sebacyl chloride and, later, glutaraldehyde) to cross-link hemoglobin (9,11,12). These bifunctional agents can cross-link hemoglobin on the surface of microscopic hemoglobin emulsions to form cross-linked hemoglobin membranes or can cross-link very small microdroplets into solid polyhemoglobin microspheres of 1 μm or less in diameter. However, these polyhemoglobins did not survive sufficiently in the circulation. Other workers cross-linked hemoglobin molecules into smaller units which stayed in solution (23). In this form, the polyhemoglobin survived significantly longer in the circulation.

Polyhemoglobins do not have 2,3-DPG and, as a result, have high oxygen affinity. Pyridoxal phosphate, when linked to hemoglobin, decreases oxygen affinity (26), but pyridoxalated hemoglobin, like hemoglobin, is rapidly converted into the dimer and is removed from the circulation (27). However, by combining the approaches of cross-linked and pyridoxalated hemoglobin, one can prepare pyridoxalated polyhemoglobin (PPHb) (28–39), and in this form the polyhemoglobin is not converted to the dimer and is therefore not excreted by the kidney (33). As a result, the survival time of polyhemoglobin in the circulation achieves a half-life of 27 hours in rats (33).

In vitro characterization of PPHb shows that it has significant oxygen carriage and acceptable oxygen affinity (34). It is effective in the resuscitation of lethal hemorrhagic shock in rats (35) and appears as effective as whole blood. It is significantly more effective than stroma-free hemoglobin, pyridoxalated hemoglobin, albumin solution, or Ringer solution (35) (Table 1). PPHb can successfully replace up to 97% of the red blood cells in rats (36,37). These studies have shown that PPHb is effective as a blood substitute for acute blood replacement.

Before using PPHb for clinical applications, it is important to further study the toxicity and safety of the material. Recent studies have demonstrated that PPHb stimulates rather than represses the RE system (38). Thus, it does not appear to have an adverse affect on this system. Another potential problem is PPHb-induced vasoconstriction due to contaminants in PPHb preparations. This problem has been minimized by the development of purification methods for PPHb (34,40).

The cross-linked nature and size of polyhemoglobin may cause adverse immunological reactions. We therefore carried out a series of immunological studies on hemoglobin and polyhemoglobin (41,42) in which we measured antibody titers were determined after administration of immunizing

Table 1 Characteristics of Modified Hemoglobin Preparations[a]

Property[b]	SFHb	PSFHb	PHb	PPHb
P_{50} (mmHg)	14	26	10	18
Half-life (h)	4	4	30	30
Long-term survival (%)	0	0	0	50
Hemorrhagic shock survival (%)	0	17	not tested	75

[a]Abbreviations: SFHb, stroma-free hemoglobin; PSFHb, pyridoxalated stroma-free hemoglobin; PHb, polyhemoglobin; PPHb, pyridoxalated polyhemoglobin.

[b]Properties: P_{50}, oxygen tension (in mmHg) at which 50% of the maximal oxygen carried is released; half-life, survival time in the systemic circulation; long-term survival, survival after exchange of 97% of the red blood cells in the animal are exchanged and replaced by the hemoglobin preparation; hemorrhagic shock survival, survival after removal of 60% of the blood in experimental animals and replacement by the hemoglobin preparation, and resulting in death in all control animals.

doses of different preparations of hemoglobin to rats. The preparations were administered in Freund's adjuvant subcutaneously on days 0, 14, and 28. On day 35, antibody titers were determined against the various hemoglobins. We found that polyhemoglobin and hemoglobin from a homologous source did not produce any significant antibody titers. Polyhemoglobin and hemoglobin from a heterogenous source resulted in significant increases in antibody titers. In nonimmunized animals, transfusions of hemoglobin and polyhemoglobin from either homologous or heterogenous sources did not result in any adverse effects (42). In animals which had been immunized, transfusion of hemoglobin or polyhemoglobin from a homologous source again produced no effect, but transfusion of either from a heterogenous source resulted in severe anaphylactic reactions.

A number of studies have been carried out to improve the P_{50} of hemoglobin preparations. The P_{50} is the oxygen tension (in mmHg) at which 50% of the maximal oxygen carried by hemoglobin is released. Thus, this parameter is an indicator of the ease that a hemoglobin preparation can release the oxygen it carries. For example, bovine hemoglobin has a P_{50} of 20–30 mmHg even without pyridoxalation (43). After cross-linking, bovine polyhemoglobin has a P_{50} of 24 mmHg, i.e., comparable to that of human red blood cells. This finding together with the ready availability of bovine hemaglobin suggests it could be a viable substitute for human hemoglobin. The potential immunological problems described above may limit the use of bovine hemoglobin to specific human applications. Other approaches to increase the P_{50} of hemoglobin include the use of ATP and other cofactors instead of pyridoxyl phosphate to prepare the hemoglobin (40,44). To date, these approaches are still experimental.

We have also cross-linked hemoglobin to polymers to make hemoglobin conjugates (7–11). A number of workers are cross-linking hemoglobin to soluble polymers to form soluble conjugated hemoglobins (24,25). The hemoglobin can also be pyridoxalated to increase its P_{50}. The results obtained in these studies are similar in many instances to those using PPHb. However, in use as a blood substitute, a PPHb hemoglobin concentration of 14 g/dl blood is possible while with conjugated hemoglobin, because of high oncotic pressure, substitution only up to 6–7 g/dl blood of conjugated hemoglobin is possible.

In summary, microencapsulation or cross-linking of hemoglobin has resulted in potentially useful blood substitutes. Their present potential is for acute blood volume replacement rather than for long-term blood replacement. Possible applications for these materials include acute blood replacement situations, emergencies, mass disasters, war, organ preservation, and some types of surgery (45).

III. THERAPEUTIC APPLICATIONS OF IMMOBILIZED ENZYMES

A. Introduction

The therapeutic potential of immobilized enzymes were known as early as the 1960s (46–48). Examples include the demostration that artificial cells containing catalase, asparaginase, or urease are effective, respectively, for replacing hereditary enzyme deficiency in acatalasemic mice (46), suppressing the growth of lymphosarcoma in mice (47), and decreasing system urea levels in animals (9,48). Since these early studies, many workers have studied this and other techniques for defining the clinical application of immobilized enzymes (1–7).

B. Inborn Errors of Metabolism

Microencapsulated catalase was successfully used for enzyme replacement in acatalasemic mice with a congenital deficiency of the enzyme catalase (46). Furthermore, unlike free heterogenous catalase, microencapsulated catalase does not cause immunological reactions (46,49). Liposomes may also be used to microencapsulate enzymes and other materials and aid in therapy (50). For example, liposomes containing sulfatide, phosphatidylcholine, and cholesterol may be incorporated into the central nervous system (51). Surface incorporation of antibodies onto liposomes may also be used to target them (50). Other possible modifications include surface attachment of lectins, glycoproteins, and ligands. While liposomes have the advantage of being easily prepared and affordable, they appear to enhance the immune response to the protein they entrap (52).

Red blood cells have been used to microencapsulate enzymes by hemolysis and resealing (53). Rh antibody-coated human red blood cells containing β-glucosidase have been used for targeting Gaucher's disease (52). Other immobilization approaches for such therapy have included cross-linkage of the enzymes with carrier proteins (54,55) or with polymers (56,57). Modifications of enzymes by elective removal of carbohydrates, coupling of a recognition marker, and selection of a specific isozyme are other useful approaches (58).

We recently reported the use of artificial cells containing bacterial phenylalanine ammonia lyase for phenylketonuria in rats (59,60). This approach addressed the problem of the availability of the human enzyme system and the requirement for cofactor recycling. By administering these artificial cells orally, we also solved the problem of in vivo accumulation of the cells following parental administration (59,60). Oral administration of the cells to the phenylketonuria rat model for 7 days resulted in a lowering of system phenylalanine levels from the control group. Oral asparaginase-glutamine-tyrosinase artifical cells deplete the respective amino acids (61) from the enterocirculation (61). Oral xanthine oxidase artifical cells used in clinical trial lowered hypoxanthine in Lesch-Nyhan disease (62,63).

C. Asparaginase and Other Enzymes in Chemotherapy

Since the initial discovery by Broome, extensive research has been carried out using asparaginase and other enzymes for the degradation of nonessential amino acids required for tumor cell growth (5). The use of these enzymes is, however, associated with toxicity problems, immunogenicity, and duration of action. Using microencapsulated asparaginase, we demonstrated the feasibility of the immobilized enzyme for suppression of lymphosarcoma growth in mice (47). Many other groups have since investigated the use of most available immobilization approaches for asparaginase (1–7,56,57,61,64–69).

Asparaginase itself can also be modified (6). For example, it can survive longer in the circulation after deamination, acylation, and/or carbodiimide treatment (which affects its primary amino groups). Certain polyamino acids have also been conjugated to enzymes to prolong their half-life in the circulation or decrease immungenicity. For example, poly(N-vinylpyrrolidone) conjugated to beta-D-N-acetyl-hexosaminidase A increased the survival and decreased the immunogenicity of the enzyme after injection into animals (6).

D. Detoxification and Removal of Waste Metabolites

The successful use of artificial cell immobilized adsorbents for the treatment of accidental or suicidal poisoning (1,70–76) has overshadowed the earlier, high expectations of immobilized enzyme use in blood detoxification. Adsorbent hemoperfusion can rapidly remove most of the toxicant in poisoning cases such as sedative, barbiturate, salicylate, and other overdoses. However, absorbent hemoperfusion is not as specific as the use of specific enzyme systems. Where it is important to have specificity, there is still a need for immobilized enzyme systems.

Renal failure results in the inability of the body to excrete waste metabolites, electrolytes, and water. The standard treatment for renal failure is based on hemodialysis where the waste metabolites are removed using dialysis equipment. It has been found, however, that hemoperfusion utilizing 100 g of artificial cells containing activated charcoal is more efficient in waste removal than hemodialysis (74–75). For example, hemoperfusion clearances for creatinine and uric acid using an artificial cell system were 250 ml/min as compared to approximately 90–150 ml/min using standard hemodialysis. Hemoperfusion in series with dialysis resulted in the reduction of time required for hemodialysis from 12 to 8 hours per week. A second-generation artificial kidney utilizing the activated charcoal-containing cells in series with a small ultrafiltrator has now been developed. When applied to renal failure patients, the only additional step required other than use of this device is the need to remove urea. This need can be addressed using other artificial cells.

We have reported that artificial cells containing urease can rapidly convert urea into ammonia in vivo in extracorporeal circulation (48). While ammonia adsorbant can be used to remove the resulting ammonia and this approach has been used in some studies for dialysate regeneration (77), the amount of adsorbant required is too large for normal extracorporeal blood recirculation in patients. Our studies have evaluated the use of microencapsulated urease together with an ammonia adsorbent for oral administration (78,79). This approach has resulted in a significant lowering of systemic urea levels in the rat and is being further developed for possible human use by Battelle Memorial Institute (80). Clinical studies utilizing our method at the Mayo Clinic have shown it to be effective for urea removal from blood (81). However, the amount of encapsulated material required for efficacy needs to be decreased. Thus, we recently demonstrated that further design of the immobilized enzyme system can achieve this volume reduction (82). Instead of using ammonia adsorbant in the formulation, we prepared artificial cells containing a multienzyme system which allows each artificial cell to convert the ammonia resulting from urea degradation to amino acids (see below) (83–85).

Hemoperfusion using artificial cells containing activated charcoal can result in the temporary recovery of consciousness in grade IV coma patients (86,87). Such treatment in the earlier grades of fulminant hepatic failure resulted in a 70% survival rate compared to a 30% survival rate in untreated patients (88,89). This treatment is, however, only useful in acute liver failure. An artificial liver support system for chronic hepatic failure (e.g., cirrhosis) requires other metabolic functions. For example, in hepatic failure, there are marked metabolic disturbances in amino acid metabolism. This includes elevation of aromatic amino acids such as tyrosine and phenylalanine and changes in amino acid ratios. Hemoperfusion using artificial cells containing tyrosinase resulted in significant lowering of tyrosine levels in rats with fulminant hepatic failure (91,92) and artificial cells containing phenylalanine ammonia lyase effectively removed phenylalanine in vivo (59,60). Another exciting approach is the immobilization of the enzymes required for the urea cycle (93). Other approaches include the microencapsulation of intracellular organelles of liver cells (94). Enzymes extracted from liver cells have also been immobilized to Sepharose for possible detoxification applications (95).

E. Multienzyme Systems

Multienzyme systems are required for most of the metabolic functions in the body, and immobilized multienzyme systems can be used to correct metabolic imbalance. For example, as discussed above, a multienzyme system containing urease, leucine dehydrogenase, glucose dehydrogenase, and a transaminase in artificial cells can convert ammonia to lysine, leucine, and valine (84,85). There are the three essential amino acids which need to be increased in hepatic coma. In another example, artificial cells containing urease, glutamate dehydrogenase, glucose dehydrogenase, and a transaminase were used to convert urea and ammonia into glutamic acid and alanine (83).

The cofactors required for immobilized multienzyme systems can be coimmobilized within the artificial cells in two ways. The first is to use a lipid polymer artificial cell membrane which can retain cofactors such as NADH and NADPH (83–85). The second approach is to cross-link cofactors such as NADH and NADPH to a soluble macromolecule such as dextran (83–85).

IV. MICROENCAPSULATED CELL CULTURES IN DIABETES MELLITUS AND OTHER APPLICATIONS

We have successfully microencapsulated biological and human cells inside artificial cells including human blood cells (1,11,12). We proposed the use of this approach for insulin-producing islets, liver cells, and other cells for in vivo replacement in order to avoid immunological rejection. This approach has now been supported by other workers. For example, Sun et al. used an improved microencapsulation method for islet cells (96) and showed that intraperitoneal injection of the formulation can successfully maintain diabetic rats for over 12 months. The microencapsulated islet cells respond to glucose level and secrete the required amount of insulin to maintain normal glucose levels. Microencapsulation of hybridoma cell cultures has also been used for the large-scale production of monoclonal antibodies and interferon (97).

We recently reported that microencapsulated hepatocytes significantly increased the survival time of fulminant hepatic fallure rats (98). Suspensions of living hepatocytes were microencapsulated inside 300-μm mean diameter alginate artificial cells. Forty-eight hours after injection of galactosamine into galactosamine fullminant hepatic failure rats, the grade II coma hepatic failure rats were divided into pairs. One of the pair was randomly chosen for the control group and the other for the treated group. Each rat in the control group received one peritoneal injection of microcapsules containing no hepatocytes while each rat in the treated group received microcapsules containing hepatocytes. The survival of the treated group was found to be significantly higher than the control group (98).

Microencapsulated rat hepatocytes also remained viable when implanted as a xenograft in mice (99). Further, after implantation, the hepatocytes could regenerate: Cell viability increased from the original 60% at implantation to nearly 100% 30 days after implantation. Microencapsulation also retained the activity of hepoatocyte stimulating factor within the artificial cells (100). Implantation into Gunn rats lowered the systemic bilirubin level (101).

V. HEMOPERFUSION: EXTRSCORPOREAL BLOOD RECIRCULATION

In the case of hemoperfusion, the immobilized proteins, enzymes, or cells are retained in an extracorporeal column (48), and blood is recirculated by a pump through the column. As a result, foreign material is not introduced into the body. Microencapsulation of a type of microorganism has been studied for converting cholesterol to carbon dioxide (102). Artificial cell immobilized absorbents (such as activated charcoal) have already become a routine procedure in clinical practice for the treatment of poisoning (1,70–76), chronic renal failure (1,74,75), removal of metals such as aluminum (76), and other applications. With the large amount of clinical experience already available using immobilized adsorbents, this approach could be developed for clinical applications of immobilized proteins, enzymes, and cells. Examples of experimental studies using this approach have already been discussed above, e.g., immobilized urease, asparaginase, tyrosinase, and others. In addition, immobilized heparinase is also being studied for potential therapeutic applications (103).

VI. CONCLUSION

Since the first demonstration of the possible medical application of immobilized enzymes and proteins (1), many different possibilities are now available (1–8). With the recent rapid progress and interest in biotechnology, one can foresee increasing progress in this area for therapeutic applications. This would include the immobilization of peptides, proteins, enzymes, hemoglobin, cell cultures, microorganisms, and other bioreactants.

REFERENCES

1. T. M. S. Chang, *Artificial Cells,* Charles C. Thomas, Springfield, IL, 1972.
2. T. M. S. Chang, ed., *Biomedical Applications of Immobilized Enzymes and Proteins,* Vols. I & II, Plenum, New York, 1977.
3. T. M. S. Chang, *Methods in Enzymology 137:*444 (1988).
4. L. Chibata, *Immobilized Enzymes,* Wiley-Interscience, New York, 1978.
5. J. S. Holcenberg and J. Roberts, eds., *Enzymes as Drugs,* Wiley-Interscience, New York, 1981.
6. H. H. Weetall and D. A. Cooney, in *Enzymes as Drugs* (J. S. Holcenberg and J. Roberts, eds.), Wiley-Interscience, New York, 1981, p. 395.
7. T. M. S. Chang, *Appl. Biochem. & Biotechnol. 10:*5 (1984).
8. T. M. S. Chang, Hemoglobin corpuscles. Report of a research project of B.Sc. honours Physiology, McGill University, 1957.
9. Chang, *Science 146*(3643):524 (1964).
10. T. M. S. Chang, F. C. MacIntosh, and S. G. Mason, Canadian Patent 873,815 (1971).
11. T. M. S. Chang, *Semipermeable aeveour microencapsulation,* Ph.D. thesis, McGill University, 1965.
12. T. M. S. Chang, F. C. MacIntosh, S. G. Mason, *Can. J. Physiol. Pharmacol. 44:*115 (1966).
13. L. Djordjevich, I. F. Miller, *Exp. Hematol. 8:*584 (1980).
14. A. C. Hunt, R. R. Burnette, in *Advances in Blood Substitute Research* R. B. Bolin, R. P. Geyer, G. J. Nemo, eds.), Alan R Liss Inc, New York, 1983, p. 59.
15. B. P. Gaber, M. C. Farmer, *Prog. Clin. Biol. Res. 165:*179 (1984).
16. C. S. Ecanow, B. Ecanow, U.S. Patent No. 4,439,424 (1987).
17. E. Tsuchida, H. Nishide, M. Yuasa, M. Sekine, *Bulletin of the Chemical Society of Japan 57:*776 (1984).
18. B. P. Gaber, M. C. Farmer, in The Red Cell: Sixth Annual Arbor Conference, New York, Alan R. Liss, Inc., 1984.
19. R. L. Beissinger, M. C. Farmer, J. L. Gossage, *Trans. Am. Soc. Artif. Inter. Organs 32:*58 (1986).
20. L. Djordjevich, J. Mayoral, A. D. Ivankovich, W. Gottschalk, *Anesthesiology 55:*A86 (1981).
21. C. A. Hunt, R. R. Burnette, R. D. MacGregor, et al. *Science 230:*1165 (1985).
22. M. C. Farmer, A. S. Rudolph, K. D. Vandegriff, M. D. Havre, S. A. Bavne, S. A. Johnson, *J. Biomat. Artif. Cells Artif. Org. 16:*289–300 (1989).
23. H. F. Bunn, J. H. Jandl, *Trans. Assoc. Am. Physicians 81:*147 (1968).
24. K. Iwasaki, Y. Iwasaki, *Artif. Org. 10:*411 (1986).
25. S. C. Tam, J. Blumenstein, J. T. Wong, *Proc. Natl. Acad. Sci. USA 73:*2128 (1976).
26. R. Benesch, R. E. Benesch, S. Yung, R. Edalji, *Biochem. Biophys. Res. Commun. 63:*1123 (1975).
27. A. G. Greenburg, R. Hayashi, I. Krupenas, *Surgery 86:*13 (1979).

28. L. R. Sehgal, A. L. Rosen, S. A. Gould, H. L. Sehgal, L. Dalton, J. Mayoral G. S. Moss, *Fed. Proc. 39:*2383 (1980).
29. R. Dudziak, K. Bonhard, *Anesthetist 29:* 181 (1980).
30. F. DeVenuto, A. I. Zegna *Surg. Gynecol. Obstet. 155:*342 (1982).
31. P. E. Keipert, J. Minkowitz, T. M. S. Chang, *Int. J. Artif. Organs 5*(6):383 (1982).
32. L. R. Sehgal, A. L. Rosen, S. A. Gould, H. L. Sehgal, G. S. Moss, *Transfusion 23*(2):158 (1983).
33. P. E. Keipert, T. M. S. Chang, *Trans. Am. Soc. Artif. Intern. Organs 23:*329 (1983).
34. P. E. Keipert, T. M. S. Chang, *Appl Biochem & Biotechnol 10:*133 (1984).
35. P. E. Keipert, T. M. S. Chang, *Biomat. Med. Dev. Artif. Organs 13:* 1 (1985).
36. P. E. Keipert, T. M. S. Chang, *Vox Sanguinis 53:*7 (1987).
37. J. Hobbhahn, H. Vogel, N. Kothe, W. Brendel, P. Keipert, F. Jesch, *Acta Anesth. Scand. 29:*537 (1985).
38. D. H. Marks, J. Patressi, I. T. Chaudry, *Circ. Shock. 16:*165 (1985).
39. R. Stabilini, G. Palazzini, G. P. Pietta, M. Pace, A. Calatroni, E. Raffaldoni, A. Ghessi, G. Aguggini, A. Agostoni, *Int. J. Artif. Organs 6*(6):319 (1983).
40. J. C. Hsia, T. M. Hayes, *J. Chromatography 303:*425 (1984).
41. C. M. Hertzman, P. E. Keipert and T. M. S. Chang, *Int. J. Artif. Organs 9:*179 (1986).
42. T. M. S. Chang, R. Varma, *Biomaterial, Artificial Cells and Artificial Organs 15:*443 (1987).
43. M. Feola, H. Gonzalez, P. C. Canizaro, D. Bingham, *Surgery Gynecology & Obstetrics 157:*399 (1983).
44. P. W. Maffuid, A. G. Greenburg, E. S. Lee, T. Velky, D. B. Hoyt, *Am. College Surg. Forum 34:*5 (1983).
45. T. M. S. Chang, R. Geyer, (eds.), *Blood Substitutes,* Marcel Dekker, New York, 1989.
46. T. M. S. Chang, M. J. Poznansky, *Nature 218*(5138):242 (1968).
47. T. M. S. Chang, *Nature 229*(528):117 (1971).
48. T. M. S. Chang *Trans. Am. Soc. Artif. Intern. Organs 12:*13 (1966).
49. M. J. Poznansky, T. M. S. Chang, *Biochim. Biophys. Acta 334:*103 (1974).
50. G. Gregoriadis in *Enzyme Replacement Therapy of Lysosomal Storage Diseases* (J. M. Tager, J. M. Hooghwinkel, and W. T. Daoms, eds.), North-Holland, Amsterdam, 1974, p. 131.
51. M. Naoi, K. Yagi, *Biochem. Int. 1:*591 (1980).
52. G. Gregoriadis *Drug Carriers in Biology and Medicine,* Academic Press, New York, 1979.
53. G. Ihler, R. Glew in *Biomedical Applications of Immobilized Enzymes and Proteins* (T. M. S. Chang, ed.), Plenum Press, New York, 1977, p. 219.
54. T. M. S. Chang, *Biochem. Biophys. Res. Commun. 44*(6):1531 (1971).
55. M. J. Poznansky, *J. Appl. Biochem. Biotech. 2:*41 (1984).
56. A. Abuchowski, F. F. Davis, in *Enzymes as Drugs* (J. S. Holcenberg and J. Roberts, eds.), Wiley-Interscience, New York, 1981, p. 367.
57. R. L. Foster, T. Wileman, *J. Pharm. Pharmacol. 31* (Suppl):37P (1979).

58. G. A. Grabowski, R. J. Desnick, in *Enzymes as Drugs* (J. S. Holcenberg and J. Roberts, eds.) Wiley-Interscience, New York, 1981, p. 167.
59. L. Bourget, T. M. S. Chang, *FEBS Lett. 180:*5 (1985).
60. L. Bourget, T. M. S. Chang, *Biochim. Biophys. Acta 883:*432 (1986).
61. T. M. S. Chang, C. Lister, *J. Biomat. Artif. Cells. Artif. Org. 16:*915 (1988).
62. T. M. S. Chang, *J. Biomat. Artif. Cells. Artif. Org. 17:*611 (1989).
63. R. M. Palmour, P. Goodyer, T. Reade, T. M. S. Chang, *Lancet* 2(8664):687 (1989).
64. T. M. S. Chang, *Enzyme 14*(2):95 (1973).
65. E. D. SiuChong, T. M. S. Chang, *Enzyme 18:*218 (1974).
66. T. Mori, T. Tosa, I. Chibata, *Biochem. Biophys. Acta 321:*653 (1973).
67. L. D. S. Hudson, M. B. Fiddler, R. J. Desnick, *J. Pharmacol. Exp. Ther. 208:*507 (1979).
68. G. Schmer, J. S. Holcenberg, in *Enzymes as Drugs* (J. S. Holcenberg and J. Roberts, eds.), Wiley-Interscience, New York, 1981, p. 385.
69. D. A. Cooney, H. H. Weetall, E. Long, *Biochem. Pharmacol. 24:*503 (1975).
70. T. M. S. Chang, J. F. Coffey, P. Barre, A. Gonda, J. H. Dirks, M. Levy, C. Lister, *Can. Med. Assoc. J. 108:*429 (1973).
71. T. M. S. Chang, J. F. Coffey, C. Lister, E. Taroy, A. Stark, *Trans. Amer. Soc. Artif. Internal Organs 19:*87 (1973).
72. T. M. S. Chang, *Clin. Toxicol. 17:*529 (1980).
73. M. C. Gelfand, J. F. Winchester, J. H. Knepshield, K. M. Hansen, S. L. Cohan, B. S. Stranch, K. L. Geoly, A. C. Kennedy, G. E. Schreiner, *Trans. Amer. Soc. Artif. Internal Organs 23:*599 (1977).
74. V. Bonomini, T. M. S. Chang, eds., *Hemoperfusion (Contributions to Nephrology Series*), S. Karger AG, Basel, Switzerland, 1982.
75. S. Sideman, T. M. S. Chang, eds. *Hemoperfusion: I. Artificial Kidney and Liver Support and Detoxification* Hemisphere, Washington, 1980, and *Hemoperfusion: II. Devices and Clinical Applications,* Samuel Neaman Inst. for Adv. Studies in Sci. and Technol., Technion, Haifa, Israel, 1981.
76. T. M. S. Chang, and N. Nicolaev, eds., *Int. J. Biomaterials, Artificial Cells and Artificial Organs 15:*1 (1987).
77. A. Gordon, A. J. Lewin, M. H. Maxwell, R. Martin, in *Artificial Kidney, Artificial Liver and Artificial Cells* (T. M. S. Chang, ed.), Plenum, New York, 1978, p. 23.
78. T. M. S. Chang, S. K. Loa, *Physiologist 13:*70 (1970).
79. T. M. S. Chang, *Kidney Intnl. 10:*S218 (1976).
80. D. L. Gardner, R. D. Falb, B. C. Kim, D. C. Emmerling, *Trans. Amer. Soc. Artif. Internal Organs 17:*239 (1971).
81. C. Kjellstrand, H. Borges, C. Pru, D. Gardner, D. Fink, *Trans. Amer. Soc. Artif. Internal Organs 27:*24 (1981).
82. E. Wolfe, T. M. S. Chang, *Int. J. Artif. Organs 10:*43 (1987).
83. T. M. S. Chang, *Methods in Enzymology 136:*67 (1987).
84. K. F. Gu, T. M. S. Chang, *Int. J. Biomaterials, Artificial Cells and Artificial Organs 15:*297.
85. K. F. Gu, T. M. S. Chang, *J. Appl. Biochem. Biotechnol. 12:*227 (1990).

86. T. M. S. Chang, *Lancet ii:*1371 (1972).
87. R. Williams, I. M. Murray-Lyon, *Artificial Liver Support,* Pitman, London, 1975.
88. T. M. S. Chang, C. Lister, E. Chirito, P. O'Keefe and R. Resurreccion, *Trans. Amer. Soc. Artif. Internal Organs 24:*243 (1978).
89. T. M. S. Chang, *Seminars in Liver Diseases Series 6:*148 (1986).
90. A. E. S. Gimson, S. Brande, P. J. Mellon, J. Canalese, R. Williams, *Lancet ii:*681 (1982).
91. C. D. Shu, T. M. S. Chang, *Int. J. Artif. Organs 4:*82 (1981).
92. Z. Q. Shi, T. M. S. Chang, *Trans. Am. Soc. Artif. Internal Organs 28:*205 (1982).
93. N. Siegbahn, K. Mosbach *FEBS Lett. 137:*6 (1982).
94. Z. Y. Yuan, T. M. S. Chang, *Int. J. Artif. Organs 9*(1):63 (1986).
95. G. Brunner, F. W. Schmidt, eds., *Artificial Liver Support,* Springer-Verlag, Berlin, 1981.
96. A. M. Sun, G. M. O'Shea, M. F. A. Goosen, *J. Appl. Biochem. & Biotechnol. 10:*87 (1984).
97. Bulletin on Tissue Microencapsulation, Damon Corporation, Needham Heights, MA, 1981.
98. H. Wong, T. M. S. Chang *Int. J. Artif. Organs 9:*335 (1986).
99. H. Wong, and T. M. S. Chang, *J. Biomat. Artif. Cells Artif. Org. 16:*731 (1988).
100. S. A. Kashani, and T. M. S. Chang, *J. Biomat. Artif. Cells Artif. Org., 16,*741 (1988).
101. S. Bruni, T. M. S. Chang, *J. Biomat. Artif. Cells. Artif. Org. 17:*403 (1989).
102. F. Garofalo, T. M. S. Chang, *J. Biomat. Artif. Cells. Artif. Org. 17:*271 (1989).
103. R. Langer, P. J. Blackshear, T. M. S. Chang, M. D. Klein, J. S. Schultz, *Trans. Am. Soc. Artif. Internal Organs 32:*639 (1986).

10

Immobilized Microbial and Animal Cells as Enzyme Reactors

Subhash B. Karkare

AMGEN, Inc.
Thousand Oaks, California

I. INTRODUCTION

The field of cell immobilization has grown over the past several years to encompass an ever-expanding array of cells ranging from single enzyme deactivated cells to living microbial cells and more recently animal cells. Consequently, some new techniques of immobilization have emerged to keep pace with the specific needs of these different types of cells, and some novel concepts in reactor design and operation have also evolved to meet these needs. In many cases, existing technology has been suitably adapted to provide for the special requirements of these systems. In this chapter, we will attempt to present an overview of this technology with an emphasis on the more recent developments, especially in the area of animal cell immobilization.

While it is true that different types of cells have distinct characteristics and require special considerations, it is still possible to make certain generalizations regarding reaction kinetics of these systems. Several aspects of immobilization techniques are also very similar. We will therefore dwell on these topics with the broad perspective of immobilized cells in general and then point out the idiosyncracies of animal cells which are more delicate than the rest and require unique design considerations. First, we will review the common techniques used for cell immobilization and discuss the suitability of each method for various types of cells. The next section will focus on different aspects of microbial cell immobilization including single enzyme systems, living cell systems, kinetics, and reactor design. Immobilized cell systems are generally used in the continuous mode for several days and even months. This necessitates stringent hardware design criteria which differ significantly from the fermentation norms. Further, the susceptibility of some mammalian cells to toxic effects of some materials severely restricts the range of materials that can be used safely in hardware design. Hence, in this chapter, we have attempted to include sections on the often neglected topic of hardware design for both microbial and animal cells. Finally, we have a section on the special considerations required for development of immobilized animal cell systems.

II. TECHNIQUES OF MICROBIAL AND ANIMAL CELL IMMOBILIZATION

Immobilization in the most general sense of the word can be taken to mean a method by which the free movement of cells is restricted. Typically, this leads to higher cell densities than those obtained in free suspension resulting in greater volumetric productivities. Perhaps an even more significant result is the fact that the dilution rate (or throughput) in the reactor can now be decoupled from the growth rate of the cells. In other words, growth is not necessary for an immobilized cell reactor to operate at steady state. Numerous techniques have been developed for cell immobilization. The technique of choice may depend on several factors including the cell type, culture kinetics, reactor design considerations, and economics. The techniques can be broken down into several categories depending on the method used to restrict the movement of the cells. These include surface adhesion, covalent attachment, entrapment in porous materials or gels, encapsulation, and cell retention by membranes. We will consider a few illustrative examples of each method to demonstrate the diversity of materials and methods employed in this field.

A. Surface Adhesion

Many microbial and animal cells naturally adhere to surfaces. This property lends itself readily to an immobilized cell system. A reactor with high surface area to volume ratio can effectively immobilize a large number of cells. The most common use of this technique can be found in the culturing of anchorage-dependent animal cells. For instance, microcarriers are commonly used to propagate mammalian cells for use in production of biologicals. Reuveny (1) describes the different types of microcarriers in use and their application in mammalian cell culture. The technique of immobilization in this case is quite simple. The cells are placed in proximity with the presterilized microcarriers under slow agitation conditions for a few hours. The cells attach to the microcarriers by virtue of their charge and grow until the surface is confluent. Nutrients are perfused through the reactor at an appropriate rate to supply the requirements of the cells. The same principle of adhesion can also be used in other cell-culturing systems such as a multidisc reactor (2), packed bed of glass beads (3), or spiral film bottles (4).

In the microbial world, the second oldest fermentation process (vinegar fermentation) utilized bacteria adhering to wood shavings for converting ethanol to acetic acid. Another traditional use of surface adhesion of bacteria can be found in wastewater treatment (5). Ion exchange resins have also been used to adsorb bacteria for the purposes of immobilization. An example is the immobilization of *Azotobacter vinelandii* on Cellex E resin for nitrogen fixation (6).

B. Covalent Attachment

In cases where bacteria do not naturally adhere to a surface, chemical cross-linking may be used to form a covalent bond between the cell wall and the immobilization support. For example, Constantinides et al. (7) used glutaraldehyde to covalently bond *Brevibacterium flavum* to a collagen matrix. Generally speaking, covalent attachment requires the use of chemicals (such as glutaraldehyde) which may be harmful to cells. Hence this type of immobilization is restricted to nonliving single enzyme systems or, as in the above case, to living but nongrowing cells (see Chapter 3). Mammalian cells are even more sensitive to toxicity from such chemicals. However, use of attachment factors such as fibronectin for promoting cell attachment may fall under this category (8).

C. Entrapment in Porous Materials or Gels

This is a very broad category comprising scores of materials and methods that vary considerably in complexity. Typically, the cells are entrapped in a porous matrix in which the pores are large enough to hold more than one cell but are small enough to severely restrict the free movement of the cells. This results in cells growing to high densities within the confines of the pores. Criteria for required pore sizes have been defined by Messing et al. (9). In some cases, the cells are simply encapsulated by a shell of semipermeable membrane such as in the Damon process (10). We can break these methods down into two classes depending on whether the cells are preinoculated during the formation of the matrix or gels or the cells are postinoculated after the matrix is sterilized. Figure 1 shows the typical steps

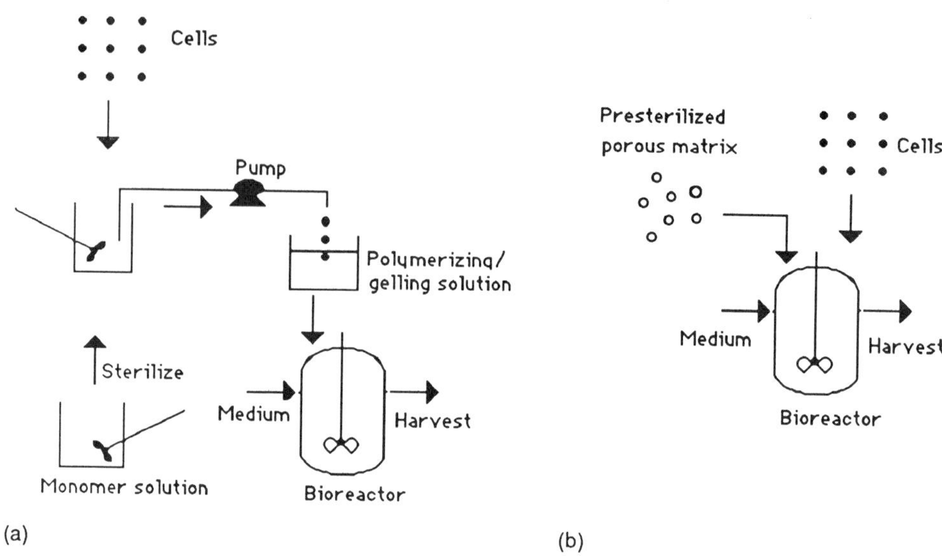

Figure 1 Common cell entrapment methods: (a) cell entrapment with preinoculation; (b) cell entrapment with postinoculation.

involved in these two methods. These methods have been used widely, both for microbial as well as animal cell immobilization.

Examples of animal cell immobilization include entrapment in agarose beads (11), in porous collagen microspheres (12), or in a ceramic matrix (13). The first example involves preinoculation of cells in the agarose solution before the matrix is formed, whereas in the latter two methods, cells are inoculated into a preformed, presterilized matrix. Microbial cells have been immobilized in matrices such as carrageenan (14), polyacrylamide (15), and alginate (16). Bacteria have also been immobilized onto controlled pore glass matrices (9). A more complete listing of materials used for microbial entrapment can be found in a recent review by Scott (17).

D. Cell Retention by Membranes

This is another broad category which includes immobilization methods such as hollow fiber technology, tangential flow filtration, and dialysis culture. The cells in all these methods are not attached to any support but are physically restrained in a given space by means of microporous (or in some cases smaller pore size) membranes. This allows the use of dilution rates in excess of the maximum specific growth rate, which results in high cell density. To compensate for cell death and accumulating cell debris, it is usually necessary to "purge" some cells out of the device either periodically

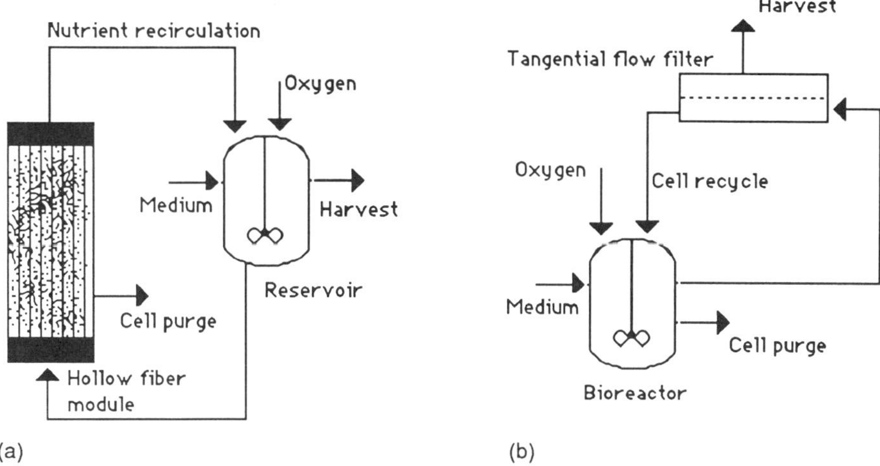

Figure 2 Cell retention by membranes: (a) static cell growth in hollow fiber reactor; (b) cell recycle by tangential flow filter.

or on a continuous basis. The very nature of this method automatically rules out attachment-dependent cell lines, which are commonly encountered in mammalian cell culture. Figure 2 shows two schemes commonly used for cell retention by membranes. In the case of hollow fibers, the cells are held static in the extra-capillary space, while the nutrients are circulated through the tubes from a reservoir which may be replenished on a periodic or continuous basis. This method is especially suitable for mammalian cells which are highly shear-sensitive. The second method utilizes a conventional fermentor coupled to a tangential flow filter. The cells are retained by the filter and are recycled back to the fermentor while the product and spent medium passes through the filter. This method is particularly suitable for bacterial cells which are not as shear sensitive but have a high oxygen demand. However, these are not hard and fast rules. In fact each method has been applied both to microbial as well as animal cells. Although Figure 2 shows two of the common techniques used for cell retention by membranes, several variations of these systems have been used. For example, oxygen may be introduced in a separate membrane oxygenator placed in the nutrient recycle loop of Figure 2a. Alternatively, oxygen may be supplied to the cells via separate tubes within the hollow fiber module.

Numerous applications have been reported for immobilization of mammalian cells in hollow fiber reactors. Altshuler et al. (18) immobilized hybridomas for the production of monoclonal antibodies in Amicon hollow fiber modules. In another variation of this system, Ku et al. (19) used the lumen of the hollow fibers to supply gaseous nutrients for oxygenation and pH control to support the immobilized mammalian cells. Hollow fiber reactors have also been employed for immobilizing microbial cells. For instance, Robertson and Kim (20) used a dual aerobic hollow fiber bioreactor to grow *Streptomyces aureofaciens*. They utilized a concentric silicone tubule to provide the oxygen transfer, which is often the limiting step for hollow fiber reactors.

Cell recycle using a tangential flow filter was used by Flickinger et al. (21) for continuous culture of a hybridoma at very low specific growth rates. The same type of system was recently utilized for ethanol production by Lee and Chang (22), who used a hollow fiber cartridge as the tangential flow filter. Since this technique provides a homogeneous cell suspension, the system lends itself to easy modeling and quantification. However, the cells are susceptible to higher shear damage both in the filter and in the fermentor. A variation of the same principle involves introducing the filtration element in the fermentor itself in the form of a rotating filter element. For example, Tolbert et al. (23) used a rapidly rotating porcelain filter in a 40 liter reactor to retain cells in the reactor, while perfusing medium at a fast rate.

In this section, we have attempted to describe the most commonly used techniques for immobilization. Needless to say, all the techniques developed so far have not been covered, nor has an attempt been made to review all the publications in this field. However, the literature cited can be a good starting point to become familiar with the state of the art.

III. IMMOBILIZED MICROBIAL CELLS

Immobilized microbial cells have been used commercially for vinegar production for centuries. However, research in this field has only recently intensified and actual commercial applications of immobilized cells are still rare. Earlier efforts in this field centered on utilizing a single enzyme from the whole cell for effecting bioconversions. For example, *Streptomyces phaechromogenes* was immobilized to produce fructose from glucose by using a single enzyme in the organism, glucose isomerase (24). This technique is an improvement over immobilizing the enzyme itself as it obviates the need to purify the enzyme and also imparts greater stability to the catalyst. However, it soon became apparent that the same process could be applied to viable (resting or growing) cells, thus utilizing entire metabolic pathways. Use of growing cells also allows the activity of the cells to be maintained indefinitely. A good example of this is the production of ethanol for extended periods of time by immobilized yeast (25). The kinetics of these two types of system have some differences, which will be pointed out in the following sections, and the consequent reactor operating strategies will be discussed.

A. Single Enzyme Systems

When an enzyme of commercial interest is derived from a microbial source, it makes sense to immobilize the whole cell, treating it in such a way that the activity of the enzyme is retained while other pathways are deactivated. High fructose corn syrup is an example of a commercial application of this process. These systems can be used with a variety of reactor types including packed bed, spiral wound, fluidized bed, and stirred tank configurations. Since only one enzyme needs to be active, there is usually no oxygen transfer requirement, which makes reactor design relatively straightforward and dependent mainly on reaction kinetics. For instance, if substrate inhibition is significant, a stirred tank geometry is beneficial, while in other cases, a plug flow packed bed reactor may yield the most cost-effective conversion of substrate to product. The enzymes do lose activity over a period of time. This necessitates periodic recharging of the reactor with fresh biocatalyst. The length of time between charges is determined usually by the minimum production capacity needed from the plant or, in the case of fixed substrate flow rate, the minimum product concentration necessary for effective downstream purification. An alternative reactor operating strategy is to continuously recharge the reactor with biocatalyst and to remove an equal amount of spent catalyst at the same rate. This leads to lower effective catalyst utilization, but costly downtimes are obviated since the process can operate continuously. Obviously, the operating strategy of choice will depend on economics. If the relative cost of the catalyst is much higher than labor costs, the former strategy makes more sense and vice versa. Venkatasubramanian and Harrow (26) have discussed the design and operating strategies for such systems in greater detail. The materials requirements for reactor design are generally not as stringent as in live cell reactors. However, it is important to make sure that the enzyme is not deactivated by any heavy metal ions that may leach out of the materials.

B. Living Cell Systems

Living cells offer the possibility of utilizing entire biochemical pathways for complex biosynthesis. Traditionally, this has been accomplished in batch reactors with cells suspended in nutrient medium. Depending on the culture kinetics, this process takes between a few hours to about a week. After this, the broth is removed for purification and the whole process is started all over again.

The fermentation process can be converted into a continuous process by constantly feeding in nutrients and removing product at the same volumetric rate. In some cases, this leads to higher volumetric productivities (especially for growth-associated products), while in others, the productivity may suffer (for non–growth associated products such as antibiotics). The continuous process (or "chemostat" as it is known) suffers from the fact that it has to depend on cell growth to maintain a steady state. Further, the rate at which nutrients can be converted into product is limited by the maximum specific growth rate of the cells. These two constraints on the chemostat can be overcome by immobilizing the live cells. Since the cells do not leave the reactor with the effluent stream, there is no upper limit on the dilution rate that can be used. Also, a maintenance medium can be used to restrict (or even stop) the cell growth. Thus, non–growth associated products can be produced at high rates in a continuous system. The cells growing in an immobilized state can reach extremely high densities. This increases the volumetric productivity of the reactor which can improve the economics for certain products.

Living cell systems do have several problems associated with them which have to be overcome. The high cell densities often lead to diffusional resistances that limit the particle size to a few micrometers (depending on the cell). For example, Shinmyo et al. (27) found that immobilized *Bacillus amyloliquefaciens* inhabited only the outer 50 μm thickness of carrageenan beads due to diffusional limitations to oxygen transfer. The matrix used for immobilization is often mechanically weak and needs gentle handling. This makes the problem of oxygen transfer worse, because high shear agitation is often used to supply oxygen in fermenters. Hence, novel oxygenation methods have to be devised to supply the increased oxygen demand of these high density cultures. Due to these limitations, the application of immobilized living cell systems has been restricted to anaerobic systems such as ethanol production.

Novel approaches are being taken, however, to solve the problems of oxygen transfer and shear sensitivity of these systems. For example, Dunn et al. (28) used a fluidized bed reator with continuous recycle of the fluid through a separate oxygenator for wastewater nitrification by bacteria adhering to sand particles. Dorr Oliver markets a commercial system called Oxitron with a similar design. Enfors and Mattiason (29) have attempted to solve the problem by increasing the solubility of the medium (by addition of perfluorocarbons) and by generating oxygen in situ (by co-immobilizing algae or catalase with the bacteria of interest). The reactor needs to be supplied with light (when algae are co-immobilized) or with hydrogen peroxide (when catalase is co-immobilized) to produce the requisite oxygen. These processes have not yet found commercial applications.

The reactor operating strategies for living cell systems are more complex than single enzyme systems, since the culture kinetics and product formation kinetics vary significantly from cell to cell. For example, ethanol can be produced continuously at a high production rate for several months by feeding the reactor with a growth-production medium at a constant feed rate (30). The growth and outgrowth of cells in this case keeps the reactor at a constant and viable cell density. Thus, the ethanol productivity remains constant throughout the run. On the other hand, when a non–growth associated product is to be produced, the operating strategy will typically involve a growth cycle followed by a production cycle in which a nutrient essential for growth is removed from the medium. This strategy was employed to increase the productivity of candicidin by immobilized *Streptomyces griseus* (31). In some cases, the growth-promoting nutrient concentration can be adjusted to a low enough level so that the slow growth essentially matches the cell deactivation rate. A constant and high level of productivity was maintained by this slow growth strategy for the case of a recombinant growth hormone production by immobilized yeast cells (32).

When living cells are immobilized in high densities in close proximity to each other, it is possible that the physiology of the cells may undergo some changes. These physiological changes and their effect on the process have been the subject of some recent investigations. Doran and Bailey (33) showed that immobilization enhanced the fermentative activity in yeast and reduced the reproductive rate of yeast. Bailey et al. (34) showed that immobilized bacteria were more prone to cell synchrony effects and modeled the synchronous growth kinetics for immobilized *Acetobacter suboxydans*. These studies indicate that major shifts in metabolism, fermentation kinetics, and yields can occur due to immobilization of cells. Hence the information derived from studies of suspended cells should be applied with caution to immobilized cell systems.

C. Kinetics of Immobilized Microbial Cell Systems

In the case of single enzyme systems, since the cells themselves are not viable, the activity of the enzyme does decay over time. The catalyst is therefore characterized by a half-life (or a decay constant k_d) and typical Michaelis-Menten constants including the half velocity constant K_m and the maximum reaction rate V_{max}. In certain cases, substrate and/or product inhibition constants may also be involved. Since the cells are immobilized inside a matrix, internal and external mass transfer kinetics have to be taken into account when designing the reactors. The treatment for reactor design is very similar to that in heterogeneous catalysis. Intrinsic kinetic parameters mentioned above are determined experimentally along with data on substrate and product diffusion kinetics. Depending on the catalyst geometry, appropriate effectiveness factors can then be determined. Venkatasubramanian et al. (35) have described the rate expressions applicable to single enzyme systems along with the various reactor performance equations.

Predictably, the kinetics of living cell systems are more complex. In addition to the reaction rate of interest, it is also necessary to consider the rates of cell growth and death as well as the kinetics of oxygen transfer and consumption. Venkatasubramanian et al. (35) have presented a rudimentary analysis of immobilized living cell systems with the simplifying assumptions of negligible mass transfer resistances. They have presented reactor performance equations for a continuous stirred tank system which demonstrate existence of dilution rate optima for the case of growth-associated product formation. In the case of non–growth associated kinetics, the objective is to achieve as high a cell density as possible and subsequently minimize the growth rate while allowing the cells to remain viable. Park et al. (36) have analyzed the performance of an immobilized cell reactor for the production of penicillin. In this case the reaction rate is dependent not only on the carbon substrate concentration, but also on the oxygen concentration near the cell. By superimposing the diffusive transport equations on the intrinsic kinetics, the authors simulated the dependence of reactor productivity on biofilm thickness in stirred tank as well as plug-flow reactors and demonstrated that the optimal biofilm thickness depends not only on the penetretion depth of a limiting nutrient but on the total reactor configuration as well. The cell specific reaction kinetics was assumed to be the same as for suspended cells. Doran and Bailey (37) have compared the fermentation rates and cell cycle operation of immobilized and suspended yeast cells and found them to be significantly different. Hence it is necessary to obtain intrinsic kinetics of the immobilized cells per se whenever possible.

D. Principles of Reactor Design

The kinetics of immobilized cells is not the only factor to be considered in the design of bioreactors. For example, the kinetics of a process may suggest that the optimal configuration is a plug flow reactor. However, the oxygen transfer considerations may force the use of several stirred tank reactors in series or the use of a three-phase fluidized bed. Kinetics usually determine the size of the reactor, but the actual configuration of the reactor may depend on several other factors including oxygen requirements, scalability, convenience of use, contamination potential, and, of course, economics. The literature in this field has concentrated primarily on the process design aspects of this problem. The equally important hardware design concepts have been largely ignored. In this section, we will cite some of the recent literature focusing on process design and then discuss the major problems associated with hardware design.

1. Process Design

The design of reactors for single enzyme systems is very similar to that of heterogeneous catalytic reactors and has been well documented (35,38,39). Process design for living cell reactors has been the subject of more recent research. Venkatasubramanian and Karkare (40) discuss the process engineering considerations involved in the design of immobilized living cell reactors. They have pointed out the relative merits and demerits of a variety of reactor configurations including packed bed, continuous stirred tank, fluidized bed, and hollow fiber reactors. The choice of the reactor type depends on considerations such as ease of operation, mixing characteristics, oxygen transfer, distribution of viable cells, catalyst attrition and replacement, ease of scale-up, reaction kinetics, process control requirements, and cost of the reactor. Obviously, the reactor type chosen will vary depending on which of these considerations are more important. For example, the hollow fiber reactors can be excellent for the production of high value small volume biochemicals. However, they are limited in their scale-up potential. Kleinstreuer and Agarwal (41) have analyzed the performance of hollow fiber reactors using a computer simulation model. Similar principles can be used to analyze a broad family of membrane reactors. Miller and Mellick (42) discuss some aspects of modeling bioreactors including immobilized living cell reactors.

The type of reactor is first chosen based on small-scale experimental results and the considerations outlined before. Next, the volume of the reactor can be determined based on reaction kinetics and the capacity requirement of the plant. It is prudent practice to opt for at least two reactors (instead of one large reactor) to ensure against possible losses due to contamination or other failures. Based on the reactor dynamics (mixing characteristics, reaction times, oxygen consumption rates etc.), the control scheme can now be worked out for the reactor. This forms the basis of the process flow sheet which quantitates the material flow and the process and instrumentation diagram (P&ID) which indicates all the components in greater detail. The next step is to translate this process into hardware that will allow reliable operation of the process for the specified amount of time.

2. Hardware Design

Most immobilized cell reactors operate in the continuous mode to take full advantage of the high cell densities obtained. This implies that the threat of contamination is worse for these reactors than for normal batch fermentors. The financial consequences due to loss of a run after several days of operation can be significant. Hence the hardware design should meet stringent criteria for sterilizability, and leakproof operation.

To ensure sterilizability, it is necessary to avoid dead legs that do not allow free passage of steam. Any vertical dead leg with an aspect ratio greater than 2 should be vented during steaming. U-shaped bends should be avoided to prevent condensate accumulation in the bend. The system design should be such that condensate drains freely towards the steam trap without forming puddles. Special attention should be paid to the proper placement of valves which can accumulate condensate in their cavities. The material of construction should be pressure rated for the steam pressure and should not be affected by repeated steamings. For instance, polycarbonate is steam sterilizable. However, any residual stresses due to machining can cause cracks to form after this material is steamed. Stainless steel is the material of choice whenever cost permits its use. If plastics (such as polycarbonate, polysulfone, etc.) must be used, proper heat treatment procedures recommended by the manufacturer should be followed after fabrication.

The choice of valves and pumps is an important consideration in ensuring leakproof operation over a long period of time. Generally, devices with static seals are preferred over rotating mechanical seals. This means diaphragm valves and pumps should be used whenever possible. However, the elastomer used for the static seal should be chosen carefully. For example, TEFLON has a tendency to "flow" and loose the sealing pressure after steaming. This often leads to leaks in valves after sterilization. If TEFLON must be used, it is important to establish proper sealing torques immediately after steaming. In cases where diaphragm pumps cannot be used, magnetic couplings should be considered. If this is not possible, the mechanical seal should be protected by live steam on the outboard side of the seal.

All entrance and exit ports (such as medium inlet, sampling line, gas inlet, etc.) to the bioreactor should be protected either by a 0.1-μm filter or a steam block. Figure 3a shows a typical arrangement of a steam block to protect a sampling line from grow-through contamination. A similar arrangement can be used to make aseptic connections between two pieces of equipment (such as the medium vessel and the bioreactor). Traditionally, the fermentation industry has depended on heating the ports where connections are made with a flame. This practice is not reliable and should be avoided if long-term reactor operation is desired. The connections must be steamed at 121°C for at least 15 minutes to ensure sterility. The foregoing hardware design considerations, although mundane, often determine the

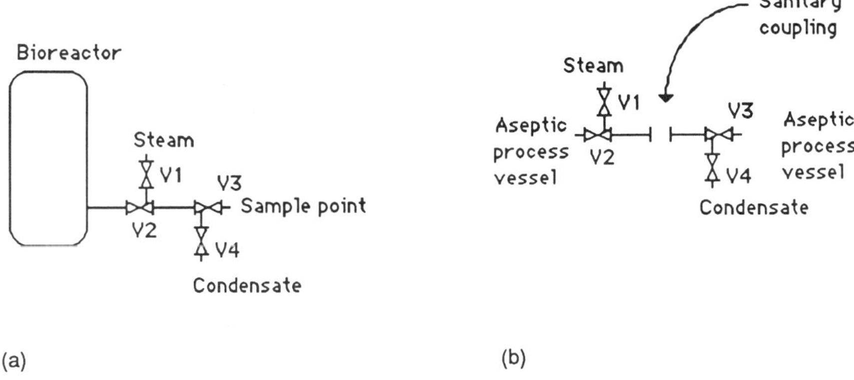

Figure 3 Hardware designs for aseptic operation: (a) steam block protection for sample line (V1, V2, V3, V4 are diaphragm valves); (b) aseptic connections between process vessels.

difference between a successful commercial process and one that appears attractive only on paper.

IV. IMMOBILIZED ANIMAL CELLS

The need to culture animal cells in large scale became evident with the difficulties faced in producing complex glycosylated proteins with recombinant bacteria. Some of the major products of biotechnology such as Tissue Plasminogen Activator (t-PA) and erythropoietin are made in mammalian cells. It is interesting that immobilized cell technology seems to have found a niche in this new industry much faster than in the traditional fermentation industry. Several immobilized cell processes have now been commercialized for the production of proteins from mammalian cells. Bioresponse, Charles River Biotechnial Services (CRBS), Damon Biotech, Endotronics, Karyon, and Verax are some of the companies that have developed various immobilization technologies for culturing of animal cells. These include hollow fiber devices (Bioresponse and Endotronics), encapsulated/entrapped cells in stirred reactors (Damon and Karyon), fixed bed ceramic matrix reactor (CRBS), and fluidized bed of porous collagen matrix (Verax). One of the reasons for this fast-paced development may be the fact that animal cells are immobilized in their natural environment and may be inherently more suitable for this technology than bacterial cells. The techniques of animal cell immobilization are similar to those used in microbial cell immobilization.

A recent review by Nilsson (43) covers the basic methods used in immobilizing animal cells. Animal cells are more fragile than their bacterial counterparts and grow much more slowly. These and other differences between the two require some special considerations when designing animal cell bioreactors. In the following sections we will describe the problems specific to animal cell culturing and ways of overcoming these problems in practice.

A. Special Considerations for Animal Cell Culturing

One of the most widely discussed problems with culturing animal cells is their shear sensitivity. However, until recently, there was no theoretical or even empirical basis for development of a rational approach to reactor design to overcome this problem. The problem of fluid dynamic shear is encountered in microcarrier culture systems and in the case of cell retention by tangential flow filters. Some recent studies have attempted to quantitate the shear sensitivity of cells by experimental measurements. For example, Stathopoulos and Hellums (44) studied the shear stress effects on human embryonic kidney cells by establishing a well-defined flow field between two parallel plates. They demonstrated that exposure to shear stresses greater than 2.6 N/m^2 for more than 2 hours led to cell death and increased urokinase production. Smith et al. (45) used a Contraves viscometer to study the shear sensitivity of murine hybridomas. They found that shear damage occurred at shear rates of about 870 s^{-1}. However, it is very difficult to apply the results of intrinsic measurements of shear sensitivity to reactor design problems. In the reactor environment, cells are subjected to varying degrees of shear for varying lengths of time.

Many techniques for scale-up based on shear sensitivity have been put forth over the years. Hirtenstein and Clark (46) correlated the shear damage to the stirrer speed. However, Sinskey et al. (47) showed that scale-up based on impeller tip speed did not work effectively. They correlated cell damage to an "integrated shear factor" defined as the ratio of impeller tip speed to the distance between the impeller tip to the vessel wall. More recently, Croughan et al. (48) have correlated the shear-related damage to the turbulent eddy length in the fermentor. The cells appear to suffer damage when the eddy length is less than the microcarrier diameter. This theory appears to have general validity and can form the basis of rational reactor design as will be discussed in the next section.

Another problem with mammalian cell culture is their sensitivity to air–liquid interfaces. Gas bubbles are reported to cause damage to cells due to bubble-induced shear, surface tension effects due to foaming, or cell entrainment in foam (49). Hence, it may not be possible in all cases to use gas sparging for oxygenating the reactors. Surface aeration or membrane aeration is often used in microcarrier reactors. Yet another problem is the vulnerability of mammalian cells to toxic effects of trace metals and other chemicals. For example, butyl rubber, which is widely used as a gasket material for fermenters, has been found to be toxic to mammalian cells. The toxic effect may be due to the rubber itself or some of the fillers or plasticizers used during manufacturing. Finally, the slow growth rates of mammalian cells make them even more vulnerable to bacterial contamination than the bacterial systems. Absolute asepsis is a must for the successful operation of mammalian cell culture systems. In the last section we will discuss principles of reactor design that are used to overcome the limitations mentioned above.

B. Principles of Reactor Design

Microcarriers are commonly used for attachment dependent mammalian cells. The most suitable reactor configuration for this is the stirred tank fermentor. The design of such fermenters has to be such that the hydrodynamic effects are not deleterious to the cells. At the same time, mixing should be adequate for control purposes and for oxygenation. Cherry and Papoutsakis (50) have described the hydrodynamic effects on cells in agitated fermenters. More recently, Nelson (51) has elaborated on the procedures used for designing and scaling up stirred reactors for mammalian cell culture. He has presented correlations for designing various types of oxygenation systems and to determine the limits on agitation power input to scale up at constant eddy length.

Another approach to designing mammalian cell reactors is to avoid the problems of shear damage and gas sparging by suitable matrix and reactor configuration. For example, the shear damage can be avoided either by maintaining the cells in a static mode (and supplying nutrients by diffusion) or by entrapping them in a porous matrix. Hollow fiber reactors are an example of the former type. The diffusion limitations can be severe in hollow fiber reactors. Kleinstreuer and Agarwal (41) have given a good theoretical basis for the analysis and design of these reactors. Tharakan and Chau (52) have described the fluid dynamics in hollow fiber reactors and have discussed the various modes of operation for these reactors. A similar method was used by Tolbert et al. (53) to develop a static maintenance reactor, which consists of cells immobilized in a packed bed of matrix. Nutrients and gases are diffused in and out of this matrix via semipermeable tubes dispersed inside the reactor. The reactor provides a shear free environment for the cells. However, the diffusion limitations may provide a nonhomogeneous environment for the cells and control of the reactor pa-

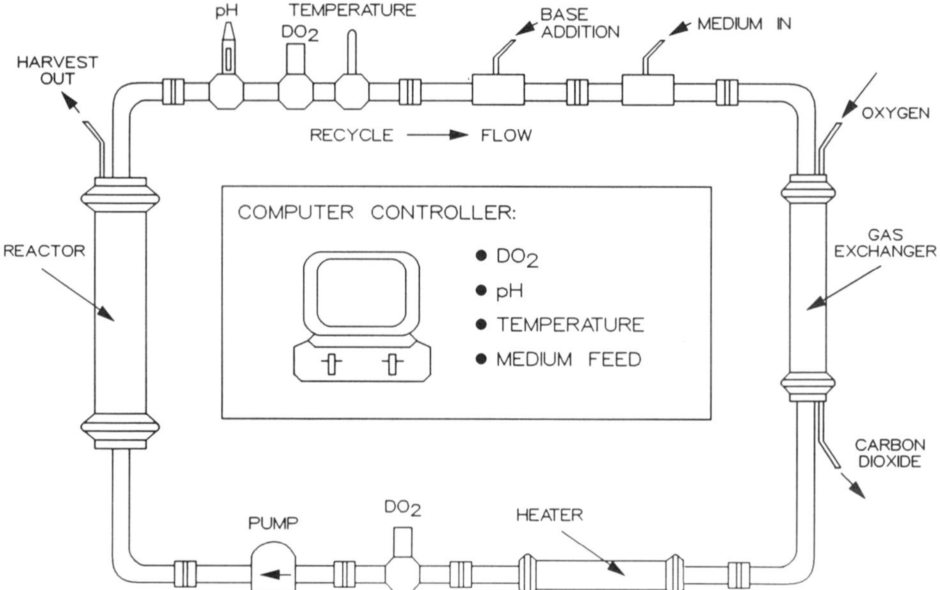

Figure 4 Schematic of a fluidized bed bioreactor. (Courtesy of Verax Corporation.)

rameters such as pH may be difficult. Fluidized bed reactors with cells entrapped in porous sponge beads (54) combine the advantages of protection from shear and homogeneous mixing. Figure 4 shows a schematic of this reactor system. The oxygenation in this case is accomplished by a membrane oxygenator to avoid the problems of gas liquid interfaces. Similar oxygenation schemes can be used for stirred reactors as well.

Hardware design principles mentioned in the previous section apply to the design of mammalian cell reactors as well. As mentioned before, mammalian cells are even more prone to the threat of contamination due to their slow growth rates. The materials of construction have to be carefully chosen to avoid any toxicity problems. Glass, 316 stainless, medical grade silicone, and teflon are some of the materials that have been proven to be effective in designing these reactors. If other materials are necessary, they should be tested for possible toxic effects before they are used.

The kinetics of mammalian cells are only poorly understood so far. Suitability of a particular reactor type is determined on a case-by-case basis depending on lab-scale results obtained. Usually, cell-specific productivities are determined under a given set of conditions. The objective of scale-up is to simulate these conditions as closely as possible in the larger scale reactors.

There is still room for major advances in the area of kinetics and optimization of immobilized animal cell reactors. A significant amount of current research is focused on kinetics of product formation and secretion and factors affecting these processes. Immobilization may help in increasing cell-specific productivities of some mammalian cells if they respond to certain autocrines produced by the surrounding cells. This is one of the rationales behind the current thrust to develop immobilized cell systems.

REFERENCES

1. S. Reuveny, *Advances in Biotechnological Processes 2:*1 (1983).
2. J. B. Schleicher, and R. E. Weiss, *Biotechnol. Bioeng. 10:*617 (1968).
3. R. E. Spier, J. Chem. Tech. Biotechnol.; *23:*304 (1982).
4. L. Nardelli, and G. F. Panina, *Develop. Biol. Stand. 37:*133 (1976).
5. N. E. Kinner, and T. T. Eighmy, *J. Water Pollut. Control Fed. 57:*536 (1985).
6. R. C. DiLuccio, and D. J. Kirwan, *Biotechnol. Bioeng. 26:*87 (1984).
7. A. Constantinides, D. Bhatia, and W. R. Vieth, *Biotechnol. Bioeng. 23:*899 (1981).
8. R. J. Klebe, K. L. Bentley, and P. C. Schoen, *Journal of Cellular Physiology 109:*481 (1981).
9. R. A. Messing, R. A. Opperman, and F. B. Kolot, in *Immobilized Microbial Cells,* ACS Symp. Ser. 106 K. Venkatasubramanian ed., American Chemical Society, Washington, D. C., 1979.
10. R. Rupp, K. Gilbride, and M. Oka, in *Large Scale Cell Culture Technology* (B. K. Lyderson ed.), Hanser, New York, 1987.
11. K. Nilsson, W. Scheirer, H. W. D. Katinger, and K. Mosbach, *Methods Enzymol. 121:*352 (1986).
12. R. C. Dean, Jr., S. B. Karkare, P. G. Phillips, N. G. Ray, and P. W. Runstadler, Jr., in *Large Scale Cell Culture Technology* B. K. Lyderson ed., Hanser, New York, 1987.
13. B. Lyderson, G. Pugh, M. Paris, B. Sharma, and L. Noll, *Biotechnology 3:*63 (1985).
14. S. B. Karkare R. C. Dean, Jr, and K. Venkatasubramanian, *Biotechnology 3:*247 (1985).
15. H. Kautola, Y. Y. Linko, and P. Linko, *Ann. N.Y. Acad. Sci. 434:*454 (1984).
16. A. Constantinides and G. K. Chotani, *Ann. N.Y. Acad. Sci. 434:*347 (1984).
17. C. D. Scott, *Enzyme Microb. Technol. 9:*66 (1987).
18. G. L. Altshuler, D. M. Dziewulski, J. A. Sowek, and G. Belfort, *Biotechnol. Bioeng. 28:*646 (1986).
19. K. Ku, M. J. Kuo, J. Delente, B. S. Wildi, and J. Feder, *Biotechnol. Bioeng. 23:*79 (1981).
20. C. R. Robertson and I. H. Kim, *Biotechnol. Bioeng. 27:*1012 (1985).
21. M. C. Flickinger, N. K. Goebel, M. Bohn, D. Bibila, and D. W. Karl, L-Glutamine stimulated 9.2.27 monoclonal antibody synthesis and evidence for

immunoglobulin heterogeneity during very slow hybridoma growth in a recycling reactor, paper presented at Engineering Foundation Conference on Cell Culture Engineering, Palm Coast, FL, Jan 31–Feb 5, 1988.
22. C. W. Lee, and H. N. Chang, *Biotechnol. Bioeng. 29:*1105 (1987).
23. W. R. Tolbert, J. Feder, and R. C. Kimes, *In Vitro 17:*885 (1981).
24. M. Kumakura, M. Yoshida, and I. Kaetsu, *Biotechnol. Bioeng. 21:*679 (1979).
25. M. Nagashima, M. Azuma, and S. Noguchi, *Ann. N.Y. Acad. Sci. 413:*457 (1983).
26. K. Venkatasubramanian and L. S. Harrow, *Ann. N.Y. Acad. Sci. 326:*141 (1979).
27. A. Shinmyo, H. Kimura, and H. Okada, *European J. Appl. Microbiol. Biotechnol. 14:*7 (1982).
28. I. J. Dunn, H. Tanaka, S. Uzman, and M. Denac, *Ann. N.Y. Acad. Sci. 413:*168 (1983).
29. S. O. Enfors, and B. Mattiason, in *Immobilized Cells and Organelles* (B. Mattiason ed.), CRC Press, Boca Raton, FL, 1983.
30. I. Chibata, in *Immobilized Microbial Cells* (K. Venkatasubramanian, ed.), ACS Symp. Series 106, Washington, D.C., 1979.
31. S. B. Karkare, K. Venkatasubramanian, and W. R. Vieth, *Ann. N.Y. Acad. Sci. 469:*83 (1986).
32. S. B. Karkare, D. H. Burke, R. C. Dean, Jr., J. Lemontt, P. Souw, and K. Venkatasubramanian, *Ann. N.Y. Acad. Sci. 469:*91 (1986).
33. P. M. Doran, and J. E. Bailey, *Biotechnol. Bioeng. 28:*73 (1986).
34. K. M. Bailey, K. Venkatasubramanian, and S. B. Karkare, *Biotechnol. Bioeng. 27:*1208 (1985).
35. K. Venkatasubramanian, S. B. Karkare, and W. R. Vieth, *Applied Biochemistry and Bioengineering 4:*311 (1983).
36. Y. Park, M.E. Davis, and D.A. Wallis, *Biotechnol. Bioeng. 26:*457 (1984).
37. P. M. Doran and J. E. Bailey, *Biotechnol. Bioeng. 28:*1814 (1986).
38. S. H. Lin, and C. K. Wei, *Chem. Eng. Sci. 34:*827 (1979).
39. S. H. Lin, *Chem. Eng. J. 17:*55 (1979).
40. K. Venkatasubramanian, and S. B. Karkare, in *Immobilized Cells and Organelles* (B. Mattiason, ed.), CRC Press, Boca Raton, FL, 1983.
41. C. Kleinstreuer, and S. S. Agarwal, *Biotechnol. Bioeng. 28:*1233 (1986).
42. R. Miller, and M. Melick, *Chemical Engineering 94*(2):112 (1987).
43. K. Nilsson, *Trends in Biotechnol. 5:*73 (1987).
44. N. A. Stathopoulos, and J. D. Hellums, *Biotechnol. Bioeng. 27:*1021 (1985).
45. C. G. Smith, P. F. Greenfield, and D. H. Randerson, *Biotechnol. Techniques 1:*39 (1987).
46. M. Hirtenstein, and J. Clark, in *Tissue Culture in Medical Research* (R. Richards and K. Rajan, eds.), Pergamon Press, Oxford, 1980.
47. A. J. Sinskey, R. J. Fleischaker, M. A. Tyo, D. J. Giard, and D. I. C. Wang, *Ann. N.Y. Acad. Sci. 369:*47 (1981).
48. M. S. Croughan, J. F. Hamel, and D. I. C. Wang, *Biotechnol. Bioeng. 29:*130 (1987).

49. R. E. Spier and B. Griffiths, *Devel. Biol. Stand.* 55:81 (1983).
50. R. S. Cherry and E. T. Papoutsakis, *Bioprocess Engineering 1:*29 (1986).
51. K. L. Nelson, *Biopharm.,* (March 1988).
52. J. P. Tharakan, and P. C. Chau, *Biotechnol. Bioeng. 28:*1064 (1986).
53. W. R. Tolbert, J. Feder, and C. Lewis, Jr., U.S. Patent 4,537,860 (1985).
54. R. C. Dean Jr.; S. B. Karkare, N. G. Ray, P. W. Runstadler, Jr., and K. Venkatasubramanian, *Ann. N.Y. Acad. Sci. 506:*129 (1988).

11

Industrial Applications of Immobilized Proteins

Ichiro Chibata, Tetsuya Tosa, and Tadashi Sato

Tanabe Seiyaku Co., Ltd.
Osaka, Japan

I. INTRODUCTION

In the field of biotechnology the biochemical industry has utilized the action of enzymes and microorganisms to effect savings of resources and energy. The production of useful chemicals by biochemical reactions requires, in most cases, the utilization of enzymes as catalysts. Enzymes are protein biocatalysts that participate in the many biochemical reactions occurring in living organisms. The characteristic that makes most enzymes far superior to ordinary chemical catalysts from a catalytic point of view is specificity. However, enzymes are not always ideal catalysts for practical applications.

Immobilization of enzymes has been extensively studied since the late 1960s as one of the methods to make enzymes more suitable for the desired purposes, and several new techniques have emerged to stimulate their application as biocatalysts. Immobilized biocatalysts are defined as enzymes, microbial cells, animal cells, or plant cells physically confined or localized in a certain defined region of space with retention of their catalytic activities, and which can be used repeatedly and continuously. In 1969, we succeeded in industrializing a continuous process for the production of L-amino acids using immobilized aminoacylase, the first industrial application of immobilized enzymes.

Although enzymes are produced by all living things, enzymes from microbial sources are the most suitable for industrial purposes. Microbial enzymes can be classified into two groups: extracellular and intracellular. For the utilization of the intracellular enzymes, it is necessary to extract the enzyme from microbial cells. However, such extracted enzymes are generally unstable. In order to avoid the extraction and separation process of enzymes and resulting enzyme instability, we have attempted the direct immobilization of whole microbial cells. In 1973, we succeeded in the industrialization of a process for the continuous production of L-aspartic acid using immobilized microbial cells, the first industrial application of immobilized microbial cells. This was followed, in 1974, by the industrial production of L-malic acid using immobilized microbial cells.

Following development of these processes, the technique of using immobilized biocatalysts was applied to other industrial processes, such as production of high fructose syrup, 6-aminopenicillanic acid, and low lactose milk.

The enzyme reactions using immobilized biocatalysts as described above are carried out by the action of single enzyme. However, many useful compounds, especially those produced by fermentation, are usually formed by multistep reactions catalyzed with more than one enzyme in living microbial cells. These reactions often require generation of ATP and other coenzymes. If immobilized cells are kept in a living state, they may be applied to such multienzyme reactions. The production of ethanol, antibiotics, amino acids, organic acids, and other chemicals has been attempted using such systems.

II. FOOD PROCESSING

A. Dairy Products

1. Hydrolysis of Lactose

Milk is a nutritionally excellent food but contains approximately 5% lactose. This lactose must be removed before feeding babies deficient in the enzyme hydrolyzing lactose, β-galactosidase, in the intestine. Direct oral administration of the enzyme to β-galactosidase–deficient babies is carried out, but this is not always satisfactory since allergic reactions can occur in some cases.

In order to resolve this problem, β-galactosidase has been immobilized by various methods and applied to milk lactose hydrolysis. For example, by using a column packed with β-galactosidase immobilized in polyacrylamide gel (1) or a cyanuric chloride derivative (2), continuous removal of lactose from milk could be achieved.

In Italy, similar experiments were carried out using purified β-galactosidase from *Escherichia coli* or yeast immobilized by entrapping in triacetylcellulose (3). This immobilized enzyme was very stable and retained initial enzyme activity even after continuous operation for 80 days. This system is used for the industrial removal of lactose in milk (3).

Immobilized β-galactosidase has also been used for the production of a sweet syrup (consisting of glucose and galactose formed by the hydrolysis of lactose) and the resulting enhancement of the sweetness of the milk for use in use for ice cream or condensed milk. Also, in order to enhance sweetness, whey was treated with immobilized β-galactosidase, and subsequently with immobilized glucose isomerase.

Since large amounts of lactose are in demand in the food industry, further applications of immobilized β-galactosidase are expected to be developed in the future.

2. Milk Processing

It is known that trypsin prevents the occurrence of an "oxidized" flavor in milk. However, the use of this enzyme has been limited because of its high cost and difficulty in inactivating the enzyme during milk pasteurization. To resolve these problems, utilization of immobilized enzyme prepared by covalent binding of trypsin to porous glass has been studied (4), although the usefulness of this method has not yet been established.

Hydrogen peroxide is sometimes used for the sterilization of milk. In order to remove residual hydrogen peroxide after sterilization, a method utilizing catalase immobilized by covalent binding with carboxychloride resin was patented in the United States (5).

3. Production of Cheese

Rennin is the milk-clotting enzyme from calf rennet used for cheese production. Recently, microbial milk-clotting enzymes have also been used in cheese production. However, rennin is in short supply and costly. In order to overcome this shortage of rennin, continuous production of milk curd using immobilized rennin has been attempted. The clotting of milk by rennin involves two reactions: hydrolysis of casein and subsequent association of the resulting micelles. If the first reaction can be performed by an immobilized enzyme, separation of the reaction product becomes easy. Thus, rennin immobilized on CNBr-activated Sepharose or by carrier cross-linking using glutaraldehyde and AE-cellulose was prepared, and basic studies on the continuous clotting of milk were carried out (6). Some problems remain regarding the stability of the immobilized rennin and the type of enzyme reactor most suitable for the recovery of curd, and this process has not yet been industrialized.

B. Liquors

1. Beer Production

The brewing of beer using immobilized yeast has been investigated (7): on passing the wort at 15°C through a column packed with yeast immobilized on polyvinylchloride and porous bricks, alcoholic fermentation occurred, and beer was produced. Further development of this kind of technique can be expected.

When beer is stored for a long period, it becomes turbid (so-called chill-haze) due to the reaction of polyphenols and polypeptides contained in the beer. In order to prevent the formation of chill-haze, papain is used at present as a clarifying agent. However, when this proteolytic enzyme remains in beer for a long time, excess proteolysis occurs, and this is undesirable. Thus, for controlled treatment to prevent chill-haze in beer, utilization of immobilized papain and immobilized polyphenol oxidase has been studied (8).

2. Hydrolysis of Starch, Proteins, and Polypeptides

Continuous preprocessing of alcoholic fermentation liquors using immobilized enzymes has been investigated. For instance, in the production of beer, attempts have been made to hydrolyze starch by using immobilized amylases from microorganisms instead of malt, or to hydrolyze proteins or polypeptides by using immobilized proteolytic enzymes have been made. To date, such processes are not used industrially.

3. Miscellaneous Applications for Food Processing

A West German patent (9) describes how glucose oxidase and peroxidase bound to cellulose membrane were used for the removal of glucose in egg white albumin. This patent claimed that when glucose oxidase and catalase immobilized in a starch matrix are coated on the inner surface of a can, decomposition of mayonnaise in the can is retarded. No decomposition or changes in flavor could be observed after storage for 5 months. Further, tannase immobilized on aminoalkylated glass beads by means of glutaraldehyde has been used for the treatment of tea cream (10). Other interesting reports include application of immobilized enzymes for the purification of sugar (11). In sugar production, dextran in the juice from pressed cane causes various problems during the purification process. Accordingly, preliminary removal of dextran is desirable. Hence, a bacterial dextran-hydrolyzing enzyme was immobilized by using CM-cellulose azide or CNBr-activated cellulose, and the hydrolysis of dextran in sugar solutions can be efficiently carried out by using this immobilized enzyme. In addition, β-galactosidase was adsorbed in nylon pellets treated with formic acid and then treated with dimethyladipimide and methylacetimidate. The resulting immobilized preparation was used for the reduction of raffinose in beet sugar molasses (12).

To remove the bitterness of orange juice, naringinase is used. Naringinase is produced by microorganisms, and it decomposes naringine, a bitter substance in oranges, to rhamnose and purunine. Naringinase was immobilized on ethylene-maleic anhydride copolymer, and its application for this purpose was studied. The removal of bitterness was efficiently carried out by this method in both batch and column systems (13). Also, we investigated immobilization of naringinase from microorganism by using DEAE-cellulose, DEAE-Sephadex, and tannin-aminohexyl cellulose (14,15).

It has been reported that palatinose, a sweetener for anti–dental caries preparations, is produced industrially from sucrose by an action of β-glucosyltransferase of *Protaminobacter rubrum* immobilized with calcium alginate.

III. PHARMACEUTICAL AND SPECIALTY CHEMICAL PRODUCTION

Applications of immobilized biocatalysts are very useful for many chemical transformations in the pharmaceutical and chemical industries. In this section, some processes utilized in the pharmaceutical and chemical industries on a pilot plant or laboratory scale are described.

A. Production of Antibiotics

In the pharmaceutical industry, applications of immobilized biocatalysts are very useful for the production of antibiotics. Immobilized biocatylysts have been utilized for the commercial production of 6-aminopenicillanic acid (6-APA), penicillins, cephalosporins, etc. For example, *E. coli* penicillin amidase entrapped in cellulose triacetate fibers was used in a continuous flow reaction to produce ampicillin or amoxycillin from 6-APA and D-phenylglycine methyl ester or D-*p*-hydroxyphenylglycine methyl ester, respectively (16). Conidia, mycelium, and protoplasts of *Penicillium chrysogenum* were immobilized with κ-carrageenan (17), polyacrylamide (18), and calcium alginate (19) for studying biosynthesis of penicillin G from glucose.

6-APA is an important intermediate for the production of semisynthetic penicillins in these reactions and is industrially produced on a large scale from penicillin G or penicillin V by deacylation using the action of penicillin amidase. These antibiotics are produced by fermentation. This enzymatic process is widely applied, and the conventional process based on soluble enzyme has been converted to immobilized biocatalyst process. For example, partially purified penicillin amidase from *E. coli* was entrapped into cellulose triacetate fibers and was used for production of 6-APA from penicillin G (20). Immobilized whole cells were used to hydrolyze penicillins by a continuous method for the production of 6-APA from penicillin G using *E coli* cells immobilized with polyacrylamide (21). Whole cells of *Pleurotus ostreatus* immobilized by entrapment into chitosan have also been used for production of 6-APA from penicillin V (22).

Another approach to 6-APA production utilized the cloning of the penicillin amidase gene of *E. coli* ATCC 11105 using multicopy plasmids. A new hybrid strain *E. coli* SK (PHM 12) resulted having high enzyme activity (23). The whole cells of the new *E. coli* strain were immobilized into calcium alginate or an epoxymatrix obtained by polycondensation of water-soluble epoxyprecursor and poly functional amine (24). This is considered to be the first promising example combining the techniques of genetic and enzyme engineering.

Cephalosporins can be synthesized from penicillins or produced by *Cephalosporium acremonium* fermentation. The cephalosporin fermentation product is cephalosporin C, which contains the nucleus 7-aminocephalosporanic acid (7-ACA) and the side chain, α-aminoadipic acid. Cephalosporin can be produced chemically from penicillins by expanding the 5-membered thiazolidine ring of penicillin to the 6-membered dihydrothiazine ring of cephalosporin containing the nucleus, 7-aminodesacetoxy cephalosporanic acid (7-APCA).

Cephalosporin amidase from various microorganisms can be immobilized by various methods, and the immobilized enzymes are used for production of various cephalosporin derivatives. For example, the enzyme from *E. coli* was entrapped in cellulose triacetate fibers and was used for production of cephalexin from 7-APCA and D-phenylglycine methyl ester (16). Commercially important cephaloglycine and cephalothin were synthesized from 7-ACA and D-phenylglycine methyl ester or 2-thiophene acetic acid by continuous passing the substrate solution through an immobilized enzyme column. Cephalosporin C was synthesized from 3-(N-morpholino) propane sulfonic acid by whole cells of *Streptomyces clavuligerus* immobilized with polyacrylamide (25).

Most studies on the enzymatic deacylation of cephalosporins have been carried out for the 7-ADCA nucleus. Substrates for the reaction are usually either 7-phenylacetamidodesacetoxycephalosporanic acid (7-phenylacetyl-ADCA) and 7-phenoxyacetamidodesacetoxy-caphalosporanic acid (phenoxyacetyl-7-ADCA). These substrates are easily obtained by ring expansion reactions from penicillin G and penicillin V.

Few reports have been published on the utilization of immobilized cells for the total production of antibiotics.

Whole cells of *Bacillus* sp., bacitracin-producing bacteria, were immobilized into a polyacrylamide gel lattice, and the immobilized living cells were used for production of bacitracin in batch and in continuous culture systems (26, 27).

The macrolide antibiotic, tylosin, and the nucleoside peptide antibiotic, nikkomycin, were produced by living cells of *Streptomyces* sp. and *Streptomyces tendae,* respectively, immobilized with calcium alginate (28).

Conidia of *Penicillum urticae* immobilized with κ-carrageenan was germinated in situ to form a patulin-producing cell mass. By incubating this immobilized condidia in a growth-supporting medium, the immobilized preparation was used for production of patulin from glucose (29,30).

B. Production of Amino Acids

Amino acids are widely used in the food, feed, medicine, and cosmetic industries, and also as starting materials for synthetic chemicals. For food and nutritional applications, only the L-isomer of amino acids is biologically active.

There are many synthetic chemical methods for amino acid production, but the products are racemic mixtures of L- and D-isomers. To obtain the L-amino acid from the chemically synthesized DL-form, optical resolution is necessary. On the other hand, biosynthesis of amino acids by microorganisms or isolated enzyme systems leads exclusively to production of the L-isomer. For this reason, the biological production method is generally the preferred one. Thus, immobilized biocatalysts have been applied to develop a more efficient process for producing optically active amino acids.

1. Optical Resolution of DL-Amino Acids

Among the many optical resolution methods, the enzymatic method using mold aminoacylase developed in our laboratories is one of the most advantageous procedures, yielding optically pure amino acids. A chemically synthesized acyl-DL-amino acid is asymmetrically hydrolyzed by aminoacylase to give L-amino acid and unhydrolyzed acyl-D-amino acid. Both products are easily separated by the difference in their solubilities. Unhydrolyzed acyl-D-amino acid is racemized and reused for the resolution procedure.

To improve the procedure, we extensively studied the continuous optical resolution of DL-amino acids using immobilized aminoacylase. Aminoacylase from *Aspergillus oryzae* was immobilized by a variety of methods and tested for enzyme activity. As a result, relatively active and stable immobilized aminoacylases were obtained by ionic binding to DEAE-Sephadex (31), covalent binding to iodoacetyl cellulose (32), and entrapment in polyacrylamide gel (33). Further, to select the most suitable prepa-

Table 1 Characteristics of Immobilized Aminoacylase

Preparation characteristics	Ionic (DEAE-Sephadex)	Covalent (iodoacetyl-cellulose)	Entrapping (polyacrylamide)
Preparation	easy	difficult	moderate
Enzyme activity	high	high	high
Cost of immobilization	low	high	moderate
Binding force	moderate	strong	strong
Regeneration	possible	impossible	impossible
Operation stability (half-life)[a]	65 days at 50°C	—	48 days at 37°C

[a]Time required for 50% of the enzyme activity to be lost.

ration for industrial purposes, the properties of these three immobilized aminoacylases were compared as shown in Table 1 (34). Finally, we chose aminoacylase immobilized by ionic binding to DEAE-Sephadex for industrial purposes, because its preparation is easy, retained activity is high and stable, and the regeneration of deteriorated preparations is possible. Further, we carried out chemical engineering studies and designed an enzyme reactor system for continuous production of L-amino acids using immobilized aminoacylase as shown in Figure 1 (34).

Since 1969 we have been industrially operating several enzyme reactors in our plants for the production of L-methionine, L-valine, L-phenylalanine, etc. This is one of the first industrial applications of immobilized enzymes in the world.

Aminoacylase from *Aspergillus* sp. has also been immobilized by covalent binding to alkylaminosilanizide porous glass with glutalaldehyde or to the diazonium derivative of arylaminosilanized porous glass. These immobilized aminoacylases have been used for the continuous preparation of L-amino acids from acetyl-DL-amino acids (35).

In addition to these immobilized enzyme methods, whole *Aspergillus ochraceus* cells having aminoacylase activity were immobilized by cross-linking with egg albumin and glutaraldehyde and used for the continuous optical resolution of acetyl-DL-methionine on a laboratory scale (36).

Wheat germ acid phosphatase was immobilized in κ-carrageenan gel, followed by cross-linking with glutaraldehyde, and used for the preparative resolution of DL-threonine (37). In this method, DL-threonine was first chemically O-phosphorylated and then asymmetrically hydrolyzed by the immobilized phosphatase. As a result, optically pure L-threonine and O-phospho-D-threonine were prepared. Further, optically active D-phenylglycine, which is an important chemical for the preparation

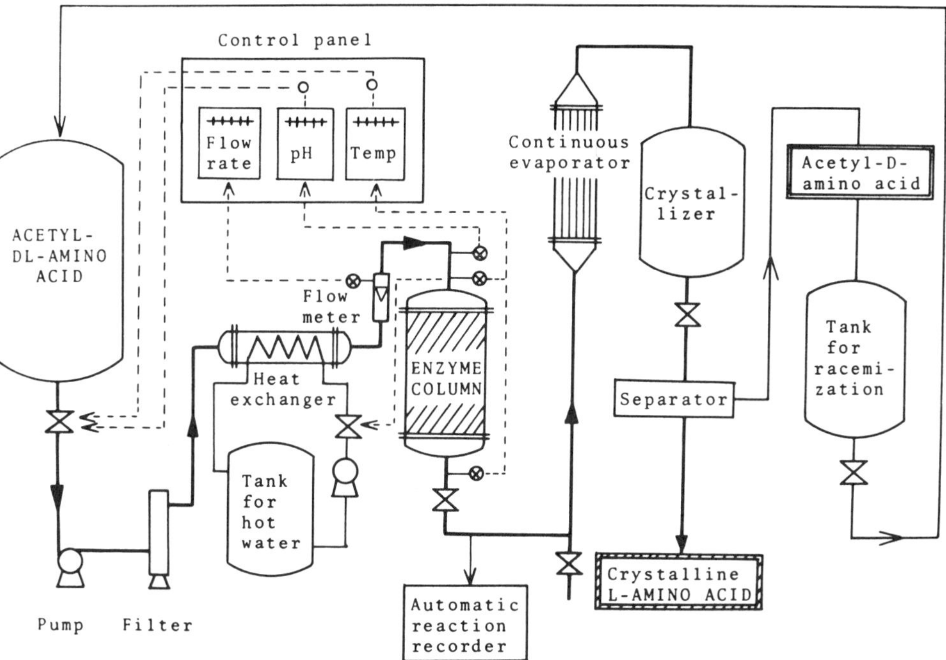

Figure 1 Flow diagram for continuous production of L-amino acid by immobilized aminoacylase (34).

of semisynthetic penicillins and cephalosporins, was obtained from the racemic mixture of its ester derivative using immobilized subtilisin (38).

2. Biosynthesis of Optically Active Amino Acids

Many papers on the biosynthesis of optically active amino acids using immobilized biocatalysts have been published, as shown in Table 2. In this section, industrial application of immobilized biocatalysts for production of L-aspartic acids and L-alanine are mainly described.

L-Aspartic Acid. L-Aspartic acid is widely used as a medicine and food additive, and demand for the amino acid is rapidly increasing as a component of the synthetic sweetener Aspartame.

Traditionally, the acid has been industrially produced by fermentative or enzymic batch process from fumaric acid and ammonia using the action of aspartase. For example, since 1960, L-aspartic acid has been industrially produced by Tanabe Seiyaku Co., Ltd., Japan, by a batchwise reaction using intact *Escherichia coli* having high aspartase activity.

Table 2 Production of Optically Active Amino Acids by Immobilized Biocatalysts

Amino acids	Biocatalysts	Matrices	Ref.
L-Alanine	Aspartase and L-aspartate 4-decarboxylase	Ultrafiltration membrane	39,40
	DL-Lactate dehydrogenase and L-alanine dehydrogenase	Ultrafiltration membrane	41
	Pseudomonas dacunhae (L-aspartate 4-decarboxylase)	Carrageenan	42
	Escherichia coli (aspartase) and *Pseudomonas dacunhae* (L-aspartate 4-decarboxylase)	Carrageenan	43,44
	Corynebacterium dismutans (multienzymes)	Polyacrylamide	45
L-Arginine	*Serratia marcescens* (multienzymes)	Carrageenan	46
L-Aspartic acid	Aspartase	Polyacylamide	47
		Duolite A-7	48,49
		Cellulose triacetate	50
	Escherichia coli (aspartase)	Polyacrylamide	51
		Carrageenan	52
		Polyurethane	53
ϵ-Aminocaproic acid	*Achromobacter guttatus* (cyclic dimer hydrolase)	Polyacrylamide	54
L-Citrulline	*Pseudomonas putida* (L-arginine deiminase)	Polyacrylamide	55
L-Glutamic acid	*Brevibacterium flavum* (multienzymes)	Collagen	56
	Corynybacterium glutamicum (multienzymes)	Polyacrylamide	57

L-Isoleucine	*Serratia marcescens* (multienzymes)	Carrageenan	58
L-Leucine	L-Leucine dehydrogenase and formate dehydrogenase	Ultrafiltration membrane	59
L-Lysine	L-α-Amino-ϵ-caprolactam hydrolase and α-amino-ϵ-caprolactam racemase	DEAE-Sephadex	60
	Microbacterium ammoniaphilum (diaminopimelic acid decarboxylase)	Polyacrylamide	61
D-α-Phenylglycine	*Bacillus* sp. (hydantoinase)	Polyacrylamide	62
D-α-Hydroxy-Phenylglycine	*Bacillus* sp. (hydantoinase)	Polyacrylamide	62
L-Tryptophan	Tryptophan synthase	Cellulose triacetate	63
	Tryptophanase	CNBr-activated Sepharose	64
	Escherichia coli (tryptophan synthase)	Polyacrylamide	65,66
		Chitosan	67
	Escherichia coli (tryptophanase)	Polyacrylamide	68
5-Hydroxytryptophan	*Escherichia coli* (tryptophan synthase)	Polyacrylamide	69
L-Tyrosine	β-Tyrosinase	CNBr-activated Sepharose	70
	Erwinia herbicola (β-tyrosinase)	Collagen and glutaraldehyde	71
	Escherichia intermedia (β-tyrosinase)	Polyacrylamide	72
L-DOPA	*Erwinia herbicola* (β-tyrosinase)	Collagen and dialdehydestarch	71
		Carrageenan	72

In order to improve the productivity of this system, we studied the continuous production of L-aspartic acid using immobilized aspartase (47). Most of the immobilization methods we tried for aspartase resulted in low activity and poor yield. Although entrapment into a polyacrylamide gel lattice gave relatively active immobilized aspertase, its operational stability was not sufficient for the industrial production of L-aspartic acid. We thus considered that if whole microbial cells having the enzyme activity could be directly immobilized, these disadvantages might be overcome. Whole cells of *E. coli* having high aspartase activity were immobilized by various methods, and the most active immobilized *E. coli* cells were obtained by entrapping the cells into polyacrylamide gel (51).

By using a column packed with the immobilized *E. coli* cells, the operational stability was investigated. We found that the immobilized cell column was very stable and its half-life at 37°C was 120 days (73,74).

For industrial application of this technique, we carried out a chemical engineering analysis of the continuous enzyme reaction using a column packed with the immobilized *E. coli* cells, and an aspartase reactor system was designed (74). As this aspartase reaction is exothermic, the column reactor used for industrial production of L-aspartic acid was designed as a multistage system with a radiator as shown in Figure 2. This immobilized cell system had been operating industrially since 1973 at Tanabe Seiyaku Co., Ltd., Japan.

For further improvement of this process, a novel technique was discovered using κ-carrageenan as an immobilization matrix for *E.coli* cells. The efficiency of *E. coli* cells immobilized with polyacrylamide or κ-carrageenan in L-aspartic acid production was compared. As shown in Table 3, it was found that *E. coli* cells immobilized with κ-carrageenan and treated with

Table 3 Comparison of Productivities of *Escherichia coli* Immobilized with Polyacrylamide and with Carrageenan for Production of L-Aspartic Acid

Immobilization method	Aspartase activity (unit/g cells)	Stability at 37°C (half-life, day)	Relative productivity
Polyacrylamide	18,850	120	100
Carrageenan	56,340	70	174
Carrageenan (GA)	37,460	240	397
Carrageenan (GA + HMDA)	49,400	680	1,498

GA: glutaraldehyde; HMDA: hexamethylenediamine.

Productivity = $\int_0^t E_0 \exp(-kd \cdot t)\, dt$.

E_0 = initial activity, kd = decay constant, t = operational period.

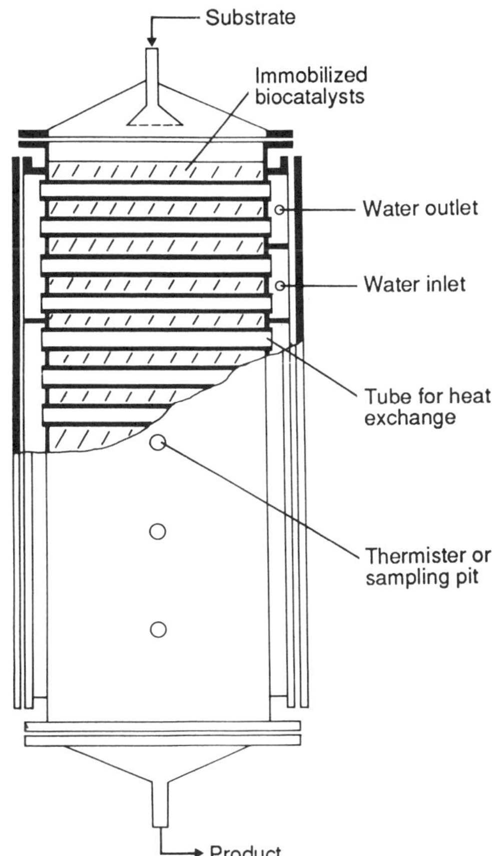

Figure 2 Heat-exchange type bioreactor for production of L-aspartic acid (74).

glutaraldehyde and hexamethylenediamine showed the highest productivity, and its half-life at 37°C was 680 days (52). We thus changed the conventional polyacrylamide gel method to the carrageenan method for industrial production of L-aspartic acid in 1978. In this method, if a 1000-liter column is used, the theoretical yield of L-aspartic acid is 3.4 ton/day or 100 ton/month.

We have also developed a strain of *E. coli* having a higher aspartase activity (75). The aspartase activity of the new strain is about seven times greater than the parent strain. However, the new strain has fumarase activity, converting fumaric acid to L-malic acid. This is disadvantageous for the production of L-aspartic acid from fumaric acid and ammonia. Thus, in order to employ this strain with its higher aspartase activity for industrial production of L-aspartic acid, it was necessary to eliminate the fumarase activity.

Several treatments for specifically eliminating fumarase activity from the bacterical strain were tested, and it was found that when the strain was treated in a culture broth (pH 4.9) containing 50 mM L-aspartic acid at 45°C for 1 h, fumarase activity was almost completely eliminated without inactivation of the aspartase (76). The treated cells were immobilized with κ-carrageenan, and the continuous production of L-aspartic acid by the immobilized preparation was investigated. We found its aspartase activity was much higher and, in addition, its stability was increased due to the inactivation of some proteases in the intact cells during the pretreatment.

In 1982, we changed from the conventional method for production of L-aspartic acid to the improved method using treated cells of the strain having higher aspartase activity. This improved method continues to give us very satisfactory results in industrial production of L-aspartic acid.

Besides our studies, several papers have been published on the continuous production of L-aspartic acid using immobilized enzyme and microbial cells. For example, the continuous production of L-aspartic acid from ammonium fumarate was investigated at a laboratory scale using *E. coli* cells immobilized with polyurethane (53). In another example, aspartase extracted from *E. coli* cells was immobilized in the presence of aspartate to a weakly basic anion exchange resin, Duolite A-7, by ionic binding, and the immobilized enzyme was used for continuous production of L-aspartic acid from ammonium fumarate (48).

L-Alanine. L-Alanine is a useful amino acid not only in medicine but also as a food additive because of its taste. It had been industrially produced from L-aspartic acid in our plant since 1965 by a batchwise enzyme reaction using L-aspartate 4-decarboxylase from *Pseudomonas dacunhae.*

To develop a more efficient process for producing L-alanine, we studied a system of continuous L-alanine production from L-aspartic acid using *P. dacunhae* cells immobilized with κ-carrageenan (43,77). In the continuous production system using immobilized *P. dacunhae,* one of the problems is the evolution of CO_2 gas during the L-aspartate 4-decarboxylase reaction. It is not efficient to perform this reaction using a conventional column system at normal pressure because the evolution of CO_2 gas makes it difficult to obtain complete plug flow of substrate solution, and there is an increase in the pH of the reaction mixture within the column as the reaction progresses. Therefore, we investigated reaction systems using immobilized *P. dacunhae* for continuous production of L-alanine and designed a closed column reactor which carries out the enzyme reaction at high pressure, such as 10 kg/cm². By using this reactor, liberated CO_2 gas is dissolved into the reaction mixture, complete plug flow of the substrate solution is obtained, and the pH of reaction mixture is not changed significantly. The efficiency of the closed column reactor was 50% higher than that of the conventional one.

In order to increase the productivity of L-alanine, we investigated its production from fumaric acid and ammonia by combining the systems for production of L-aspartic acid using immobilized *E. coli* as described above and the system for production of L-alanine using immobilized *P. dacunhae* as shown in Figure 3 (43,44). In these studies, we carried out chemical engineering studies and designed an enzyme reactor system for production of L-alanine from fumaric acid and ammonia using the two immobilized cell systems (78,79). In 1982, Tanabe Seiyaku Co., Ltd., Japan, industrialized successfully this continuous production system of L-alanine. This is

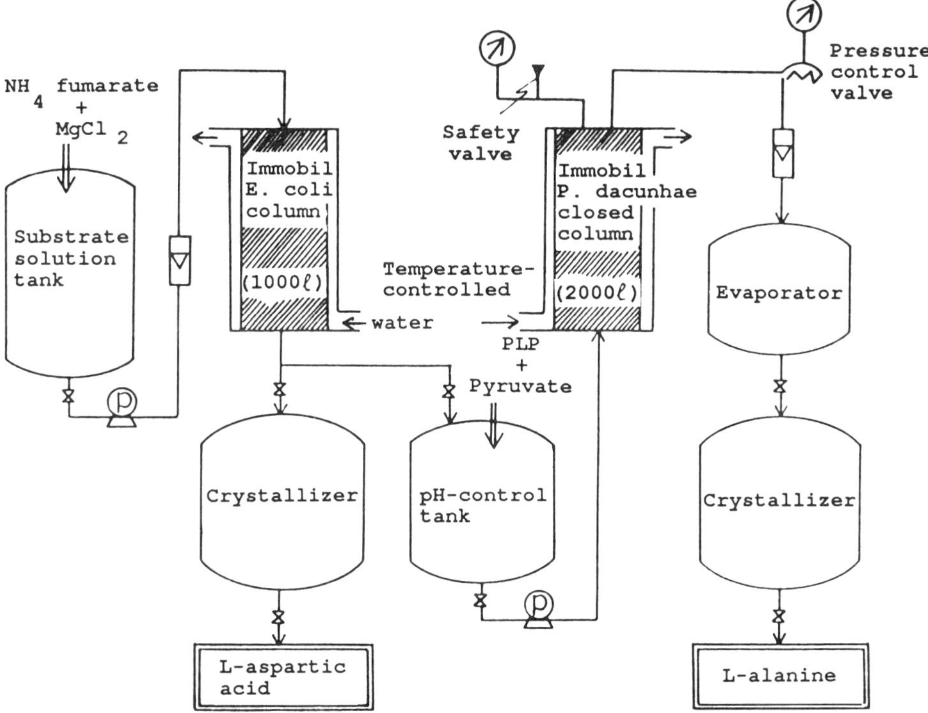

Figure 3 Flow diagram for continuous production of L-aspartic acid and L-alanine by immobilized cells (43).

considered to be the first industrial application of sequential enzyme reactions using two different immobilized cells.

Continuous production of L-alanine from ammonium fumarate has also been carried out in a two-stage membrane reactor containing partially purified aspartase and L-aspartate 4-decarboxylase (39,40). In addition, the production of L-alanine from DL-lactate using a multienzyme system with cofactor regeneration was carried out in a membrane reactor (41). In the process, a form of the coenzyme NAD(H) covalently bound to water-soluble polyethylene glycol 20000 together with three enzymes—L-lactate dehydrogenase, D-lactate dehydrogenase, and L-alanine dehydrogenase—was retained by an ultrafiltration membrane.

L-Lysine. In 1975 an interesting report on the preparation of L-lysine using two kinds of immobilized enzymes was publicized by Tore Co., Ltd., Japan (60). Cyclohexane, a byproduct of nylon production, was used as a starting material, and was converted to DL-α-amino-ϵ-caprolactam by a synthetic chemical method. L-α-Amino-ϵ-caprolactam hydrolase and α-amino-ϵ-caprolactam racemase were immobilized by ionic binding to DEAE-Sephadex, and both immobilized enzymes were allowed to react with DL-α-amino-ϵ-caprolactam at the same time. L-α-amino-ϵ-caprolactam was hydrolyzed to give L-lysine, and the remaining D-α-amino-ϵ-caprolactam was racemized to the DL-form and again hydrolyzed to L-lysine. Essentially, the DL-form was completely converted to L-lysine.

In another study, *Microbacterium ammoniaphilum* having diaminopimeric acid decarboxylase activity was immobilized into a polyacrylamide gel, and the immobilized cells were used for production of L-lysine from diaminopimelic acid (61).

L-Tryptophan and L-Tyrosine. L-Tryptophan can be produced from indole, pyruvate, and ammonia by a biosynthetic method involving tryptophanase. This enzyme was immobilized by covalent binding to CNBr-activated Sepharose, and the immobilized enzyme was used for production of L-tryptophan from indole, pyruvate, and ammonia (64,65).

A partially purified tryptophan synthase from *E. coli* was immobilized by entrapment in cellulose triacetate fibers for production of L-tryptophan from indole and L-serine in batch and continuous processes (63).

Analogous to the case of tryptophanase, tyrosine can be synthesized from phenol, pyruvate, and ammonia by the action of β-tyrosinase. Also, the enzyme can produce 3,4-dihydroxy-L-phenylalanine (L-DOPA), a drug used to treat Parkinson's disease, from pyrocatechol, pyruvate, and ammonia. β-Tyrosinase from *Escherichia intermedia* was immobilized by covalent binding to CNBr-activated Sepharose (70), and the immobilized enzyme was used for the continuous production of L-tyrosine and L-DOPA. Whole cells of *E. intermedia* were immobilized with polyacrylamide gel and used for production of L-tyrosine from phenol, pyruvate, and ammonia (72). *Erwinia herbicola* having β-tyrosinase activity was immobilized by cross-linking with collagen and glutaraldehyde or dialdehyde starch, and the immobilized cells were used for production of L-tyrosine from phenol, pyruvate, and ammonia, and production of L-DOPA from pyrocatechol, pyruvate, and ammonia, respectively (71). The same cells immobilized with κ-carrageenan were used for production of L-DOPA from pyrocatechol and DL-serine (72).

Other Amino Acids. In addition to those described above, other amino acids can also be produced by immobilized biocatalysts as shown in Table 2. For example, L-leucine has been produced continuously from α-ketoisocaproate in a membrane reactor containing L-leucine dehydrogenase, formate dehydrogenase, and NAD(H) covalently bound to polyethylene glycol (MW 10000) (59). In this case, formate dehydrogenase was used for the regeneration of NADH.

L-Glutamic acid is industrially produced by chemical synthesis or fermentation. To develop a more efficient method, whole cells of *Corynebacterium glutamicum* (a glutamic acid–producing bacterium) were immobilized by entrapping into polyacrylamide gel, and the production of L-glutamic aid from glucose by using the immobilized cells was investigated (57). Living cells of *Brevibacterium flavum* immobilized into collagen were used for the continuous production of L-glutamic acid in a recycle reactor system (56). We studied continuous production of L-citrulline from L-arginine using immobilized *Pseudomonas putida* having L-arginine deiminase activity (55). In addition, we presented a study on continuous production of L-isoleucine and L-arginine from glucose using immobilized living cell systems (46,58).

D-α-Phenylglycine and D-α-hydroxyphenylglycine, which are important components of the semisynthetic penicillins and cephalosporins, were produced from phenylhydantoin and ρ-hydroxyphenylhydantoin, respectively, by immobilized *Bacillus* sp. containing hydantoinase activity (62).

C. Organic Acids

Organic acids are widely used for food and medical applications, and some are produced by conventional fermentation. For improvement of productivities, immobilized biocatalysts have been utilized in organic acid production as shown in Table 4.

In 1974 we developed a continuous commercial process for producing L-malic acid from fumaric acid using the fumarase activity of *Brevibacterium*

Table 4 Production of Organic Acids by Immobilized Biocatalysts

Organic acid	Biocatalysts	Matrices	Ref.
Acetic acid	*Acetobacter aceti* (multienzymes)	Porous ceramic	80,81
	Acetobacter sp (multienzymes)	Hydrous titanium	82
Citric acid	*Aspergillus niger* (multienzymes)	Calcium alginate	83,84
α,ω-Dodecanedioic acid	*Candida tropicalis* (oxidase)	Carrageenan	85
Gluconic acid	*Aspergillus niger* (glucose oxidase)	Calcium alginate	83
α-Kato acid	*Trigonopsis variabilis* (D-amino acid oxidase)	Calcium alginate	86
	Chlorella vulgaris and *Anacystis nidulans* (L-amino acid oxidase)	Agarose	87
12-Ketocheno-deoxycholic acid	*Brevibacterium fuscum* (reductase)	Carrageenan	88
2-Ketogluconic acid	*Serratia marcescens* (multienzymes)	Collagen	89
2-Ketoglonic acid	*Gluconobacter melanogenus* (L-sorbose dehydrogenase) and *Pseudomonas syringus* (L-sorbosone oxidase)	Polyacrylamide	90
Lactic acid	*Lactobacillus casei* (multienzymes)	Polyacrylamide	91
	Lactobacillus lactis (multienzymes)	Calcium alginate	83
	Lactobacillus delbruekii (multienzymes)	Calcium alginate	92
	Lactobacillus sp (multienzymes)	Calcium alginate	93
	Mixed culture of *Lactobacilli* and *yeasts* (multienzymes)	Gelatin	94
L-Malic acid	*Brevibacterium ammoniagenes* (fumarase)	Polyacrylamide	95,96
	Brevibacterium flavum (fumarase)	Carrageenan	97,98
	Thermus rubens nov sp (fumarase)	Duolite A-7	99
D-Tartaric acid	*Achromobacter* sp (α-tartaric epoxydase)	Polyacrylamide	100
α,ω-Tridecane-dioic acid	*Candida tropicalis* (oxidase)	Carrageenan	85
Urocanic acid	*Achromobacter liquidum* (L-histidine ammonia-lyase)	Polyacrylamide	101
	Micrococcus luteus (L-histidine ammonia-lyase)	Carbodiimide activated CM-cellulose	102
	Pseudomonas fluorescens (L-histidine ammonia-lyase)	Hollow fiber	103

ammoniagenes cells immobilized with polyacrylamide gel (95,96). Subsequently we used the carrageenan method to improve L-malic acid productivity (97). By screening various microorganisms for higher fumarase activity, *Brevibacterium flavum* was found and shown to exhibit higher enzyme activity after immobilization with κ-carrageenan than *B. ammoniagenes* (98) as shown in Table 4. Therefore, the *B. ammoniagenes* method was changed to the *B. flavum* method for commercial production in 1977. In other studies, *Thermus rubens* nov sp., a thermophilic bacterium, was immobilized ionically on Duolite for production of L-malic acid from fumaric acid (99).

Achromobacter liquidum cells having high L-histidine ammonia-lyase activity were immobilized into a polyacrylamide gel. The immobilized cells were used for the production of urocanic acid from L-histidine (101).

Trigonopsis variabilis having high D-amino acid oxidase was immobilized with calcium alginate, and the immobilized cells were used for production of α-keto acid from D-amino acid (86). The enzyme showed relatively high activity towards the following D-amino acids: valine, leucine, isoleucine, methionine, phenylalanine, tyrosine, tryptophan, and histidine.

2-Ketogluconic acid, an important intermediate for synthesis of vitamin C, was produced by using *Serratia marcescens* cells immobilized into a collagen membrane (89).

Immobilized living cell systems can also be applied to aerobic multistep reactions such as production of acetic acid, citric acid, etc. For example, whole cells of *Acetobacter aceti* were immobilized onto porous ceramic, and the immobilized living cells were used for production of acetic acid in a simple medium containing glucose (80,81). Citric acid was produced from glucose by using living cells of *Aspergillus niger* immobilized with calcium alginate (83,84), and 12-ketochenodeoxycholic acid (used chemotherapeutically for the solubilization of cholesterol gallstones) was produced from dehydrocholic acid by using living cells of *Brevibacterium fuscum* immobilized with κ-carrageenan (88).

Recently, whole cells of *Pseudomonas chlororuphis* having nitrilehydratase were immobilized with polyacrylamide gel, and the immobilized cells were used for industrial production of acrylamide from acrylonitrile (104).

IV. WASTE TREATMENT

The application of immobilized enzymes and microbial cells for waste treatment has been studied for a number of years. For example, conversion of ϵ-aminocaproic acid cyclic dimer in wastewater from a nylon plant to a biodegradable substance was studied using *Achromobacter guttatus* immobilized in polyacrylamide gel (54). Humphrey and co-workers at the University of Pennsylvania studied the continuous measurement of specific substances in wastewater by using immobilized enzymes and also studied the reduction of phenol content in wastewater by using immobilized polyphenol oxidase (105).

Cyanide in wastewater is usually decomposed by the action of microorganisms, i.e., by the so-called active sludge method. To improve this method, the cyanide-decomposing enzyme was immobilized by the entrapping method using polyacrylamide gel, packed into a column, and removal of cyanide from wastewater using the resulting enzyme column was inverstigated.

Decomposition of ammonia by *Nitrosomonas europhaea* entraped with alginate (106) and hydrolysis of palatione by palatione-hydrolyzing enzyme immobilized on porous glass (107) has been investigated.

Though many kinds of harmful substances in wastewater cannot be simultaneously decomposed, as they can be in the case of the active sludge method, these immobilized cell and enzyme systems are advantageous for the treatment and removal of specific substances. Thus, further development of systems utilizing immobilized enzymes and microbial cells in this field may be expected.

V. CONCLUSION

At present, examples of immobilized cell and enzyme systems used for industrial purposes are limited to the following nine cases:

1. Production of L-amino acid from acetyl-DL-amino acids using immobilized amino acylase.
2. Production of 6-aminopenicillanic acid using immobilized penicillin amidase.
3. Production of high fructose syrup using immobilized glucose isomerase.
4. Hydrolysis of lactose in milk using immobilized β-galactosidase.
5. Production of L-aspartic acid using immobilized microbial cells.
6. Production of L-malic acid using immobilized microbial cells.
7. Production of L-alanine using immobilized microbial cells.
8. Production of acrylamide using immobilized microbial cells.
9. Production of palatinose using immobilized microbial cells.

Besides the use of immobilized biocatalysts for the chemical processes stated above, immobilized biocatalyst systems are applied in a variety of fields and play a very important role in enzyme engineering and in the field of biotechnology. For example, immobilized living cell systems are now being developed not only for microbial cells but also for plant and animal cells, and it is expected that these systems will be an important technology for biotechnology.

If novel microorganisms with desired characteristics are produced by genetic engineering, that is, recombinant DNA technique, and are used as immobilized biocatalysts, this will be a very promising technology in the pharmaceutical and chemical industries.

REFERENCES

1. K. Ohmiya, C. Terao, S. Shimizu, and T. Kobayashi, *Agric. Biol. Chem. 39:*491 (1975).
2. H. Samejima and K. Kimura, *Enzyme Engineering 2:*131 (1974).
3. F. Morisi, M. Pastore, and A. Viglia, *J. Dairy Sci. 56:*1123 (1973).
4. W. F. Shipe, G. Senyk, and H. H. Weetall, *J. Dairy Sci. 55:*647 (1972).
5. H. R. Schreiner, U.S. Patent No. 3282702 (1966).
6. M. L. Green, and G. Crutchfield, *Biochem. J. 115:*183 (1969).
7. G. Corrieu, H. Blachere, A. Ramirez, J. M. Navarro, G. Durand, I. N. S. A. Toulouse, B. Duteurtre, and M. Moll, Fifth International Fermentation Symposium, Berlin, 1976, p. 294.
8. P. R. Witt, Jr., R. A. Sair, T. Richardson, and N. F. Olson, *Brewers Dig.*, October:70 (1970).
9. F. Leuschner, British Patent No. 953414 (1964).
10. H. H. Weetall and C. C. Detar, *Biotechnol. Bioeng. 16:*1095 (1974).
11. N. W. H. Cheethan, and G. N. Richards, *Carbohydrate Res. 30:*99 (1973).
12. J. H. Reynolds, *Biotechnol. Bioeng. 16:*135 (1974).
13. L. Goldstein, A. Lifshitz, and M. Sokolovsky, *Int. J. Biochem. 2:*448 (1971).
14. M. Ono, T. Tosa, and I. Chibata, *J. Ferment. Technol. 55:*493 (1977).
15. M. Ono, T. Tosa, and I. Chibata, *Agric. Biol. Chem. 42:*1847 (1978).
16. W. Marconi, F. Batoli, F. Cecere, G. Galli, and F. Morisi, *Agric. Biol. Chem. 39:*277 (1975).
17. Y. M. Deo and G. M. Gaucher, *Biotechnol. Bioeng. 26:*285 (1984).
18. Y. Morikawa, I. Karube, and S. Suzuki, *Biotechnol. Bioeng. 21:*261 (1979).
19. W. Kurzatkowski, W. Kurytowicz, and A. Paszkiewicz, *Eur. J. Appl. Microbiol. Biotechnol. 15:*211 (1982).
20. F. Giacobbe, A. Iasonna, and F. Cecere, *Enzyme Engineering 4:*245 (1978).
21. T. Sato, T. Tosa, and I. Chibata, *Eur. J. Appl. Microbiol. 2:*153 (1976).
22. M. Kluge, J. Klein, and F. Wagner, *Biotechnol. Lett. 4:*293 (1982).
23. H. Mayer, H. Collins, and F. Wagner, *Enzyme Engineering 5:*61 (1980).
24. J. Klein and F. Wagner, *Enzyme Engineering 5:*335 (1980).
25. A. Freeman and Y. Aharonowitz, *Biotechnol. Bioeng. 23:*2747 (1981).
26. Y. Morikawa, K. Ochiai, I. Karube, and S. Suzuki, *Antimicrob. Agents Chemother. 25:*126 (1979).
27. Y. Morikawa, I. Karube, and S. Suzuki, *Biotechnol, Bioeng. 22:*1015 (1980).
28. M. Veelken and H. Pape, *Eur. J. Appl. Microbiol. Biotechnol. 15:*206 (1982).
29. Y. M. Deo, and G. M. Gaucher, *Biotechnol. Lett. 5:*125 (1983).

30. A. Jones, D. Berk, B. H. Lesser, L. A. Bahie, and G. M. Gaucher, *Biotechnol. Lett. 5:*785 (1983).
31. T. Tosa, T. Mori, N. Fuse, and I. Chibata, *Enzymologia 31:*214 (1966).
32. T. Sato, T. Mori, T. Tosa, and I. Chibata, *Arch. Biochem. Biophys. 147:*788 (1971).
33. T. Mori, T. Sato, T. Tosa, and I. Chibata, *Enzymologia 43:*213 (1972).
34. I. Chibata, T. Tosa, T. Sato, T. Mori, and Y. Matsuo, *Proc. IV. IFS: Ferment Technol. Today* 383 (1972).
35. H. H. Weetall and C. C. Detar, *Biotechnol. Bioeng. 16:*1537 (1974).
36. K. Hirano, I. Karube, and S. Suzuki, *Biotechnol. Bioeng. 19:*311 (1977).
37. M. P. Schollar, B. Sigal, and A. M. Klibanov, *Biotechnol. Bioeng. 27:*247 (1985).
38. H. Schutt, G. Schmidt-Kastner, A. Arens, and M. Preiss, *Biotechnol. Bioeng. 27:*420 (1985).
39. A.-S. Jandel, H. Husted, and C. Wandrey, *Eur. J. Appl. Microbiol. Biotechnol. 15:*59 (1982).
40. C. Wandrey, R. Wichmann, and A.-S. Jandel, *Enzyme Engineering 6:*61 (1982).
41. C. Wandrey, E. Friolitakis, and R. Wichmann, *Enzyme Engineering 7:*91 (1984).
42. K. Yamamoto, T. Tosa, and I. Chibata, *Biotechnol. Bioeng. 22:*2045 (1984).
43. T. Sato, S. Takamatsu, K. Yamamoto, I. Umemura, T. Tosa, and I. Chibata, *Enzyme Engineering 6:*271 (1982).
44. S. Takamatsu, I. Umemura, K. Yamamoto, T. Sato, and I. Chibata, *Eur. J. Appl. Microbiol. Biotechnol 15:*147 (1982).
45. J. M. Sarkar and J. Mayaudon, *Biotechnol. Lett. 5:*201 (1983).
46. M. Fujimura, J. Kato, T. Tosa, and I. Chibata, *Appl. Microbiol. Biotechnol. 19:*136 (1984).
47. T. Tosa, T. Sato, T. Mori, Y. Matuo, and I. Chibata, *Biotechnol. Bioeng. 15:*69 (1973).
48. Y. Yokote, S. Maeda, H. Yabushita, S. Noguchi, K. Kimura, and H. Samejima, *J. Solid-phase Biochem. 3:*247 (1978).
49. K. Kimura, K. Takayama, T. Ado, T. Kawamoto, and I. Masunaga, Japanese Patent 81-75097 (1981).
50. F. Pittalis, F. Bartoli, and F. Morisi, *Enzyme Microb. Technol. 1:*189 (1979).
51. I. Chibata, T. Tosa, and T. Sato, *Appl. Microb. 27:*878 (1974).
52. T. Sato, Y. Nishida, T. Tosa, and I. Chibata, *Biochim. Biophys. Acta 570:*179 (1979).
53. M. C. Fusee, W. E. Swann, and G. J. Calton, *Appl. Environ. Microbial. 42:*672 (1981).
54. S. Kinoshita, M. Muranaka, and H. Okada, *J. Ferment. Technol. 53:*223 (1975).
55. K. Yamamoto, T. Sato, T. Tosa, and I. Chibata, *Biotechnol. Bioeng. 16:*1589 (1974).
56. A. Constantinides, D. Bhatia, and W. R. Vieth, *Biotechnol. Bioeng. 23:*899 (1981).

57. W. Slowinski and S. E. Charm, *Biotechnol. Bioeng. 15:*973 (1973).
58. M. Wada, T. Uchida, J. Kato, and I. Chibata, *Biotechnol Bioeng. 22:*1175 (1980).
59. R. Wichmann, C. Wandrey, A. F. Buckmann, and M.-R. Kula, *Biotechnol. Bioeng. 23:*2789 (1981).
60. T. Fukumura, Japanese Patent 74-15795 (1974).
61. O. Kanemitsu, Japanese Patent 75-132181 (1975).
62. H. Yamada, S. Shimizu, H. Shimada, Y. Tani, S. Takahashi, and T. Ohashi, *Biochimie 62:*395 (1980).
63. P. Zaffaroni, V. Vitobello, F. Cecere, E. Giacomozzi, and F. Morisi, *Agric. Biol. Chem. 38:*1335 (1974).
64. S. Fukui, S. Ikeda, M. Fujimura, H. Yamada, and H. Kumagai, *Eur. J. Biochem. 51:*155 (1975).
65. S. Fukui, S. Ikeda, M. Fujimura, H. Yamada, and H. Kumagai, *Eur. J. Appl. Microbiol. 1:*25 (1975).
66. W.-G. Bang, U. Behrendt, S. Lang, and F. Wagner, *Biotechnol. Bioeng. 25:*1013 (1983).
67. K.-D. Verlop, and J. Klein, *Biotechnol. Lett. 3:*9 (1981).
68. P. D.-L. Marechal, R. Calderon-Seguin, J. P. Vandecasteele, and R. Azerad, *Eur. J. Appl. Microbiol. Biotechnol. 7:*33 (1979).
69. I. Chibata, T. Kakimoto, and K. Nabe, Japanese Patent 74-81590 (1974).
70. R. Axen, and S. Ernback, *Eur. J. Biochem. 18:*351 (1971).
71. H. Yamada, K. Yamada, H. Kumagai, T. Hino, and S. Okamura, *Enzyme Engineering 3:*57 (1978).
72. G. Para, S. Rifai, and J. Baratti, *Biotechnol. Lett. 6:*703 (1984).
73. T. Tosa, T. Sato, T. Mori, and I. Chibata, *Appl. Microbiol. 27:*886 (1974).
74. T. Sato, T. Mori, T. Tosa, I. Chibata, M. Furui, K. Yamashita, and A. Sumi, *Biotechnol. Bioeng. 18:*1797 (1975).
75. N. Nishimura and M. Kisumi, *Appl. Environ. Microbiol. 48:*1072 (1984).
76. I. Umemura, S. Takamatsu, T. Sato, T. Tosa, and I. Chibata, *Appl. Microbiol. Biotechnol. 20:*291 (1984).
77. S. Takamatsu, K. Yamamoto, T. Tosa, and I. Chibata, *J. Fermet. Technol. 59:*489 (1981).
78. S. Takamatsu, T. Tosa, and I. Chibata, *Nippon Kagaku Kaishi 9:*1369 (1983).
79. T. Tosa, S. Takamatsu, M. Furui, and I. Chibata, *Enzyme Engineering 7:*450 (1984).
80. C. Ghommidh and J. M. Navarro, *Biotechnol Bioeng. 24:*1991 (1982).
81. C. Ghommidh, J. M. Navarro, and G. Durand, *Biotechnol. Lett. 3:*93 (1981).
82. J. F. Kennedy, S. A. Barker, and J. D. Humphreys, *Nature 261:*242 (1976).
83. P. Linko, *Advances in Biotechnology:*711 (1981).
84. H. Eikmeier and H. J. Rehm, *Appl. Microbiol. Biotechnol. 20:*365 (1984).
85. Z.-H. Yi and H. J. Rehm, *Eur. J. Appl. Microbiol. Biotechnol. 16:*1 (1982).
86. P. Brodelius, B. Hagerdal, and K. Mosbach, *Enzyme Engineering 5:*383 (1980).
87. P. Wikstrom, E. Szwajcer, P. Brodelius, K. Nilsson, and K. Mosbach, *Biotechnol. Lett. 4:*153 (1982).

88. H. Sawada, S. Kinoshita, T. Yoshida, and H. Taguchi, *J. Ferment. Technol. 59:*111 (1981).
89. K. Venkatasubramanian, A. Constantinides, and W. R. Vieth, *Enzyme Engineering 3:*29 (1978).
90. C. K. A. Martin and D. Perlman, *Eur. J. Appl. Microbiol. Biotechnol. 3:*91 (1976).
91. A. Tuli, R. P. Sethi, P. K. Khanna, and S. S. Marwaha, *Enzyme Microb. Technol. 7:*164 (1985).
92. S.-L. Strenroos, Y.-Y. Linko, and P. Linko, *Biotechnol. Lett. 4:*159 (1982).
93. P. Tipayang and M. Kozaki, *J. Ferment. Technol. 60:*595 (1982).
94. A. L. Compere and W. L. Griffith, *Dev. Ind. Microbiol. 17:*241 (1975).
95. K. Yamamoto, T. Tosa, K. Yamashita, and I. Chibata, *Eur. J. Appl. Microbiol. 3:*169 (1976).
96. K. Yamamoto, T. Tosa, K. Yamashita, and I. Chibata, *Biotechnol. Bioeng. 19:*1101 (1977).
97. I. Takata, K. Yamamoto, T. Tosa, and I. Chibata, *Eur. J. Appl. Microbiol. Biotechnol. 7:*161 (1979).
98. I. Takata, K. Yamamoto, T. Tosa, and I. Chibata, *Enzyme Microb. Technol. 60:*431 (1980).
99. Y. Ado, T. Kawamoto, I. Masunaga, T. Takayama, S. Takasawa, and K. Kimura, *Enzyme Engineering 6:*303 (1982).
100. Y. Kawabata and S. Ichihara, Japanese Patent 77-102496 (1977).
101. K. Yamamoto, T. Sato, T. Tosa, and I. Chibata, *Biotechnol. Bioeng. 16:*1601 (1974).
102. T. R. Jack and J. E. Zajic, *Biotechnol. Bioeng. 19:*631 (1977).
103. J. K. Kan and M. L. Shuler, *Biotechnol. Bioeng. 20:*217 (1978).
104. I. Watanabe, K. Sakashita, and Y. Ogawa, Japanese Patent 83-35078 (1983).
105. A. E. Humphrey and E. K. Pye, *Chem. Eng. News,* (Jan. 3):25 (1972).
106. C. G. V. Ginkel, J. Tramper, K. C. A. M. Luyben, and A. Klapwijk, *Enzyme Microb. Technol. 5:*297 (1983).
107. D. M. Munnecke, *Biotechnol. Bioeng. 21:*2247 (1979).

Index